Riverbank Filtration for Water Security in Desert Countries

NATO Science for Peace and Security Series

This Series presents the results of scientific meetings supported under the NATO Programme: Science for Peace and Security (SPS).

The NATO SPS Programme supports meetings in the following Key Priority areas: (1) Defence Against Terrorism; (2) Countering other Threats to Security and (3) NATO, Partner and Mediterranean Dialogue Country Priorities. The types of meeting supported are generally "Advanced Study Institutes" and "Advanced Research Workshops". The NATO SPS Series collects together the results of these meetings. The meetings are co-organized by scientists from NATO countries and scientists from NATO's "Partner" or "Mediterranean Dialogue" countries. The observations and recommendations made at the meetings, as well as the contents of the volumes in the Series, reflect those of participants and contributors only; they should not necessarily be regarded as reflecting NATO views or policy.

Advanced Study Institutes (ASI) are high-level tutorial courses intended to convey the latest developments in a subject to an advanced-level audience

Advanced Research Workshops (ARW) are expert meetings where an intense but informal exchange of views at the frontiers of a subject aims at identifying directions for future action

Following a transformation of the programme in 2006 the Series has been re-named and re-organised. Recent volumes on topics not related to security, which result from meetings supported under the programme earlier, may be found in the NATO Science Series.

The Series is published by IOS Press, Amsterdam, and Springer, Dordrecht, in conjunction with the NATO Public Diplomacy Division.

Sub-Series

A.	Chemistry and Biology	Springer
B.	Physics and Biophysics	Springer
C.	Environmental Security	Springer
D.	Information and Communication Security	IOS Press
E.	Human and Societal Dynamics	IOS Press

http://www.nato.int/science
http://www.springer.com
http://www.iospress.nl

Series C: Environmental Security

Riverbank Filtration for Water Security in Desert Countries

edited by

Chittaranjan Ray
University of Hawaii
Honolulu, HI, USA

and

Mohamed Shamrukh
Minia University
Egypt

Published in cooperation with NATO Public Diplomacy Division

Proceedings of the NATO Advanced Research Workshop on
Water Security in Desert Countries
Qena, Luxor, Egypt
21-24 February 2009

ISBN 978-94-007-0039-0 (PB)
ISBN 978-94-007-0025-3 (HB)
ISBN 978-94-007-0026-0 (e-book)

Published by Springer,
P.O. Box 17, 3300 AA Dordrecht, The Netherlands.

www.springer.com

Printed on acid-free paper

All Rights Reserved
© Springer Science + Business Media B.V. 2011
No part of this work may be reproduced, stored in a retrieval system, or transmitted
in any form or by any means, electronic, mechanical, photocopying, microfilming,
recording or otherwise, without written permission from the Publisher, with the
exception of any material supplied specifically for the purpose of being entered and
executed on a computer system, for exclusive use by the purchaser of the work.

CONTENTS

Preface xiii

Acknowledgment xv

Chapter 1 Riverbank Filtration Concepts and Applicability
to Desert Environments 1
   Chittaranjan Ray

Chapter 2 Water Pollution and Riverbank Filtration for
Water Supply Along River Nile, Egypt 5
   Mohamed Shamrukh and Ahmed Abdel-Wahab

   1. Introduction 6

   2. Methodology 7

   3. Drinking Water Sources and Pollution 7
      3.1. Pollution of River Nile 8
      3.2. Pollution of Groundwater 10

   4. Situation of Drinking Water in Egypt 13
      4.1. Quantity of Supplied Water 13
      4.2. Quality of Supplied Water 14

   5. Riverbank Filtration (RBF) 17

   6. RBF for Water Supply in Nile Valley 18
      6.1. Nile Water and RBF 18
      6.2. Nile—Aquifer Interaction 19
      6.3. RBF Wells at Naga Hamadi 22
      6.4. RBF Benefits and Scheme in Nile Valley 24

   7. Conclusion 25

Chapter 3 A Combined RBF and ASR System for Providing
Drinking Water in Water Scarce Areas 29
   Laxman Sharma and Chittaranjan Ray

   1. Introduction 29

   2. Methods and Procedures 33

   3. Results and Discussion 41
      3.1. The RBF Subsystem 41
      3.2. The ASR Subsystem 44
      3.3. Pyrite Oxidation and Arsenic Mobility 45

   4. Conclusions and Future Research Recommendations 46

CONTENTS

Chapter 4 Behavior of Dissolved Organic Carbon During Bank Filtration Under Extreme Climate Conditions 51
Dagmar Schoenheinz and Thomas Grischek

1. Introduction 51
2. Climate Projections and Their Relevance to the
   Abstraction of Bank Filtrate 52
3. Temperature Patterns of Water Resources 53
4. Effects of Low Flow Periods on Bank Filtrate Quality 55
5. Effects of Flood Events on River Bank Filtration 56
6. Evaluation Based on the Sum Parameter DOC 56
   6.1. Boundary Conditions 56
   6.2. Materials and Methods 57
   6.3. Results and Discussion 60
7. Adaptation Measures 64

Chapter 5 Risk Assessment for Chemical Spills in the River Rhine 69
Paul Eckert

1. Introduction 69
2. Site Description 70
3. Development of River Water Quality at the Rhine 71
4. Risk Assessment 74
   4.1. Rhine Alarm Model 74
   4.2. Well Management 75
   4.3. The Chloracetophenon Case in 2003 76
5. Conclusions 78

Chapter 6 Fluorescent Microspheres as Surrogates in Evaluating the Efficacy of Riverbank Filtration for Removing *Cryptosporidium parvum* Oocysts and Other Pathogens 81
Ronald Harvey, David Metge, Rodney Sheets, and Jay Jasperse

1. Introduction 82
2. Comparisons of Microsphere and Microbial Transport
   Behaviors in the Subsurface 83
3. Microspheres for Assessing Vulnerability of RBF Wells
   to *Cryptosporidium* Contamination 86
   3.1. Comparison of Oocyst and Microsphere Properties 86
   3.2. Bench-Scale Comparisons of Microsphere and
        Oocyst Transport 87
   3.3. Field Transport Studies Using Oocyst-Sized Microspheres 89
4. Limitations of Microspheres as Surrogates 92
5. Conclusions 93

CONTENTS

Chapter 7 Hydrogeochemical Processes During Riverbank Filtration and Artificial Recharge of Polluted Surface Waters: Zonation, Identification, and Quantification 97
Pieter J. Stuyfzand

  1. Introduction 98

  2. RBF and AR Types 98

  3. Compartments and Processes 101

    3.1. The Surface Water Compartment 103

    3.2. The Recharge Proximal Aquifer Zone 104

    3.3. The Distant Aquifer Zone 104

    3.4. The Recovery System and the Discharge Proximal Zone 105

  4. Redox Zonation 105

    4.1. Definition and Measurement 105

    4.2. Typical Zonation Patterns 108

    4.3. Redox Dependent Removal Efficiencies 112

  5. Identification and Quantification of Reactions by Mass Balancing 113

    5.1. Introduction 113

    5.2. REACTIONS+6 113

    5.3. Examples of Application 121

  6. Reactive Transport Modeling 122

  7. Concluding Remarks 125

Chapter 8 Potential of Riverbank Filtration to Remove Explosive Chemicals 129
Chittaranjan Ray, Weixi Zheng, Matteo D'Alessio, Joseph Lichwa, and Rico Bartak

  1. Introduction 129

  2. Experimental Methods 131

  3. Results 132

  4. Conclusions 134

Chapter 9 Framework for Assessment of Organic Micropollutant Removals During Managed Aquifer Recharge and Recovery 137
Sung Kyu Maeng, Saroj K. Sharma, and Gary Amy

  1. Introduction 138

  2. Methods 139

    2.1. Guidelines for Estimating Removal Efficiencies 139

    2.2. QSAR 141

  3. Results and Discussion 143

viii CONTENTS

3.1. Guidelines 143
3.2. QSAR 143
4. Conclusions 148

Chapter 10 Dissolved Organic Carbon as an Indicator Parameter for Groundwater Flow and Transport 151
Dagmar Schoenheinz

1. Introduction 152
2. Requirements of Indicator Parameters 153
3. Review of DOC Characterization and Transport Research 154
4. Modeling DOC Degradation 159
   4.1. Mathematical Model 159
   4.2. Model Assumptions 161
   4.3. Conceptual Model 161
5. Application of the Conceptual Model to the Evaluation of Anthropogenically Impacted Aquifers 163
6. Conclusion 165

Chapter 11 Planning, Design and Operations of Collector 6, Sonoma County Water Agency 169
Jay Jasperse

1. Introduction 169
2. Purpose of Collector 6 Project 170
3. Background 171
   3.1. Hydrology of the Russian River 171
   3.2. Hydrogeologic Setting 172
   3.3. SCWA Water Supply Facilities and Operations 173
4. Collector 6 Site Selection 175
5. Planning and Site Characterization Activities 176
   5.1. Phase 1—Initial Site Characterization 177
   5.2. Phase 2—Detailed Site Assessment 178
6. Development of Collector 6 Conceptual Design 179
7. Detailed Design and Construction of Collector 6 183
   7.1. Permitting 183
   7.2. Caisson 183
   7.3. Installation of Conventional Laterals 185
   7.4. Development of Conventional Laterals 187
   7.5. Performance Testing 187
   7.6. Design and Installation of Long, Large Diameter Laterals 187
   7.7. Successful Installation of Lateral A5-1 192

CONTENTS ix

7.8. Analytic Element Modeling and Installation of Lateral A5-2 193
7.9. Development of LLDLs 193
7.10. Embankment 194
7.11. Pumphouse Facility and Associated Equipment 194
8. Startup Testing 196
8.1. Pre-Production Monitoring 197
8.2. Post-Production Monitoring 197
9. Sustainable Operations 198
10. Capacity Analysis of Collector 6 199
11. Conclusions and Recommendations 200

Chapter 12 Evaluation of Bank Filtration for Drinking Water
Supply in Patna by the Ganga River, India 203
Cornelius Sandhu, Thomas Grischek, Dagmar Schoenheinz,
Triyugi Prasad, and Aseem K. Thakur
1. Introduction 203
2. Study Area 205
2.1. Physiography and Hydrogeology 205
2.2. Drinking Water Supply 210
3. Data and Methods 210
3.1. Field and Laboratory Investigations 210
3.2. Groundwater Flow Model Geometry and Boundary Conditions 212
4. Results and Discussion 214
4.1. Ground and Surface Water Levels 214
4.2. Ganga River Morphology 214
4.3. Water Quality 215
4.4. Groundwater Flow Modeling 218
4.5. Isotope Analyses 219
5. Conclusion 220

Chapter 13 Minimizing Security Risks Beyond the Fence-Line:
Design Features of a Tunnel-Connected Riverbank Filtration System 223
S. Hubbs, K. Ball, and D. Haas
1. Introduction 223
2. Design Considerations 224
2.1. Extraction Systems 224
2.2. Design Issues 225
3. Construction 226
4. Summary 234

CONTENTS

Chapter 14 Removal of Iron and Manganese Within the Aquifer Using Enhanced Riverbank Filtration Technique Under Arid Conditions 235
Kamal Ouda Ghodeif

   1. Introduction 235

   2. Iron and Manganese Distribution and Associated
      Problems in Egypt 237

   3. Hydrogeological Conditions of the Nile Delta System 238

   4. Research Assumptions, Materials and Methods 239

   5. Design of the Site and Detailed Hydrogeology 241

   6. Results and Discussion 244

      6.1. Hydrogeology and RBF Wells Productivity 244

      6.2. Water Quality and Quality Changes 245

   7. Conclusion 251

Chapter 15 Riverbank Filtration as an Alternative Treatment Technology: AbuTieg Case Study, Egypt 255
Fathy A. Abdalla and Mohamed Shamrukh

   1. Introduction 256

   2. Statement of the Problem and Objectives 257

   3. Examination Site 258

      3.1. Nile Aquifer 259

      3.2. Quaternary Aquifer 260

   4. Materials and Methods 261

   5. Results and Discussions 262

      5.1. Quality of Nile Water 264

      5.2. Behavior of Biological Contaminants 265

      5.3. Quality of Bank Filtered Water 265

   6. Conclusion 266

Chapter 16 Quality of Riverbank Filtrated Water on the Base of Poznań City (Poland) Waterworks Experiences 269
Józef Górski

   1. Introduction 269

   2. Characteristic of the Mosina-Krajkowo Waterworks 270

   3. Investigations 271

   4. Results of Investigations 273

   5. Discussion 277

   6. Conclusions 278

CONTENTS xi

Chapter 17 Riverbank Filtration as an Alternative to Surface
Water Abstraction for Safe Drinking Water Supply to the
City of Khabarovsk, Russia 281
V.V. Kulakov, N.K. Fisher, L.M. Kondratjeva, and T. Grischek

1. Introduction 281

2. Hydrology of the Amur River...................................... 283

3. Pollution of the Sungari River 284

4. Pollution in the Amur River 285

5. Accident in the Sungari River Basin in 2005 286

6. Amur and Sungari Rivers Pollution After the Accident.............. 289

7. An Alternative Means of Drinking Water Production 291

8. Conclusion 295

Index 299

PREFACE

Riverbank filtration has been used for more than a century for providing drinking water to riparian communities in Europe, particularly in Germany. In the United States, it has been used to provide municipal and industrial water for more than half a century. Despite these successes and its appeal as a low-cost technology, it has not been widely used as a viable means for water supply in developing regions of the world. Limited numbers of systems are in operation for potable water supply and some new systems are being built to pre-treat surface water that has been severely impacted by wastewater discharge. However, use and success of riverbank filtration in desert environments is rare to find in literature. Many of the communities located on perennial rivers in desert countries (e.g., Nile, Niger, etc.) directly abstract the water from rivers, treat, and supply. Riverbank filtration can provide the added benefit of natural filtration and enhance treatment efficiency even if treatment plants are used in polluted reaches of these rivers. Also, potential applications of riverbank filtration along with aquifer storage and recovery can be envisaged for ephemeral rivers or Wadis. In this book, we examine some of the systems in the western countries and a few systems that are beginning to operate on the River Nile. We address the issue with pathogens and climate change. Also, we take a holistic approach to present that for water security in desert environments (both from pollution and source), riverbank filtration alone or in combination with aquifer storage and recovery have significant relevance.

xiii

ACKNOWLEDGMENT

Organizing committee:

C. Ray, University of Hawaii, USA
M. Shamrukh, El-Minia University, Egypt
K. Ghodeif, Suez Canal University, Egypt
T. Grischek, University of Applied Sciences, Dresden, Germany
Ahmed Khaled, Sohag University, Egypt

Official Host of the Workshop:

Suez Canal University (at Luxor, Egypt)

Financial support:

NATO SPS
with some in-kind clerical and editorial support from University of Hawaii and
El-Minia University

Chapter 1 Riverbank Filtration Concepts and Applicability to Desert Environments

Chittaranjan Ray*

Professor of Civil & Environmental Engineering and Interim Director, Water Resources Research Center, University of Hawaii at Manoa, Honolulu, Hawaii 96822, USA

Abstract   Riverbank filtration (RBF) is considered a "natural" filtration technique in which the bed and bank areas of a river serve as "treatment" zones for the river water. When wells are placed adjacent to a river and pumped, the treatment zones remove most surface water pollutants. This technology has been in operation for more than a century in Europe and for more than half a century in the United States. In many areas of the world, particularly in the populated regions of India and China, RBF has significant potential. Riverbank filtration can be used along perennial as well as ephemeral rivers in desert countries. However, studies are needed to evaluate the connectivity of the surface water and ground water, dynamic variation of the redox zone(s), flash flood impacts, and climate changes for the sustainable operation of RBF systems in desert environments.

Keywords: Riverbank filtration, collector wells, water supply, desert countries

Although surface water has been the prime source of drinking water for communities along the rivers, pollution of the rivers from municipal and industrial waste discharges as well as runoff waters from non-point sources (such as agricultural land) is causing significant concern to these communities to supply water that meets regulatory requirements. Riverbank filtration (RBF) is an alternative to traditional surface water treatment process in which RBF uses the riverbed and the underlying aquifer as "natural" filters to remove pollutants present in the surface water. The process is simple in the sense that when a well is placed in an alluvial aquifer adjoining the river and pumped steadily, the drawdown cone around the pumping well induces part of the river water to infiltrate and move towards the pumping well (Figure 1.1). As the infiltrating water moves through the porous media, many dissolved contaminants and suspended matter as well as pathogens are removed. A combination of straining, colloidal filtration, sorption, and microbial

---

* Chittaranjan Ray, Professor of Civil & Environmental Engineering and Interim Director, Water Resources Research Center, University of Hawaii at Manoa, Honolulu, Hawaii 96822, USA, e-mail: cray@hawaii.edu

C. Ray and M. Shamrukh (eds.), *Riverbank Filtration for Water Security in Desert Countries*, DOI 10.1007/978-94-007-0026-0_1, © Springer Science+Business Media B.V. 2011

degradation contribute to the removal/attenuation of these contaminants. It should be pointed out that many of these dissolved contaminants that are typically hard to remove in surface water treatment plants can be removed by RBF.

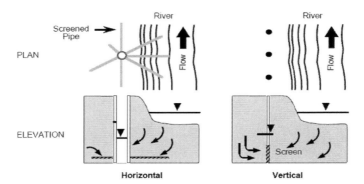

Figure 1.1. Conceptual representation of an RBF system with horizontal collector and vertical wells (after Ray et al. 2002).

RBF is a cost effective and yet efficient treatment method that can be suitable for a wide variety of conditions if managed properly. In Europe, particularly in Germany, riverbank filtration has been in operation for more than a century along the Rhine and Elbe rivers. In Düsseldorf along the River Rhine, RBF systems have been operating since 1870 (Schubert 2002). At Dresden along the Elbe River, RBF systems have been operating since 1875 (Fischer et al. 2006). Many large cities (e.g., Belgrade, Budapest, Bratislava, Köln, Heidelberg, Bonn, etc.) along major European waterways use RBF for water supply. In the United States, RBF systems have supplied water to industries and municipalities for more than half a century (Ray et al. 2002). Without such systems, many communities cannot comply with the current drinking water regulations using surface water and treating it in conventional water treatment plants. Examples of such systems are cities in the Midwestern United States: Lincoln, Nebraska (on the Platte River); Des Moines, Iowa (on the Raccoon River); and Cedar Rapids, Iowa (on the Cedar River). Nitrate and pesticides in spring runoff from agricultural runoff often exceed the maximum contaminant levels (MCLs).

While designing RBF systems, one takes into consideration the distance of the well from the river, pumping rate, well type (vertical or horizontal), well depth, and riverbed conditions. These factors affect the travel time of a water particle from the river to the well. For wells that are far from the river, the travel time of water can be long. Similarly, long travel times expected for riverbeds with fine grained materials, as the rate of percolation can be slow. In Europe, most communities use vertical wells that are typically located some distance away from the river. In the United States, the preference is for horizontal collector wells, which are large-capacity wells placed very close to the bank of the rivers. Consequently, the travel time of water in an RBF system employing vertical wells is longer than that employing horizontal collector wells. A longer travel time enables the removal of pathogens and many chemicals, including nitrate. In Europe, travel times on the order of 30 days or more are considered necessary for the removal of pathogens from surface water. However, an extremely long travel time can induce the dissolution of iron, manganese, and other redox-sensitive chemicals. These chemicals then need to be removed through a conventional water treatment plant.

Design of RBF systems heavily depends on site geology and hydrology. For example, rivers in desert countries can be flashy with very little flow in dry season. In comparison, rivers in western countries are often regulated with locks and dams with dependable flows. Thus, the siting of the RBF systems can be a problem when the flow varies significantly. In dry season, the yield from wells can be low. In some cases, communities and industries have built horizontal collector wells inside the main course of the river so that the well can be surrounded by water even in the dry season (Figure 1.2). Raising inflatable dams around the well sites could pond water and enhance infiltration in dry seasons. Low dams built and maintained in Wadis can store storm water after rain events. Wells placed around such systems can function as RBF systems. However, such shallow reservoirs can dry out during long summer months. If the local aquifers are not contaminated, the water produced from the wells on the banks of the Wadis can be stored underground. Although it involves the use of energy for injection and pumping, evaporation losses from surface impoundments are reduced. Water quality issues involving a combination of RBF systems and aquifer storage recovery systems have been examined in Chapter 3 of this book.

Figure 1.2. A horizontal collector well located within the main course of the Sabarmati River, Ahmedabad, Gujrat, India. (Photo: Courtesy of C. Sandhu, University of Applied Sciences, Dresden)

Water quality in desert environment can be different compared to temperate regions because of the higher temperature of the source water and the reactions involving organic carbon. If the source water contains significant amounts of organic carbon (typically expected in developing countries), dissolved oxygen in the subsurface can be depleted quickly, causing an anoxic environment. Subsequently, the system can face nitrate iron/manganese and sulfate reducing conditions, respectively, depending on organic load and travel time of water. Removal of trace organics and pharmaceutical compounds can be affected by subsurface redox conditions. In desert environments, the organic pollution load can be high during low-flow conditions because of low dilution. This period can be critical for water quality.

RBF systems can be implemented in desert areas on the banks of perennial rivers such as Nile, Niger, Indus, and along numerous Wadis. While RBF systems on perennial rivers like the Nile have the full potential to produce potable water, detailed studies are needed for marginal rivers or non-perennial rivers or streams. For Wadis, a completely different design option may be necessary.

If the water level in the aquifer is at significant depth below the riverbed, then a partially saturated condition exists between the aquifer and the riverbed. If an RBF system is to be built under such conditions, the quantity of filtrate that can be extracted from such systems needs to be calculated using three-dimensional two-phase flow models. The associated water quality also needs to be predicted using mass transport and biogeochemical models.

Several research issues must be undertaken for desert environments: (1) examining the connectivity between the ground water and surface water and how that would affect the operation of RBF systems, (2) dynamic variation of the redox zones at the RBF sites as functions of time and the resulting impacts on water quality, (3) dealing with flash floods in Wadis and small rivers and the resulting impact on filtrate quality, and (4) impacts of projected climate change on the sustainable operation of RBF systems.

References

Ray C, Melin G, Linsky RB (2002) Riverbank filtration – Improving source water quality. Kluwer, Dordrecht, Boston, London

Schubert J (2002) German experience with riverbank filtration systems. In: Ray C, Melin G, Linsky RB (eds) Riverbank filtration – Improving source water quality, Kluwer, Dordrecht, Boston, London, pp 35–48

Fischer T, Day K, Grischek T (2006) Sustainability of riverbank filtration in Dresden, Germany. In: Recharge systems for protecting and enhancing groundwater resources. Proc. Int. Symp. Management of Artificial Recharge, 11.-16.06.2005, Berlin, UNESCO IHP-VI Series on Groundwater No. 13, pp 23–28

Chapter 2 Water Pollution and Riverbank Filtration for Water Supply Along River Nile, Egypt

Mohamed Shamrukh[1]* and Ahmed Abdel-Wahab[2]

[1] Department of Civil Engineering, Faculty of Engineering, Minia University, 61111 Minia, Egypt.

[2] Chemical Engineering Program, Texas A&M University at Qatar, P.O. Box 23874, Doha, Qatar. E-mail: ahmed.abdel-wahab@qatar.tamu.edu

Abstract In a developing country such as Egypt, there are growing challenges for providing water supply of good quality. Committing to the Millennium Development Goals (MDG) by providing access to clean drinking water supply is an additional challenge. This is primarily attributed to treatment costs, especially when large quantities of water are treated. The two sources of potable water supply in Egypt are groundwater and surface water, either from the River Nile or from the main irrigation canals. In 2008, the total drinking water production in Egypt was about 7.5 billion m^3/year, the contributions from the Nile and groundwater being about 60% and 40%, respectively. Nile water in Egypt is facing rising sources of pollution despite all the programs for pollution control. Discharging industrial and domestic wastewater, return drainage of irrigated water, and flash flood into the River Nile represent the major sources of pollution. There are also widespread problems of iron, manganese, nitrate, and fecal coliform bacteria in the groundwater used for drinking water supply. Riverbank filtration (RBF) is a water treatment technique that can improve surface water quality. Current and previous results of water quality produced from RBF have proven its potential to treat Nile water and to avoid quality problems associated with source water. This paper illustrates the benefits of using RBF, the ability to resolve a broad range of water quality problems in an economic manner and to provide clean and safe drinking for the residents of a desert country such as Egypt.

Keywords: Riverbank filtration, Egypt, Nile valley, groundwater, water supply, water quality, potable water, water pollution

---

* Mohamed Shamrukh, Department of Civil Engineering, Faculty of Engineering, Minia University, 61111 Minia, Egypt, e-mail: mshamrukh@hotmail.com

C. Ray and M. Shamrukh (eds.), *Riverbank Filtration for Water Security in Desert Countries*, DOI 10.1007/978-94-007-0026-0_2, © Springer Science+Business Media B.V. 2011

1. Introduction

Access to good quality and safe water, makes a tremendous difference to our quality of life. As we step into the twenty-first century, it is realized that the trend towards urbanization is posing ever-increasing challenges with respect to water supply. The rate of growth of population (~1.8%), especially in the urban areas, is far exceeding that of the rural areas in Egypt and most developing countries which put an increasing stress on treatment capacities in many cities. Statistics indicate that over one billion of the world population lack access to safe water, and nearly two billion lack safe sanitation worldwide. It is also reported that more than three million die every year from water-related diseases (UNICEF 2005). A growing number of water-related diseases such as diarrhea, schistosomiasis (bilharziasis), intestinal parasites, lymphatic filariasis, and trachoma are responsible for major health problems in the majority of rural and urban residents. One of the greatest challenges facing Egypt today is the growing number of rural and urban households who need access to basic infrastructure, mainly water supply and sewage. If lacking, this can have a significant adverse impact on human health, productivity and the quality of life (UNDP 2005).

During the last 2 decades, the Egyptian governments have invested massively in providing water supply and sewage networks to both rural and urban communities. Today, all Egyptian cities and more than 90% of villages are provided with water supply services. In the new Five-Year Plan (2007–2012), the Egyptian government planned to allocate more than 72 billion EGP to expand the water supply service to secure water access for all citizens. However, even with these unprecedented investments, 20% of Egyptian villages have inadequate potable water (IDSC 2009). Confronted with this challenge, the community is looking for innovative and cost-effective technologies for potable water supply (UNDP 2008). Great efforts are being made to increase water availability for the whole Egyptian population. These efforts, however, are not being felt by the users, neither in the suburban and rural areas, nor in the cores of the capital Cairo. The main reason for this is that the rate of population increase and the city's expansion outpaces these efforts.

The major problem of water supply in Egypt and other developing countries is not a lack of technology availability. Rather, it is due to the fact that stakeholders are largely unaware of the available alternatives and the complexity of the suitability of one technology over the other applicable in their situation. The major challenge is therefore to select an appropriate technology considering the multi faceted issues including technical feasibility, affordability, customs and practices, preferences, and available institutional support. The potential goal is, presenting no single and absolute solution, but offering a comparative analysis of various options, and encouraging the decision makers and communities to adopt the one that is best suited to their needs.

This paper highlights the qualitative and quantitative problems related to drinking water supply in Egypt, summarizes drinking water resources and their pollution level, and investigates the feasibility of applying riverbank filtration (RBF) technique for water purification in the Nile valley of Egypt. Therefore, the main objective of the current study is to address the feasibility of RBF to tackle the water quality problems of drinking water supply in Egypt with a fraction of total cost (i.e., capital cost and O&M cost).

2. Methodology

The purpose of the current investigation is to present the economical, technical, and environmental benefits of using RBF for drinking water supply in the Nile valley. For this purpose, previous reported data and current measurements for Nile water and groundwater were conducted. The study is carried out in four steps. First, the pollution sources and extent of pollution in the River Nile and groundwater in Nile valley based on reported data were evaluated. Second, the quantitative and qualitative aspects of domestic water supply scheme in Nile valley were investigated. Third, section of Nile valley located between Qena and Assiut was investigated for RBF applicability to treat Nile water. In this part, additional measurements were made to supplement the available data. Measurements of flow hydraulics and alluvial sediments of River Nile were conducted. These measurements were conducted in the field or in the laboratory of New Naga-Hamadi Barrage. Physicochemical and microbiological quality analysis was performed on water samples. Temporal and spatial variations of quality parameters were monitored. These parameters include pH, TDS, BOD_5, hardness, chloride, iron, manganese, sulfate, orthophosphates, nitrate, ammonia, and bacteria. Quality measurements of samples were done according to the procedures laid down in the standard methods of the examination of water and wastewater (APHA 1998). Fourth, feasibility of using RBF technique to treat Nile water for water supply in Nile valley of Egypt was presented.

3. Drinking Water Sources and Pollution

Egypt can be classified as an arid climate with 95% of its area as desert. A narrow strip of fertile land exists along the main stem of River Nile and within a relatively small delta in the north. It became the base for economic and social life of one of the most distinguished ancient civilization where agriculture was the main human activity. The average rainfall is about 25.7 mm/year and the evapotranspiration rate ranges from about 0.7 mm/day in winter to about 15.7 mm/day in summer. The relative humidity ranges from 45% to 75% and the average daily temperature ranges from 13°C to 38°C. The total population of Egypt is about 77 million and increasing at a rate of about 1.8% annually (CAPMAS 2007).

The conventional water resources for drinking water supply in Egypt are limited to the River Nile and the groundwater in Nile valley, Nile delta, deserts and Sinai. Limited rainfall and flash floods are also available. The Nile is the main and almost the exclusive source of fresh water in Egypt. The country relies on the available water stored in Lake Nasser to meet needs within Egypt's annual share of water, which is fixed at 55.5 billion m^3/year. Each resource has its limitations on use. These limitations relate to quantity, quality, location, and cost of development. Protecting this limited amount of fresh water is crucial to sustain the development of the nation (EEAA 1999).

Water quality problems of these sources vary depending on flow, pattern of use, population density, extent of industrialization, availability of sanitation systems and the social and economic conditions. Discharge of untreated, or partially treated, industrial and domestic wastewater, leaching of pesticides and residues of fertilizers and navigation are often factors that affect the quality of water. In general, ranking pollutants according to their severity to public health and the environment puts pathogenic microorganisms on the top. This is followed by biodegradable organic compounds which deplete dissolved oxygen, affecting water suitability for many purposes. This is followed by pesticide residues and heavy metals. However, little precise information is available to the magnitude of the problem (MWRI 2002). Furthermore, in Egypt there is no national monitoring program concerning the identification and determination of organic micropollutants in drinking water resources (Badawy 2009).

3.1. Pollution of River Nile

Egypt is one of the ten countries that share the 6,650 km long basin of River Nile. The last 1,600 km of it goes through Egypt from Aswan to Mediterranean Sea. Since 1968, Aswan High Dam (AHD) is used to regulate the flow of Nile and to provide multi-year storage. There are four main barrages, Esna, Nag-Hamada, Assiut, and Delta which control the flow and divert water into the main irrigation canals without lifting stations (Abdel-Dayem et al. 2007). Only River Nile and the main irrigation canals branching from it are used to supply drinking water in Nile valley and Nile delta (Figure 2.1).

Pollution load in the Nile system (River Nile, canals, and drains) has increased in the past few decades due to population increases, several new irrigated agriculture projects, new industrial projects and other activities along the River Nile. Consequently, quality of Nile water worsened dramatically in the past few years (Abdel-Satar 2005; Abdel-Dayem et al. 2007). It is anticipated that the dilution capacity of the River Nile system will diminish as the program to expand irrigated agriculture moves forward and the growth in industrial capacity increases the volume of pollutants discharged into the Nile (MWRI 2002). The major pollution sources of Nile and main canals are effluents from agricultural drains and treated or partially treated industrial and municipal wastewaters, including oil and wastes from passenger and river boats. The most polluted part of Nile is the part located between Cairo and Mediterranean Sea within the two branches of Nile, Damietta and Rosetta (Abdo 2004; NAWQAM 1998, 2003).

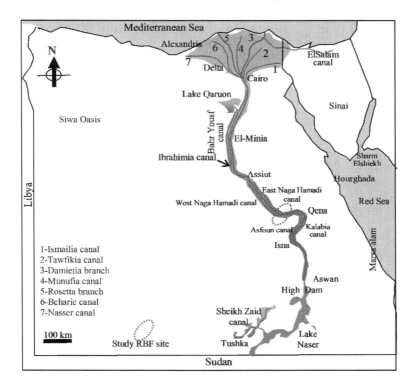

Figure 2.1. Map of Egypt with River Nile, main canals and study site, 2008.

Discharging the drainage water from drains into Nile, represent the primary source of Nile pollution. The volume of drainage water is about 12 billion m^3/year, 25% of it is in upper Egypt and the rest in Nile delta. There are 76 drains discharging drainage water into Nile system with annual volume of about the half of the total drainage water (World Bank 2005). The part of Nile from Aswan to Delta Barrage receives discharges from 67 agricultural drains of which 43 are considered major drains. Previous reported measurements indicated that out of those 43 drains, only ten are complying with the Egyptian standards regulating the quality of drainage water allowed to be discharged into Nile systems (MWRI 2002). This drainage water contains dissolved salts washed from agricultural lands as well as residues of pesticides and fertilizers. Impact of this drainage water on Nile quality has been reported by several authors (Abdo 2004; Abdel-Dayem et al. 2007).

Industrial wastewater is considered the second of the main sources of Nile water pollution because of the toxic chemicals and organic loading in this wastewater. Egyptian industry uses about 7.8 billion m^3/year of water, of which 4,050 million m^3/year are discharged into the River Nile system. The River Nile supplies 65% of the industrial water needs and receives more than 57% of its effluents (MSEA 2007). There is an increasing trend in industrial consumption of water for new development activities. There are about 129 factories discharging their wastewater into the River Nile system. Effluent wastewater is often partially treated. In spite of all official efforts to prevent this pollution source, there are 34 factories still not complying with Egyptian regulation of water disposal into Nile systems (NBI 2005).

The third major source of pollution of River Nile is the effluent of municipal wastewater. There are about 239 wastewater treatment plants with an annual effluent of 4.5 billion m^3, of which 1,300 million m^3/year are discharged into River Nile system. Though the plants offer secondary level of treatment, the real treatment removal efficiency of these are lower than their design in many sites.

All aforementioned major pollution sources deteriorate Nile water. Indication of this pollution has been reported by chemical and microbiological quality measurements along many sections of the River Nile (MWRI 2002, 2003; MSEA 2005; NBI 2005). There are elevated levels of organic matter (measured as COD or BOD_5), heavy metals (Pb, Cr, Hg, and Cd), and fecal coliform. Sixteen organochlorine pesticides were detected in the drains and to less degree in canals, including total BHC and total DDT (El-Kabbany et al. 2000). These elevated values are higher than natural occurring and at some spots of Nile are higher than allowable limits for healthy water streams. Also, low levels of DO at many sections of River Nile were recorded (Ismail and Ramadan 1995; Mohamed et al. 1998; Wahab and Badawy 2004). The two branches of Nile, Rosetta Branch and Damietta Branch, downstream Delta Barrage, represent the worst quality of River Nile. Some quality parameters of Nile were reported by NBI (2005).

3.2. Pollution of Groundwater

The Nile system comprises the Valley and Delta regions, including Cairo. These are morphologic depressions filled with Pliocene and Quaternary sediments. The hydrogeological framework of Egypt comprises six main aquifer systems (Fadlelmawla et al. 1999). The aquifer of interest of this study is the Nile aquifer system that is assigned to the Quaternary and Late Tertiary, which occupies the Nile floodplain region, including Cairo, and the desert fringes. The width of the aquifer is about 20 km and bounded by the carbonates (Figure 2.2). Almost 90% of Egypt's population lives on the Nile aquifer (Shamrukh 1999).

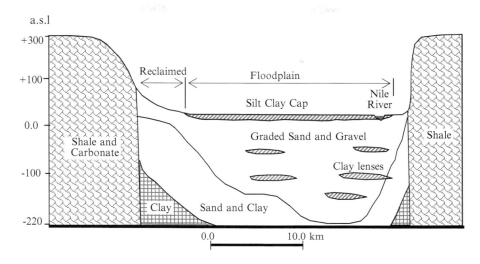

Figure 2.2. General hydrogeological cross section of Nile aquifer in Nile valley.

Nile aquifer consists of a thick layer of graded sand and gravel underlined by the Pliocene impermeable clays and covered by a silt-clay layer (5–20 m thickness) in its major part. The aquifer thickness varies from 350 m (at Sohag) to only a few meters (at Cairo and Aswan). North of Cairo, the aquifer thickness increases gradually until it reaches more than 800 m along the Mediterranean coastline. The transmissivity of the aquifer ranges on the higher side between 5,000 and 20,000 m^2/day (Shamrukh 1999; Ahmed 2009a). The main recharge source is the infiltration from the excess water application for agriculture, seepage from the irrigation canals and from sewerage system. Discharge from the aquifer is through seepage to the River Nile and groundwater extraction through wells. Thus, the main sources of pollution into Nile aquifer are agrochemicals, domestic wastewater, and natural dissolution of soil minerals such as iron and manganese.

Agrochemicals and concentrated salts are the main sources of groundwater for water supply. There have been extensive applications of chemical fertilizers (nitrogen, phosphors, sulfur, and potassium) to enhance crop production after the construction of the High Aswan Dam (Shamrukh 1999). Elevated concentrations of nitrate and other agrochemicals such as sulfate and potassium are reported in many locations of the aquifer especially in Nile delta (Abdel-Dayem and Abdel-Ghani 1992; El-Fouly and Fawzi 1994; Awad et al. 1995; Shamrukh and Abdel-Lah 2004). In addition, measurements of elevated concentrations of pesticides in Nile aquifer were reported (Abdel-Dayem and Abdel-Ghani 1992). Furthermore, percolating water into the groundwater table contains concentrated salts due to evaporation of irrigation water and due to water use with high dissolved salts in reclaimed areas in eastern and western Nile floodplain (Soltan 1998).

Pollution from sewerage systems especially in rural areas is another source. In Egypt, water supply and sewage services are not implemented simultaneously. In the rural areas, where half of the population lives, 90% of the people have no access to sewer systems or wastewater treatment facilities (UNDP 2008). The "sewage room" is the most common disposal facility where its bottom has direct contact with the ambient groundwater. This method of local sewage disposal makes it a point source of pollution with pathogens and nitrate (Figure 2.3). Therefore, adverse water quality as a result of pathogen contamination occurs locally at a number of different locations. Sampling of shallow groundwater close to this septic room shows that the risks associated with pathogen contamination in such supplies is high (Abdel-Lah and Shamrukh 2001).

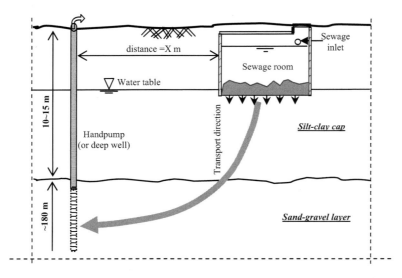

Figure 2.3. Schematic diagram shows the groundwater pollution from sewage room in Nile aquifer.

The last major source of groundwater pollution is the natural dissolution of soil minerals. The main attribute of these areas is high formation-inherited iron and manganese concentrations as a result of the highly reduced environment of the confined aquifer of the Nile basin. Therefore elevated contents of Fe and Mn in groundwater are reported at many locations by Ministry of Health (MoH 2009). Arsenic was also reported in few groundwater locations. In addition to those sources of pollution, other sources such as elevated TDS in desert fringes of Nile and salt water intrusion in coastline areas were reported (Zaghloul and Abdallah 1985).

Finally, it is clear that the main challenge for the sustainability of water resources is the control of water pollution. The Ministry of the Environment in Egypt is observing the enforcement of the legislation regarding the treatment of industrial and domestic wastewater. It is also advocating organic farming and limiting the use of chemical fertilizers and pesticides to reduce water pollution. In addition, the present policy is to minimize the use of herbicides and to depend mainly on the mechanized control of submerged weeds and water hyacinths (MSEA 2005).

4. Situation of Drinking Water in Egypt

The municipal water supply in the urban and sub-urban areas in the valley and delta of Nile includes: (i) water treated from the Nile; (ii) a mixture of water treated from the Nile and groundwater; and (iii) groundwater alone. Almost all major towns setting at the banks of Nile system are supplied from Nile water through conventional treatment plants or compact units. Towns away from Nile and small villages located at the Nile banks are supplied from aquifer groundwater. The following treatment methods are commonly used:

- Conventional surface treatment plants to treat Nile water applying conventional treatment process: coagulation/flocculation with alum, sedimentation, filtration, and disinfection with chlorine gas
- Compact units to treat the water of main canals using the same processes applied in the conventional treatment plants but in compact design
- Direct groundwater pumping using deep wells (60–100 m) without any further treatment processes, not even disinfection
- Aeration units to treat groundwater rich in Fe and Mn followed by filtration and chlorine disinfection

In 2004, the Holding Company for Water and Wastewater (HCWW) and its affiliated 23 companies was established to take over the responsibility of operating the drinking water supply schemes all over Egypt. There are 153 large surface treatment plants, 615 compact surface treatment plants, 1,684 wells, and 21 desalination plants for producing water. Quality of produced water is monitored by Ministry of Health for compliance and by HCWW for operation and quality and quality control purposes.

4.1. Quantity of Supplied Water

Currently, total drinking water produced by HCWW is about 22 million m^3/day. Drinking water reaches to about 97% of the population, however supply is intermittent, a few hours per day for 25% of the population. In summer of 2007, there was a water shortage due to limited capacity of water treatment plants. It is anticipated that this problem will be continue due to budget constraints to construct/expand treatment plants. It is difficult for the government with limited financial resources to provide enough quantity with rising demand and to keep the water in good quality. The price of consumed water, about $0.06/$m^3$ is less than its actual cost as the government subsidizes the drinking water sector. Figure 2.4 shows the number of drinking water plants and water production in the last years. It is estimated that water demand will continue increasing by 2% per year as generated by rapid population growth, urbanization, and industrialization (CAPMAS 2008).

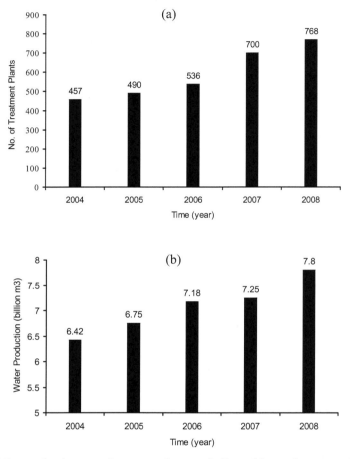

Figure 2.4. Recent development of water supply sector in Egypt (**a**) no. of treatment plants and (**b**) drinking water production.

4.2. *Quality of Supplied Water*

In Egypt, supply shortfalls exist in terms of quantity, physical, chemical and biological quality, and reliability. It is anticipated that the quality of drinking water will be the biggest challenge of water supply system in Egypt in the coming decades. Water-borne diseases from pathogens contamination is one of the biggest public health concerns (UNDP 2005). Typhoid, paratyphoid, infectious hepatitis, and infant diarrhea are some endemic diseases indicating deterioration of water quality. According to a World Bank report (2005), the cost of diarrhea and mortality due, in large part, to water pollution was estimated at US$800 million/year. Rural areas are often the most affected areas by those appreciable variations in water quality.

Quality of drinking water supplied from Nile using conventional treatment technology complies with Egyptian standards (Donia 2007; EHCW 2007). The normal water quality measurements carried out by Ministry of Health do not include pollutants of organic, microorganic or trace metals. As mentioned in previous sections, the pollution of Nile water is mainly chemical and biological. However, it is known that the conventional treatment processes does not have the ability to effectively remove dissolved materials such as pesticides, chlorinated organics, micropollutants and heavy metals. Very few research works have been carried out on such pollutants as these require extensive time and money. Elevated contents of heavy metals and pesticides in drinking water supplied from Nile have been reported (Mohamed et al. 1998; MWRI 2002). Therefore, it is imperative to carry out a precise and detailed investigation of the impact of Nile pollution on the quality of water obtained from it.

The presence of organic matter in the Nile is about 2–5 mg/L. The recommended total chlorine dose for disinfection in conventional normal or compact treatment units of Nile water is 5–8 mg/L. Chlorine is added to both the raw water (pre-chlorination) and the filtered water (post-chlorination). The applied chlorine dose is higher than what is actually needed for disinfection (Shamrukh and Abdel-Lah 2004). This higher dose is applied to maintain enough chlorine residual in order to prevent water quality deterioration in the distribution pipelines. Applying this high chlorine dose with dissolved organic matter is expected to form disinfection by-products (DBPs) formation, such as trihalomethanes (THMs) and haloacetic acids (HAAs). THMs are suspected carcinogens and/or mutagenic compounds (WHO 1996). A maximum contamination level of 100 μg/L is accepted by Egyptian standards. EI-Dib and Ali (1992, 1995) reported high concentrations of THMs in treated Nile water at Cairo. Measured concentrations of THM in winter and summer ranged 41.8–247.1, and 18.1–80.1 μg/L, respectively.

Elevated levels of chlorinated organics in Cairo drinking water have also been identified in one study (Smith 2009). These compounds have the effect of "chronic toxicity" which is usually characterized by long-term exposure to contaminants at relatively low concentrations. Adverse health symptoms may not appear for years but it may manifest later as cancerous tumors (Smith 2009). Polycyclic aromatic hydrocarbons (PAHs) measured at four locations were present in extremely high concentrations of 1,112.7–4,351.2 μg/L in raw Nile water but were not detected in treated water (Badawy and Emababy 2010).

Drinking water supplied from groundwater is also facing growing quality problems. A significant one is the high concentrations of Fe and Mn in many drinking wellfields. Table 2.1 shows the number of the villages declared by Ministry of

Housing and Utilities to have quality problems (Al-Ahram 9/28/2009). Treatment of groundwater polluted by Fe and Mn in Egypt is carried out using an aeration tower. A study carried out by Abdel-Lah et al. (2002) indicated that oxidation of Mn by aeration is not effective. Many health problems can be caused by elevated Fe and Mn in drinking water (WHO 1996).

TABLE 2.1. No. of villages with polluted drinking groundwater (mostly iron and manganese).

Governorate	No. of villages
Qena	4
Sohag	114
Assiut	168
Minia	31
Helwan	10
6th October	36
Menofia	51
AlGharbia	73
AlShargia	44

Problem of agrochemicals in groundwater is getting worse due to extensive application of chemical fertilizers and pesticides. Elevated concentrations of chemical fertilizers in drinking and irrigation wells have been reported (Shamrukh and Abdel-Lah 2004; Awad et al. 1995; El-Fouly and Fawzi 1994). In these studies, elevated concentrations higher than drinking water standards were reported for NO_3, SO_4, and K and up to 120, 530, and 45 mg/L, respectively. A recent study (Shamrukh 2010) indicated that nitrate pollution in drinking wells is increasing with time.

In addition to chemical fertilizers, elevated concentrations of pesticides and herbicides in groundwater wells were also reported (MWRI 2003). Concentrations higher than allowable limits were detected in few shallow wells. For instance, shallow groundwater in Delta was highly contaminated with organochlorine (OC) and organophosphorus (OP) residues such as BHC and DDT (Abd-Allah and Gaber 2004).

Bacterial contamination of drinking handpumps and municipal wells from leaching wastewater of sewage rooms is another problem. As shown in Figure 2.3, wastewater can easily move into handpumps and municipal wells due to the direct contact between sewage and ambient groundwater. Water quality from private handpumps that partially supply villages is not routinely sampled. However, occasional sampling shows that the risk associated with pathogen contamination in such supplies is high (Abdel-Lah and Shamrukh 2001). Their measurements indicated that drinking wells at depth of 60 m and close to sewage rooms by 40 m in rural areas were also at high risk of bacterial contamination. Drinking water associated with pathogenic pollution can cause a growing number of water-related diseases, especially in infants (WHO 1996).

5. Riverbank Filtration (RBF)

For more than 100 years, RBF has been used in Europe to supply drinking water to communities along the Rhine, Elbe, Danube, and Seine rivers. In Berlin, RBF contributes to about 70% of total drinking water demands. Potable water abstracted using RBF is about 50% in Slovak republic, 45% in Hungary, 16% in Germany and 5% in The Netherlands (Dash et al. 2008). In USA, RBF systems have also been supplying drinking water to several communities for nearly 50 years. In Europe, post-World War II when the rivers were significantly polluted with municipal and industrial effluents, RBF was the most efficient method of producing high quality drinking water (Ray et al. 2002). Recently, many countries around the world have started to evaluate RBF feasibilities for water treatment including India, South Korea, China and Jordan.

In RBF, which similar to some extent to slow-sand filtration, river water contaminants are attenuated from a combination of processes such as filtration, microbial degradation, sorption to sediments and aquifer sand, and dilution with background groundwater. According to previous published work by numerous researchers, RBF has proven its effectiveness in water treatment (Shamrukh and Abdel-Wahab 2008; Dash et al. 2008; Massmann et al. 2008; Weiss et al. 2003, 2005; Schmidt et al. 2003a, b; Hiscock and Grischek 2002; Tufenkji et al. 2002; Stuyfzand 1998; Doussan et al. 1997; Cosovic et al. 1996). Reported data indicated that RBF can effectively remove many water major pollutants and micropollutants including particulates, colloids, algae, organic and inorganic compounds, microcystins, pathogens and even heavy metals (Sontheimer 1980). Furthermore, bank filtration is able to attenuate concentration or temperature peaks and can provide protection against shock loads. Compared to traditional water treatment plants, RBF could have more advantages especially in improving the removal capacity and reducing the total cost. In addition, RBF reduces the concentrations of disinfection by-products due to its ability to remove organic matter. There is no waste generated from RBF which gives it an environmental advantage.

However in RBF, when the surface water is low in dissolved oxygen then conditions during underground passage will likely become anaerobic, which can cause iron and manganese to become soluble and therefore be drawn into the groundwater well (Hiscock and Grischek 2002). This can have the undesirable effect of degrading the water quality to unacceptable drinking water standards. In spite of that, under anoxic conditions, nitrates are reduced to nitrogen and thereby provide oxygen for organics removal and ammonia oxidation (Sontheimer 1980).

RBF is typically conducted in alluvial valley aquifers, which are complex hydrologic systems that exhibit both physical and geochemical heterogeneity. The performance of RBF systems depends upon well type and pumping rate, travel time of surface water into wells, source water quality, site hydrogeologic conditions, biogeochemical reactions in sediments and aquifer, and quality of ambient groundwater (Ray 2001).

6. RBF for Water Supply in Nile Valley

In Egypt, there is no designed RBF that has all the facilities to be investigated in depth. But there are few sites where the vertical wells are very close to Nile banks which can be considered RBF sites. One of the first studied such RBF sites is the wellfield of Sidfa city, which is south of Assiut city in upper Egypt (Abdel-Lah and and Shamrukh 2006). Sidfa RBF site includes seven vertical wells that are 70 m in depth and placed 50 m from the Nile bank. Results of Sidfa site indicated that Nile filtrate contributes about 70% of total wells production. The following sections illustrate that RBF that can be applied widely in Nile valley of Egypt.

6.1. Nile Water and RBF

In the current study, a section of Nile between Qena and Assiut was selected to evaluate the quality of Nile water and RBF applicability for water supply. Most measurements and sampling were done at Qena and Naga Hamadi barrage.

After constructing Aswan High Dam (AHD), the flow regime of Nile has been changed to dam-regulated flow. Figure 2.5 shows the discharge from AHD and level (i.e., stage) of River Nile at Qena. It is clear that the discharge in summer is double of winter. Also, discharge is minimum in December because of winter closure of irrigation systems and minimum gate openings of AHD. Lowest water level of Nile occurs during winter closure starting at the end of Dec and for 1 month long. Nile stage is almost consistent allover the year. The difference between water level in summer and winter is about 2.50 m.

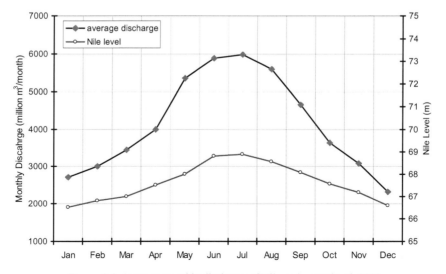

Figure 2.5. Average monthly discharge of Nile and water level, 2008.

The water quality parameters of Nile water at Naga Hamadi are shown in Figure 2.6. Temperature, turbidity, nitrate, DO and BOD_5 data are given for 3 months in 2009. In general, there are no significant variations of Nile quality between summer and winter. The suspended matter (i.e., turbidity) of the river is low compared to European and US rivers (Caldwell 2006). This low content of suspended solids is due to the sedimentation in Naser Lake, upstream of AHD. This could result in reducing the clogging of riverbed but the effect of low suspended solids on RBF performance needs a separate investigation. Dissolved oxygen of Nile water is good for RBF which is anticipated to make it aerobic and reduce the probability of iron and manganese mobilization in pumped water. Nitrate and BOD_5 can be considered as indication of Nile pollution from irrigation drains and industry. Nitrate and BOD_5 are low and illustrate that the Nile surface water is good at the studied section.

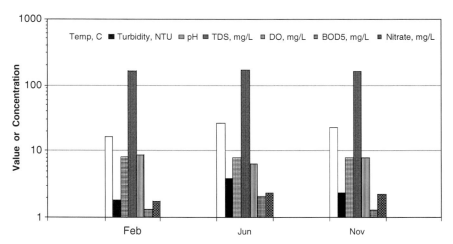

Figure 2.6. Some of quality parameters of the Nile at 10 km downstream Naga Hamadi barrage, 2008.

6.2. Nile — Aquifer Interaction

For RBF to work effectively, there should be a hydrogeologic connection between river and wells. Figure 2.7 shows the hydrogeologic cross section of Nile at new Naga Hamadi barrage. This section was drawn using the data of more than 30 boreholes made for the design of new Naga Hamadi barrage. The significant feature is that Nile cuts through the silt-clay cap forming a good hydraulic connection between the river and the main aquifer in absence of any low-permeable deposits below the riverbed. The estimated hydraulic conductance for the bed layer of River Nile ranges from 2 to 8 days. This will enhance the movement of bank filtrate from Nile into any abstraction wells installed at the Nile banks. The general stratification of Nile aquifer system at Naga Hamadi area from top to bottom is:

- Quaternary
  - Holocene silt clay or clayey silt with fine sand
  - Pleistocene sand with gravel and broken stone fragments
  - Pleistocene lenses of silt and clay
  - Plio-Pleistocene deposits

- Tertiary
  - Pliocene clay
  - Paeleocene shale
  - Eocene limestone (also east and west of the floodplain)

In general, the water table is located in the silt-clay cap 2–3 m below the ground surface, which is higher than the Nile water surface in both summer and winter. The only exception is about 50 km upstream of the four Nile barrages due to back water curve. Therefore, Nile works as a drain for the Nile aquifer. For about 50 km upstream of any Nile barrage, the flow is from Nile into the aquifer (i.e., Nile recharges the aquifer). However, at new Naga Hamadi barrage, level of water table is 63.90 m (MSL) and Nile stage is 62.8 m, in summer. In addition, water table is higher than the piezometric surface of sand-gravel layer indicating that vertical recharge from silt-clay cap into the aquifer.

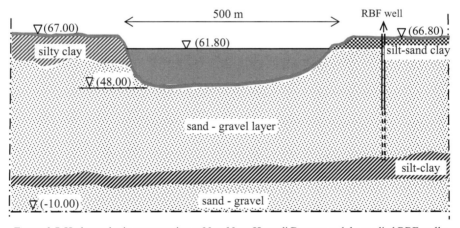

Figure 2.7. Hydrogeologic cross section at New Naga Hamadi Barrage and the studied RBF wells.

Particle size distribution of the main sand layer that form the main aquifer layer of Nile valley at Naga Hamadi barrage is shown in Figure 2.8. About 112 samples were used to draw the size distribution of this sand-gravel layer. Three significant sizes can be extracted from sieve analysis graph for each layer, they are effective diameter (d_{10}), main diameter (d_{50}), and 60% passing diameter (d_{60}). Hydraulic conductivity from the current sieve analysis using d_{10}, d_{50}, and d_{60} and applying the empirical formulas of Hazen (1911) and Shepherd (1989) are given in Table 2.2. Other aquifer characteristics that could help in assessing the applicability of RBF in Nile valley taken from reported data are also given (Ahmed 2009b). It is clear that the hydraulic conductivity in the horizontal direction, K_h, for the main aquifer layer is very good and supporting water movement from the streambed and the banks into the pumping wells.

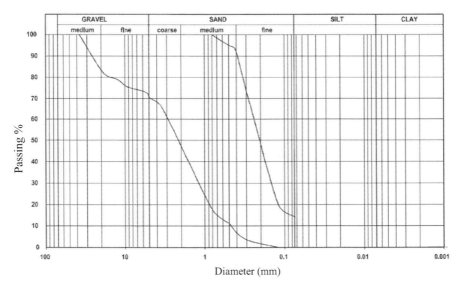

Figure 2.8. Particle size distribution of Pleistocene sand layer at the new Naga Hamadi Barrage.

TABLE 2.2. Hydrogeologic characteristics of the Nile aquifer at Naga Hamadi Barrage.

Aquifer layer	Silt-clay cap	Sand layer (Above 27.00 MSL)	Sand layer (Below 27.00 MSL)	Sand-gravel lenses	Clay lenses (At 15.00 MSL)
Thickness (m)	7–16	25–40	100–200	7–12	6–18
d_{10} (mm)	0.04	0.15–0.6	0.09–3.0	0.13–0.50	0.002
d_{50} (mm)	0.02–0.15	0.2–2.0	0.4–0.9	0.80–15.0	0.09
d_{60} (mm)	0.02–0.17	0.25–3.0	0.5–1.1	3.0–17.0	0.24
K_h (m/day) (Ahmed 2009)	0.1–0.3	40–120	40–120	40–120	na
K_v (m/day) (Ahmed 2009)	0.02–0.04	7–20	7–20	7–20	na
Hazen (1911)	–	20–260	10–80	30–310	Very low
Shepherd (1989)	–	15–300	35–110	75–330	Very low

na = not available

In addition, the ability of Nile aquifer sediments to adsorb cyanobacterial toxins supports the application of bank filtration. Zakaria et al. (2007) investigated the Nile sediments and reported that RBF can be used effectively in Nile valley of Egypt for removing these cyanobacterial toxins (microcystin) from drinking water.

6.3. RBF Wells at Naga Hamadi

There are two abstraction wells installed about 70 m away from the right bank of the River Nile and 50 m deep to supply potable water to the camp of new Naga Hamadi barrage (Figure 2.9). The supplied community is about 3,000 persons. Measurements of water quality from the abstraction wells in September 2009 were carried out. Collected samples were analyzed using Hach DR2000 spectrophotometer for physio-chemical measurements. Other meters for pH and TDS measurements were used. Instrument startup and analysis were carried out as detailed in the operating manual and each measurement was made in duplicate. Microbial measurements for pathogens were carried out at the laboratories of

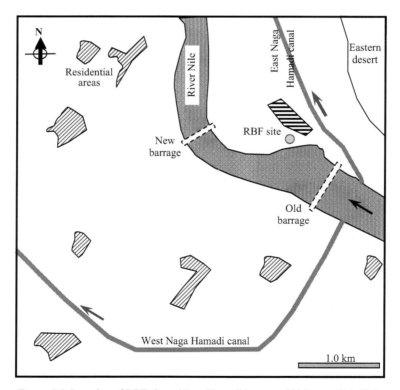

Figure 2.9. Location of RBF site at Naga Hamadi barrage, 500 km south to Cairo.

Ministry of Health in Egypt. The results of riverbank filtrates at the abstraction wells were compared with those of the River Nile water, and the ambient groundwater. Ambient groundwater was analyzed from a deep well located at 1 km from the current RBF site. The effectiveness of the RBF process is evaluated based on this comparison.

The RBF site at Naga Hamadi is the fourth RBF study site after Sidfa, AbuTieg, and ElMutia. All the three previous studied site are located on the western bank of Nile where the width of the floodplain is about 20 km. The quality of the ambient groundwater of the previous three RBF sites is different than the current investigated case. The most important feature of the current ambient groundwater is higher concentrations of almost all ions of concern. Table 2.3 shows the water quality at the RBF site at Naga Hamadi.

TABLE 2.3. Water quality at the three locations of RBF site, Naga Hamadi barrage.

Parameter	Ambient GW	Nile water	RBF
Physicochemical			
pH	7.9	7.7	7.8
Turbidity (NTU)	0.3	4	0.2
TDS (mg/L)	622	168	320
Total Hardness (mg/L as $CaCO_3$)	272	119	168
Major cations (mg/L)			
Ca^{2+}	58	26	39
Mg^{2+}	32	12	21
Na^+	57	22	38
Fe	0.2	0.15	0.16
Mn	0.6	0.07	0.09
Major anions (mg/L)			
Cl^-	42	19	27
SO_4^-	32	20	24
HCO_3^-	336	142	244
Nutrients (mg/L)			
NO_3^-	8	1.7	2.4
NH_4^-	nd	0.03	0.05
PO_4^-	1.8	0.5	1.0
Organic matter (mg/L)			
COD	nd	8.5	nd
BOD_5	nd	2.7	nd
Bacteriological parameters			
Total coliform (fcu/100 mL)	0	1,100	0
Fecal coliform (fcu/100 mL)	0	290	0

nd = not determined

Results indicate that infiltrating water from Nile dilutes the ambient groundwater moving into the RBF wells. Results show that Nile pollutants were effectively removed through the RBF process as the pumped water meets the Egyptian standards. These results of RBF effectiveness are in agreement with the results of the previous two RBF studies in Nile valley (Abdalla and Shamrukh 2009; Shamrukh and Abdel-Wahab 2008). However, the ambient groundwater at this site has elevated values of most chemical species. Therefore, this should be considered when evaluating the RBF removal effectiveness of Nile water. This situation of ambient groundwater with higher constituents than adjacent surface water is similar to another RBF study reported in India (Dash et al. 2006, 2008).

Quality results indicated that distance and location of the RBF wells from Nile are the key parameters of the RBF performance. If the RBF wells are very close to Nile then the microbial removal will be reduced. On the other hand, if the RBF is far away from Nile then the problems of Fe and Mn as well as other contaminants will be detected. Organic carbon is consumed during biodegradation, mediated by microbial activities in presence of other compounds that are electron acceptors Therefore, four zones can be defined according to the geochemical and transformation processes between Nile and RBF wells. Those four zones that influence the quality of bank filtrates from Nile are as follows (Wett et al. 2002):

1. Zone I: reduction of dissolved O_2 and NO_3
2. Zone II: reduction of Mn(IV) and Fe(III)
3. Zone III: reduction of Fe(III) and precipitation of $MnCO_3$
4. Zone IV: reduction of SO_4

In Nile aquifer, iron and manganese in drinking wells represent a major problem. To avoid the Fe and Mn dissolution and contamination, RBF wells should be installed within Zone I. Monika et al. (2009) proposed a quantitative approach which is called the *electron trapping capacity* (ETC). It is calculated using dissolved O_2 and NO_3 concentrations in bank filtrates. ETC represents the quantity of electrons that O_2 and NO_3 are capable of trapping as groundwater moves from oxidizing conditions to Mn and Fe reducing environment. The greater the ETC, the less reductive the conditions become and a lower probability of dissolution of manganese or iron oxyhydroxides exist. Therefore, ETC should not be lower than 0.2 mmol/L (Monika et al. 2009).

6.4. RBF Benefits and Scheme in Nile Valley

Current RBF results and previous work reported three RBF sites located in Nile valley improved the quality of drinking water production. In addition, most of the Egyptian towns are located within 10 km from River Nile which makes RBF physically and topographically suitable. The other advantages of using RBF among Egyptian water supply schemes are:

- Low capital and O&M costs
- Absence of or minimal addition of chemicals (i.e., coagulants)
- High rates of water-recovery
- Reducing or absence of disinfection by-products
- No sludge production or generation of hazardous waste stream
- Destruction, rather than sequestration or concentration of contaminants
- Simultaneous removal of multiple contaminants
- Protection against shock loads from flash floods and ship accidents in Nile
- Robustness over a wide range of operating conditions and water qualities

These attractive benefits of RBF have received the attention of decision makers in Egypt. The implementation of RBF application in Nile valley could be through the following three scenarios:

1. New installation of RBF vertical wells followed by online chlorine disinfection and elevated storage tank for new drinking water demands
2. Moving the current municipal wells with Fe and Mn or bacteriological problems closer to Nile
3. Stop the costly expansion of current surface treatment plants and integrate RBF wells that can be discharge water to the existing filtered water tank (i.e., Cl_2 contact tank)

7. Conclusion

This study illustrates that the River Nile and its aquifer as the main drinking water sources in Egypt is facing serious pollutant problems. River Nile receives considerable amounts of pollution from municipal and industrial effluents as well as agricultural drains. Problems of iron and manganese contaminations in the Nile aquifer are widespread. Fertilizers and septic rooms in the areas with no piped sewage system present another challenge to improving water quality of aquifers. There is an increasing demand for drinking water due the growing population. Conventional treatment plants used to treat Nile water are not capable of removing micropollutants in the River Nile. Riverbank filtration as natural treatment method could be the optimal solution to overcome Egyptian quantitative and qualitative challenges of drinking water. Previous research work and data of the present investigation shows that RBF can be an effective and economical means of drinking water production from Nile. Many of the large and small cities of Egypt are located along the banks of River Nile. RBF can be used as stand alone or as an alternative to direct Nile abstraction followed by conventional treatment process.

Acknowledgments Authors are grateful to the Assiut Company for Water and Wastewater and to new Naga-Hamadi Barrage for technical support during this research work. We are greatly indebted to the two reviewers who provided helpful comments on earlier drafts of this paper.

References

Abdalla FA, Shamrukh M (2009) Riverbank filtration as a new treatment technology in Nile Valley: Case study, AbuTieg City, Assiut, Egypt. NATO Advanced Research Workshop (*Riverbank Filtration for Water Security in Desert Countries*), Luxor, October 24–27, 2009

Abd-Allah SM, Gaber HM (2004) Monitoring of pesticide residues in different sources of drinking water in some rural areas. Alexandria J Agric Res 49:112–124

Abdel-Dayem S, Abdel-Gawad S, Fahmy H (2007) Drainage in Egypt: A story of determination, continuity, and success. Irrig Drain 56:S101–S111

Abdel-Dayem S, Abdel-Ghani M (1992) Concentration of agricultural chemicals. *Proceedings of 6th Int. Symposium on Drainage and Water Table Control*, TN, USA, pp. 353–360

Abdel-Lah A, Shamrukh M (2001) Impact of septic on ground water quality in a Nile Valley Village, Egypt. *6th Int. Water Technology Conference (IWTC)*, Alexandria, Egypt, March 2001, pp. 237–245

Abdel-Lah A, Shamrukh M (2006) Riverbank filtration: A promise method for water supply from Nile, Egypt. In: *Proceedings of 7th International Symposium on Water Supply Technology*, Yokohama, Japan, November 2006, pp. 385–395

Abdel-Lah A, Shamrukh M, Shehata A (2002) Efficiency of iron and manganese removal from groundwater using aeration tower in Nile Valley, Egypt. *3rd International Symposium on Environmental Hydrology, American Society of Civil Engineers-Egypt Section* (ASCE-EGS), Cairo, Egypt, April 2002, 7pp

Abdel-Satar AM (2005) Water quality assessment of River Nile from Idfo to Cairo. Egypt J Aqua Res 31(2):200–223

Abdo MH (2004) Environmental studies on the River Nile at Damietta Branch region, Egypt. J Egypt Acad Soc Environ Dev 5(2):85–104

Ahmed AA (2009a) Using lithologic modeling techniques for aquifer characterization and groundwater flow modeling of the Sohag area, Egypt. Hydrogeol J 17(5):1189–1201

Ahmed AA (2009b) Using generic and pesticide DRASTIC GIS-based models for vulnerability assessment of the Quaternary Aquifer at Sohag, Egypt. Hydrogeol J 17(5):203–1217

APHA AWWA, WEF (1998) Standard methods for the examination of water and wastewater 20th edn. American Public Health Association, American Water Work Association, Water Environment Federation, Washington, DC

Awad MA, Nada AA, Hamza MS, Froehlich K (1995) Chemical and isotopic investigation of groundwater in Tahta region, Sohag-Egypt. Environ Geochem and Health 17(3):147–153

Badawy MI, Emababy MA (2010) Distribution of polycyclic aromatic hydrocarbons in drinking water in Egypt. Desalination 251(1-3):34–40

Caldwell TG (2006) Presentation of data for factors significant to yield from several riverbank filtration systems in the U.S. and Europe. Stephen A. In: Hubbs S (ed) *Riverbank Filtration Hydrology*, Springer, pp. 299–344

CAPMAS (2007) Bulletin of collection, purification and distribution of water. Central Agency for Public Moblisation And Statistics, Egypt

CAPMAS (2008) Water rationalization in Egypt. Central Agency for Public Moblisation And Statistics, Egypt

Cosovic D, Hrsak V, Vojvodic V, Krznaric D (1996) Transformation of organic matter and bank filtration from a polluted stream. Water Res 30(12):2921–2928

Dash RR, Mehrotra I, Kumar P (2006) Natural bank / bed filtration: Water supply schemes in Uttaranchal, India. *World Environmental and Water Resources Congress*, ASCE

Dash RR, Mehrotra I, Kumar P, Grischek T (2008) Lake bank filtration at Nainital, India: Water-quality evaluation. Hydrogeol J 16:1089–1099

Donia N (2007) Survey of potable water quality problems in Egypt. *11th International Water Technology Conference*, Sharm El-Sheikh, Egypt, pp. 1049–1058

Doussan C, Poitevin G, Ledoux E, Detay M (1997) Riverbank filtration: Modeling of the changes in water chemistry with emphasis on nitrogen species. J Contam Hydrol 25:129–156

EEAA (1992) Environmental action plan. Egyptian Environmental Affairs Agency (EEAA), Cairo

EHCW (2007) Egyptian standards for drinking and domestic uses. Egyptian Higher Committee for Water, Egyptian Governmental Press, Egypt

El-Dib MA, Ali RK (1992) Trihalomethanes and halogenated organic formation in water treatment plant. Bull Environ Contam Toxicol 49:381–387

El-Dib MA, Ali RK (1995) THMs formation during chlorination of raw Nile River water. Water Res 29(I):375–378

El-Fouly MM, Fawzi AF (1994) Higher and better yields with less environmental pollution in Egypt through balanced fertilizers use. *Proceedings of the International Symposium, Fertilizers and Environment*, Salamanca, Spain, pp. 19–22

El-Kabbany S, Rashed MM, Zayed MA (2000) Monitoring of the pesticide levels in some water supplies and agricultural land, in El-Haram, Giza (A.R.E). J Hazard Mater A72:11–21

Fadlelmawla A, Abdel-Halem M, Vissers M (1999) Preliminary plans for artificial recharge with reclaimed wastewater in Egypt. *9th Biennial Symposium on the Artificial Recharge of Groundwater*, Tempe, AZ, June 10–12, 1999

Hazen A (1911) Discussion: Dams on sand foundations. Trans Am Soc Civ Eng 73:199–203

Hiscock KM, Grischek T (2002) Attenuation of groundwater pollution by bank filtration. J Hydrol 266:139–141

IDSC (2009) Monthly report no 30. *Information and Decision Support Center*, Egypt, June 2009 (http://www.idsc.gov.eg/News/recentnews.aspx)

Ismail SS, Ramadan A (1995) Characterisation of Nile and drinking water quality by chemical and cluster analysis. Sci Total Environ 173/174:69–81

Massmann G, Sultenfuß J, Dunnbier U et al (2008) Investigation of groundwater residence times during bank filtration in Berlin: A multi-tracer approach. Hydrol Processes 22:788–801

MoH (2009) Results of water quality examinations. Central Water Laboratory in Cairo, Ministry of Health, Egypt

Mohamed MA, Osman MA, Potter TL, Levin RE (1998) Lead and cadmium in Nile River water and finished drinking water in Greater Cairo, Egypt. Environ Int 24(7):767–772

Monika AM, Kedziorek A, Bourg CM (2009) Electron trapping capacity of dissolved oxygen and nitrate to evaluate Mn and Fe reductive dissolution in alluvial aquifers during riverbank filtration. J Hydrol 365:74–78

MSEA (2005) The annual report on the quality of water of the Nile River. Ministry of State for Environmental Affairs, Egypt

MWRI (2002) Survey of Nile system pollution sources. *APRP-Water Policy Activity*, Ministry of Water Resources and Irrigation (MWRI), EPIQ Report No. 64

MWRI (2003) Nile River water quality management study. *WPRP-Water Policy Reform Program*, Ministry of Water Resources and Irrigation (MWRI), IRG Report No. 67

NAWQAM (1998) Present status of water quality in Egypt. *National Water Quality and Availability Management Project (NAWQAM)*, National Water Research Centre, Canadian Executive Agency, Assessment Interim Report

NAWQAM (2003) Nile Research Institute data. *National Water Quality and Availability Management Project (NAWQAM)*, National Water Research Centre. Egypt

NBI (2005) Nile Basin water quality monitoring baseline report. Transboundary Environmental Action Project, *Nile Basin Initiative*

Ray C, Melin G, Linsky RB (2002) Riverbank filtration: Improving source water quality. Kluwer, The Netherlands

Ray C (2001) Riverbank filtration: An analysis of parameters for optimal performance. *AWWA annual conference*, CD Proceedings

Schmidt CK, Lange FT, Brauch HJ, Kühn W (2003a) Experiences with riverbank filtration and infiltration in Germany, *DVGW-Water Technology Center,* TZW, Germany

Schmidt CK, Lange FT, Sacher F et al (2003b) Assessing the fate of organic micropollutants during riverbank filtration utilizing field studies and laboratory test systems. Geophys Res Abstr 5:08595

Shamrukh M (1999) Effect of chemical fertilizers on groundwater quality in Upper Egypt. PhD thesis, Department of Civil Engineering, Minia University, Egypt

Shamrukh M (2010) Seasonal and long-term changes in nitrate content of water supply wells in Upper Egypt. *The 1st IWA Malaysia Young Water Professionals Conference (IWAYWP2010),* Kuala Lumpur, Malaysia, 2–4 March 2010

Shamrukh M, Abdel-Lah A (2004) Monitoring of agricultural nutrients in water supply wells, Upper Egypt. J Eng Appl Sci Cairo Univ 51(6):1135–1150

Shamrukh M, Abdel-Wahab A (2008) Riverbank filtration for sustainable water supply: Application to a large-scale facility on the Nile River. Clean Technol Environ Policy 10(4):351–358

Shepherd RG (1989) Correlations of permeability and grain size. Groundwater 27(5):633–638

Smith E (2009) Faculty Profile. *AUC Bulletin,* American University in Cairo, December 2009

Soltan ME (1998) Characterization, classification, and evaluation of some ground water samples in Upper Egypt. Chemosphere 37(4):735–745

Sontheimer H (1980) Experiences with riverbank filtration along the Rhine River. J Am Water Works Assoc 72:386–390

Stuyfzand PJ (1998) Fate of pollutants during artificial recharge and bank filtration in the Netherlands. *Artificial Recharge of Groundwater* (J.H. Peters, editor), Water Research Association, Medmenham, England, pp 119–125

Tufenkji N, Ryan JN, Elimelech M (2002) The promise of bank filtration. Environ Sci Technol 36(21):422A–428A

UNDP (2005) Egypt Human Development report, choosing our future: Towards a new social contract. United Nations Development Programme

UNDP (2008) Egypt Human Development report, Egypt's social contract: The role of civil society. United Nations Development Programme

UNICEF (2005) Water for life: Making it happen. WHO/UNICEF Joint Monitoring Programme for Water Supply and Sanitation, United Nations Children's Fund, New York

Wahab RA, Badawy MI (2004) Water quality assessment of the River Nile system: An overview. Biomed Environ Sci 17:87–100

Weiss WJ, Bouwer EJ, Aboytes R et al (2005) Riverbank filtration for control of microorganisms: Results from field monitoring. Water Res 39(10):1990–2001

Weiss WJ, Bouwer EJ, Ball WP et al (2003) Study of water quality improvements during riverbank filtration at three Midwestern United States drinking water utilities. Geophys Res Abstr 5:04297

Wett B, Jarosch H, Ingerle K (2002) Flood induced infiltration affecting a bank filtrate well at the River Enns, Austria. J Hydrol 266(3–4):222–234

WHO (1996) Guidelines for Drinking Water Quality, 2nd edition, World Health Organization, Geneva

World Bank (2005) Arab Republic of Egypt: Country environmental analysis 1992–2002. The World Bank, Washington DC

Zaghloul ZM Abdallah AM (1985) Underground water pollution in the Nile Delta area. Egypt J Geol 29(1–2):1–10

Zakaria AM, El-Sharouny HM, Ali WS (2007) Microcystin concentrations in the Nile River sediments and removal of microcystin-LR by sediments during batch experiments. Arch Environ Contam Toxicol 2(4):489–495

Chapter 3   A Combined RBF and ASR System for Providing Drinking Water in Water Scarce Areas

Laxman Sharma[1] and Chittaranjan Ray[2]*

[1] Department of Civil & Environmental Engineering, University of Hawaii at Manoa, Honolulu, Hawaii 96822, USA. E-mail: laxman@hawaii.edu

[2] Professor, Department of Civil & Environmental Engineering; Researcher, Water Resources Research Center; University of Hawaii at Manoa, Honolulu, Hawaii 96822, USA

Abstract   A novel combination of riverbank filtration (RBF) and aquifer storage recovery (ASR) in the Albany region of Georgia (USA) was investigated in order to study possible changes in water quality. In areas where there are seasonal changes in water availability, seasonal excesses can be stored underground to meet short-term demands. Using RBF as a source water, rather than obtaining water directly from the surface water, would reduce treatment costs. The RBF site taps the Flint River through the Upper Floridan Aquifer producing water that can be injected into the deeper Clayton Aquifer for storage and subsequent recovery. This study tests the conceptual framework of having such RBF and ASR schemes coupled together and, more importantly, looks at the hydrogeochemical changes that are likely to occur. It was seen, in the scenarios considered, through numerical modeling, that acceptable water can be obtained from such coupled systems. Injection of the RBF water in an aquifer with arseniferous pyrite did not mobilize any significant arsenics.

Keywords: Riverbank filtration, aquifer storage recovery, drinking water, water scarcity, PHT3D, redox modeling

1. Introduction

Spatial and temporal variations in availability of fresh water resources necessitate diversion and storage of fresh water to meet demand. The increasing and competing

---

* Chittaranjan Ray, Professor, Department of Civil & Environmental Engineering; Researcher, Water Resources Research Center; University of Hawaii at Manoa, Honolulu, Hawaii 96822, USA, e-mail: cray@hawaii.edu

C. Ray and M. Shamrukh (eds.), *Riverbank Filtration for Water Security in Desert Countries*, DOI 10.1007/978-94-007-0026-0_3, © Springer Science+Business Media B.V. 2011

demands of the residential, industrial, agricultural and environmental sectors make water a limiting or a scarce resource, which often constrains the development or economic growth of a region. The scarcity of water can exist not only in arid zones, but also in other climatic zones where there exist significant seasonal variations in water availability and use, and the present resources are already (over)allocated. A unique combination of riverbank filtration (RBF) and aquifer storage and recovery (ASR) is presented herein, where RBF is used to obtain a relatively better quality water to inject into a deeper aquifer for temporary storage, and then recover it to be used in times of water stress.

RBF (Kuehn and Mueller 2000; Ray et al. 2002; Hubbs 2006) is an efficient, yet low-cost water treatment technology for drinking water production. RBF wells for public water supply have been widely used in Europe for more than a century (Schubert 2002) and more recently in the United States (Ray et al. 2002). While it has been shown that the quality improvement of the filtrate is significant compared to the source water (Stuyfzand et al. 2004; Stuyfzand 1998; Tufenkji et al. 2002; Wang et al. 2002), very few studies have examined the feasibility of RBF for water banking or aquifer storage and recovering it later.

For example, in desert areas, rain occurs for a small duration of the year and the rivers run dry the rest of the year. Surface storage is not feasible due to excessive evaporation. Increased evaporation limits the suitability of surface impoundments in these dry areas. In monsoon-driven climates, such as in India or Nepal, the river flows diminish greatly in post-monsoon seasons. River flows can be so low that storage impoundments must be constructed in order to allow water to be pumped to treatment plants. Even in wetter areas of the United States, high consumptive uses such as irrigation and municipal demands severely limit environmental flows. For example, in the southeastern United States (particularly in South Georgia), several rivers cut into the large Floridan Aquifer – a supraregional limestone aquifer in the area. There occurs significant pumping for irrigation during these low flow periods causing the groundwater table to drop so low that it even causes depletion of flows in the rivers. This adversely affects the ecology of the downstream reaches. In an effort to maintain environmental flows, the state offers cash incentives to farmers to stop using the groundwater for irrigation so that the downstream impacts on the river can be minimized (Wilson 2007).

In Santa Rosa, California, the Sonoma County Water Agency (SCWA) pumps its water from the Russian River where an inflatable dam is raised in summer months to impound the water so that its six collector wells can pump adequate amount of water to meet summer demand (Su et al. 2007). However, the operation of this dam and a reduction of flow affect threatened species of fish (salmon) and other biota. One of the potential solutions for South Georgia or the SCWA is to bank the water in aquifers during periods of high river flows and to release the stored water from the aquifer during times of high demand. This may enable SCWA to meet regulations regarding the minimum flow in the river. For Georgia, it will not only enable the state to meet the minimum flow requirements in the lower reaches of the rivers, it will also allow economic expansion for cities that are located in southern parts of the state thus attracting industry.

ASR is increasingly becoming a popular technique to augment drinking water supplies as well as to enhance the recharge of aquifers (Pyne 2005). ASR is considered as a useful water management option in areas of water scarcity or where the seasonal demands fluctuate widely. The purpose of ASR is to store water in a suitable aquifer during times when water is available, and recover water from the same aquifer when it is needed (Dillon 2005; Pyne 2005). A large volume of water can be stored underground in suitable aquifers, reducing or eliminating the need to construct surface reservoirs and minimizing evaporation losses, saving resources (Khan et al. 2008) and without most of the undesirable environmental consequences associated with large surface reservoirs.

Using RBF water over surface water as a source for ASR has several other advantages too. RBF water is generally better than directly tapping surface water of polluted rivers. In general, it reduces turbidity, biodegradable compounds, bacteria, viruses, parasites, persistent organic contaminants and heavy metals, as well as attenuate shock loads to yield water of a relatively consistent quality as it is forced through the river bed and the bank.

There are several quality concerns when RBF and ASR systems are run in series. The quality of the filtrate must comply with local regulations before it can be injected into an aquifer. This is true for any source water for injection into potable aquifers. This paper looks at the possibility of a novel combination of RBF and ASR to address water shortage issues with reference to the Albany region of Georgia. The Georgia State Water Plan (EPD 2008) outlines surface water storage, interbasin transfer, and ASR as three main water supply management practices to address water scarcity issues. RBF is sought to be used as source water for an ASR system which injects water into the deeper aquifer during times of excess flows in the Flint River and extract it during the drier seasons. Numerical modeling is done to investigate the possibility of such a scheme and model the possible water quality changes occurring therein.

Review of Past Relevant Work

The surface water, as it flows through the river bed and the porous media, is subjected to a combination of physical, chemical, and biological processes such as filtration, dilution, sorption, and biodegradation that can significantly alter the filtrate water quality by the time it reaches the production wells (Stuyfzand 1998; Kuehn and Mueller 2000; Tufenkji et al. 2002). The passage of water through aquifers can introduce a number of water quality changes, such as attenuation or removal of organic carbon, microbes, pesticides, nitrate, and other contaminants or even leaching of minerals (Stuyfzand, this volume; Hiscock and Grischek 2002; Tufenkji et al. 2002; Ray 2004; Kuehn and Mueller 2000; Doussan et al. 1997; Miettinen et al. 1994; Massman et al. 2008; Petrunic et al. 2005).

There are currently more than 300 ASR systems in operation (NRC 2007). The injected water undergoes a complex set of geochemical reactions before it is recovered. The recovered water quality can change substantially during the cyclical processes of injection, storage within the aquifer, and subsequent withdrawal. Some of these changes are beneficial whereas others are adverse. The reactions may even alter the hydraulic properties (permeability and porosity) of the aquifer due to mineral precipitation/dissolution (Meyer 1999) and/or by biological clogging (Rinck-Pfeiffer et al. 2000). Injecting oxygen-rich potable water and nutrient-rich reclaimed water into an anaerobic aquifer leads to a variety of water quality changes (Greskowiak et al. 2005; Vanderzalm et al. 2006). This process leads to the production or consumption of protons and other reactants which in turn triggers the precipitation and dissolution of minerals, ion exchange, and surface complexation reactions (Eckert and Appelo 2002). If dissolved organic carbon (DOC) is present in the injected water, microbial reactions will degrade the carbon and deplete oxygen in the process, making the aquifer more reducing. This result can affect processes such as denitrification and other redox reactions and lead to higher concentrations of redox sensitive species such as iron, manganese, arsenic (As), and SO_4^{2-} in the recovered water.

In field settings, quality changes during the injection of high quality water have been observed by several authors. Stuyfzand (1998) gives an important overview of quality changes of injection water based on 11 deep well recharge experiments in The Netherlands; and the dissolution of calcite, dolomite and amorphous silica was reported by Mirecki et al. (1998). Effects of deep well recharge of oxic water into an anoxic pyrite-bearing aquifer was investigated and modeled by Saaltnik et al. (2003). However, only a limited number of studies report water quality changes during the injection of reclaimed water. Australia and some western states in the U.S., most notably, such as California and Arizona employ ASR systems that use recycled waters exclusively. Greskowiak et al. (2005) simulated carbon cycling and biogeochemical changes during the operation of an ASR system at Bolivar, South Australia. Their models predict that dynamic changes in bacterial population during the storage phase can affect the local geochemistry around the injection/extraction wells. Farther away from the injection wells, breakthrough of cations was strongly affected by exchange reactions. Calcite dissolution substantially increased calcium concentrations in the recovered water.

Seasonal variations in redox reactions have also been looked at in a few cases in the context of ASR. Temperature-dependent pyrite oxidation in a deep (1,000 ft) anaerobic aquifer in the Netherlands was simulated by Prommer and Stuyfzand (2005). Greskowiak et al. (2006) examined the variability in the degradation of pharmaceutical phenazone (present in the wastewater of Berlin, Germany) as a function of season. They found that the degradation was redox sensitive and breakthrough of phenazone in monitoring wells occurred in warmer summer months when anaerobic conditions developed.

In terms of benefits, ASR has the potential to remove pathogens, disinfection byproducts (from chlorinated water), DOC, and pharmaceutical residues (NRC 2007). The NRC report also points to case studies where tri-halomethanes have been formed by injecting chlorinated drinking water to aquifers that have some amount of DOC. Recently, arsenic dissolution from the Floridan Aquifer has been identified as a major problem inherent to ASR application there (Arthur et al. 2002, 2005). At the Punta Gorda ASR site in Florida, these authors showed that arsenic concentrations significantly increased during the recovery periods, and exceeded the current EPA standards for presence of arsenic in water. Similar mobilization of arsenic has also been observed at the Peace River ASR site in Florida. The limestone matrix in the Floridan Aquifer contains small amounts of arsenic, mostly associated with arseniferous pyrite along with other trace metals (Price and Pichler 2005). Under normal conditions, arsenic is in equilibrium with the native ground water. However, during ASR operation, especially when the system experiences iron-reducing conditions and the iron oxides dissolve, the adsorbed arsenic is released (Smedley and Kinniburg 2002). Jones and Pichler (2007) recently showed in the lab that the arsenic was immobile in ambient deep anoxic groundwater conditions, but became mobile as recharge water increased the redox potential of groundwater. EPA has lowered the maximum allowable limit of arsenic in drinking water supply/sources from 50 to 10 μg/L. Thereby about 13 ASR systems in Florida are in violation of the drinking water standards (NRC 2007) and are reportedly banned from operating. This underscores the importance of screening tools, such as numerical modeling, that can yield critical information a priori when it is backed by proper site investigations.

2. Methods and Procedures

The study was carried out by collating information, analyzing data and literature from various sources, estimating parameters, building conceptual models, carrying out numerical modeling using the parameters for different scenarios.

For this study, we examined various areas of Georgia for ASR feasibility. The coastal zones were excluded as the current regulations do not allow for ASR systems in the immediate future (until the prohibition for ASR expires). Areas north of the "fall line" that runs northeast from Columbus, GA to Augusta, GA were eliminated because of geological constraints. Most recently, the Flint and Chattahoochee basins experienced drought conditions that reduced stream flows significantly. In order to preserve base flow of the Flint River, the Georgia Environmental Protection Division (EPD) asked the farmers to stop withdrawing groundwater from this upper aquifer for irrigation in return for a cash rebate (Wilson 2007).

Cities that are located in the area also face limitations of growth because of restrictions of additional withdrawal from surface or ground water sources. To mitigate this, a possible solution was to consider withdrawing water from the Flint or the Chattahoochee rivers during periods of high flows and then treating this water and storing it in a deeper aquifer for future use. The needed treatment before injection could be regular treatment of surface water in a conventional water treatment plant, natural filtration such as occurring during RBF without additional treatment, or a combination of both. Such systems would allow the cities to capture excess runoff at a period of no or minimal restrictions and store that water in deep aquifers for later use.

An area in the vicinity of the City of Albany, GA, USA was selected for the study. The Water Planning and Policy Center (http://www.h20policycenter.org) had previously developed a conceptual stage feasibility assessment of an ASR system for the extraction and injection scenario (Water Resource Solutions 2006). It identified an area east/north-east of the City of Albany for a RBF and ASR scheme. It suggested extracting water from the Flint River and treating it before storing in a deeper aquifer so that the industry could pump it out at a sustained rate of 10 million gallons per day (MGD). Water for injection was to be obtained by directly pumping surface water or obtaining it from RBF schemes, the latter being the more attractive alternative. It was expected that the RBF scheme would have overall lower treatment costs, assuming that heavy metal concentrations do not increase substantially during subsurface passage. The proposed area is shown in Figure 3.1.

This area lies on the Floridan Aquifer system, having multiple aquifers and confining layers, and the geo-hydrology is well documented (Hicks et al. 1981). A generalized stratigraphy and the water-bearing properties of formations underlying the study area are shown in Figure 3.2. The aquifers included are (i) the Upper Floridan Aquifer of the Ocala Limestone formation, and (ii) the Clayton Aquifer of the Midway Group, which is a deeper aquifer. Water from the shallower aquifer will be extracted when the surface water in the Flint River is in excess of demands. This extracted water will be used to recharge the lower aquifer. This in turn can be pumped to augment water supply during dry summer months.

For the simplified case of this study, water is extracted from the RBF part of the system for the 6 months of October to March and recharged into the deeper aquifer, and subsequently withdrawn from this storage for the remaining 6 months of April to September when the water demand is higher. The water from storage could be directly used for municipal or industrial uses, or released into the river which could augment the environmental flows and recharge the depleting aquifer.

Study of the available materials and maps indicate that suitable areas for the project would be located adjacent to the river, north of the City of Albany. A linear parcel along the river would facilitate placing about ten RBF wells, each of about 1 MGD capacity, at a lateral spacing of 200 ft. This could draw water from Upper Floridan Aquifer (also called the Ocala aquifer; Figure 3.2) which is hydraulically connected to the Flint River.

A preliminary 3-D numerical model using MODFLOW (Harbaugh et al. 2000) was set up to explore the flow conditions in the aquifers and the information was used to prepare computationally more efficient 2-D models using MODFLOW. Separate models were set up for obtaining water from the river (named the RBF model) and for injection, storage and retrieval of the pumped water (named ASR model). Geochemical transport of different species in the models with reactions was carried out using the PHT3D (Prommer et al. 2003) which couples the transport simulator MT3DMS (Zheng and Wang 1999) in MODFLOW with the geochemical model PHREEQC-2 (Parkhurst and Appelo 1999).

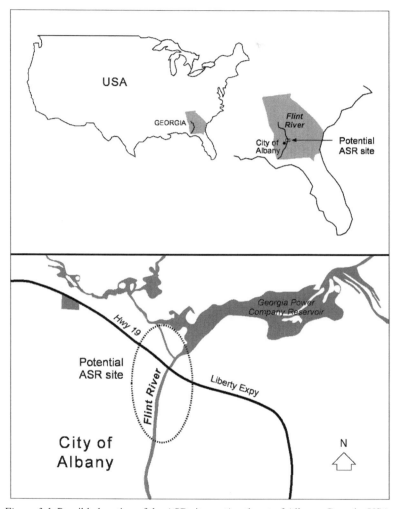

Figure 3.1. Possible location of the ASR site east/northeast of Albany, Georgia, USA.

The river water quality and stage data is obtained for the USGS Station ID 02352560 (Flint River at Albany) from the United States Geological Survey (USGS) surface water data inventory. The groundwater quality data were determined from the nearest wells, Well ID 12K129 in the Upper Floridan Aquifer and Well ID 12L020 for the Clayton aquifer. Water quality and hydrology data of the other adjacent wells (12M002, 13L002 and others) were also looked into to determine the regional flow and the required water chemistry.

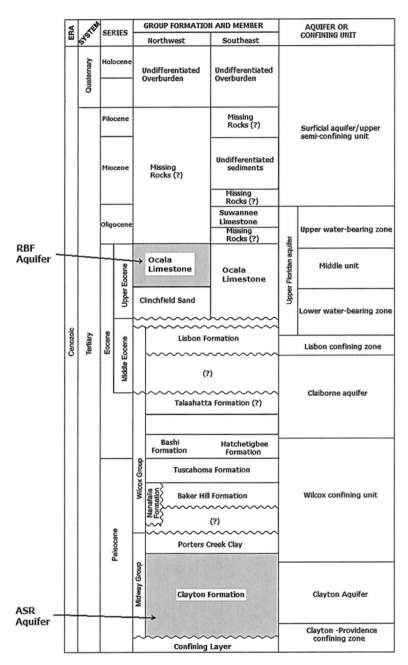

Source: http://ga.water.usgs.gov/projects/albany/stratcol.html

Figure 3.2. Aquifers in the Albany area.

The 2-D RBF model grid is set up 520 ft long and 160 ft deep with a unit thickness as shown in the upper part of Figure 3.3. The grid is divided into a total 14 layers and refined at the river portion so that the river hydrograph can be properly represented and the boundary head cells do not go dry at any time during simulations.

The pumping well is placed 120 ft away from the river bank. The river stage and aquifer data were obtained from USGS sites for surface water and ground-water data. The model is subjected to a time variant hydraulic boundary of the river defined by varying water stages, and with general head boundaries on either side defined with a hydraulic head of 154 ft at a distance of 6,000 ft from the boundary corresponding to regional groundwater levels. For simulating aqueous species composition, the river boundary is defined with the observed river water chemistry; and the left and right boundaries are set as constant concentration boundaries – the aqueous chemistry being defined with that of the ambient aquifer water quality.

The ASR aquifer, 125 ft thick, is located between the depths of 625 ft and 750 ft below the ground surface level and is described by a grid as shown in the lower part of Figure 3.3. The confined aquifer is differentiated into five layers. It is depicted by a 2-D symmetrical model, set up with the well at the left end and a general head boundary at the right side. The left boundary cells are set as point sources for injecting and extracting water (i.e., a well with reversible pumps). The length of the model domain is selected in an iterative manner so that the right boundary chemistry remained relatively unchanged from ambient conditions. The vertical grids are closer spaced near the well and placed further apart towards the right boundary as more rapid reactions are assumed to take place in the vicinity of the wells.

Detailed site specific multi-species time series data was not available and only a few parameters such as river stage, temperature and dissolved oxygen in the river water and others for the groundwater were found to be recorded intermittently; and only one set of complete aquatic chemistry data could be constructed for other species. These data are used to emulate annual time series data. Errors in this data set was minimized by carrying out a charge balance of the ions using PHREEQC-2 (Parkhurst and Appelo 1999) adjusting the small errors to chloride concentrations. The aqueous chemistry data sets for the Flint River, Upper Floridan Aquifer (the RBF aquifer) and the Lower Floridan Aquifer (the ASR aquifer) are given in Table 3.1 and are used for the initial and boundary conditions in constraining the numerical model.

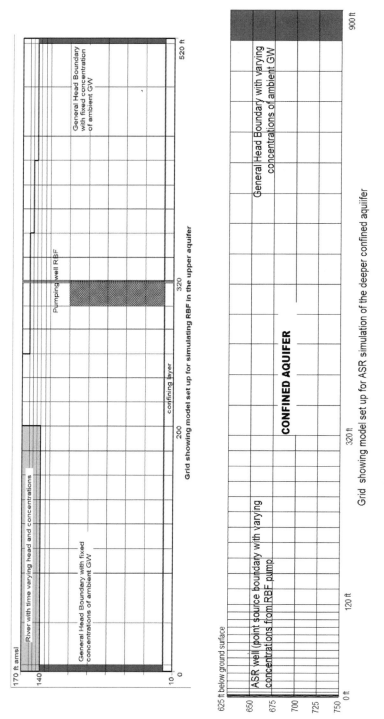

Figure 3.3. Two dimensional grid models for depicting RBF (top) and ASR (bottom).

A COMBINED RBF AND ASR SYSTEM FOR PROVIDING DRINKING WATER

TABLE 3.1. Aqueous components as model inputs from charge balanced data set.

Component	Flint River (mol/L)	Upper Floridan Aquifer (mol/L)	Lower Floridan Aquifer (mol/L)
DOC	0.00035	0	0
Temperature	9.8–30.6	20.4	23
Dissolved oxygen	4.4e-4 to 6.9e-4	5.44e-04	0
Ca(+2)	5.24e-04	9.68e-04	2.99e-04
Mg(+2)	5.76e-05	2.17e-05	2.47e-04
Na(+)	2.94e-04	9.18e-05	1.44e-03
K(+)	4.60e-05	8.19e-06	7.16e-05
Fe(+3)	1.38e-06	5.37e-08	3.22e-06
Mn(+3)	7.81e-07	1.82e-08	9.10e-08
Si	1.09e-04	1.33e-04	3.33e-04
Cl(−)	7.48e-04	1.44e-04	1.23e-03
C(+4)	5.98e-04	1.76e-03	1.34e-03
S(+6)	9.68e-05	1.87e-06	1.35e-04
N(+5)	2.71e-05	2.04e-04	2.86e-06
N(+3)	2.14e-07	7.14e-08	0
Ammonium (+)	1.61e-06	0	0
pH	7.3	7.9	7.2

Units: mol/L except temperature (°C) and *pH* is dimensionless.

The extraction rate per well is 1 MGPD (million gallons per day). These wells are placed linearly along the river bank every 200 ft so a linear approximation is made to represent the system with a 2 D vertical model. The extraction rate per unit width is 5,000 gallons per day (GPD). A simplified scenario of extraction from the RBF well for 6 months, starting October and lasting until March, is used to approximate the season when there would be excess flow in the river. The pump is turned off for the rest of the year.

The average hydraulic conductivity is assumed to be 450 ft/day in the upper aquifer, and 66 ft/day in the storage aquifer estimated from regional transmissivity values (Hicks et al. 1981; Miller 1986). The ASR wells are placed 350 ft apart with reversible pumps operating at 2,900 GPD. Equal injection and extraction rates are assumed for the ASR for a simple analysis. The model set up utilizes symmetry of the system simulating one half of the system imposing half the extraction or injection rates in the model. The actual pumping schedule is shown in Table 3.2. All simulations are run with a 50 day spin up period so as to remove any biases from arbitrarily chosen starting conditions. The ambient aquifer head in the upper layers is taken as 154 ft below the ground surface. It is important to note that the hydraulic performance of the wells and the aquifer are assumed adequate in terms of aquifer storage, recovery and hydraulic heads on the aquifer and wellheads.

TABLE 3.2. Pumping schedule.

Start time (days)	End time (days)	RBF well (GPD)	ASR well (GPD)	Remarks
0	50	−5,000	2,900	Start up period
50	230	−5,000	2,900	To storage
230	410	0	−2,900	To external supply
410	590	−5,000	2,900	To storage
590	770	0	−2,900	To external supply

The MODFLOW/MT3DMS-based multi-component reactive transport model PHT3D was used. Both the upper aquifer and the lower aquifers are carbonate aquifers, so the model was first set up to run equilibrium and dissolution reactions with oxidation of the organic carbon and other species using PHT3D with sorption described by linear isotherms and the reaction module defined by the PhreeqC-2 database (Parkhurst and Appelo 1999).

The DOC in river water as it infiltrates into the aquifer is initially degraded by the dissolved oxygen in presence of microbes and electron acceptors. Redox sensitive species such as O_2, NO_3^-, Mn^{4+}, Fe^{3+}, SO_4^{2-} are present in the aquatic environment and these also sequentially act as electron receptors for degradation of dissolved organic matter.

Key reactions involved in RBF and ASR

Mineralization of organic matter: $CH_2O + O_2 \rightarrow CO_2 + H_2O$
Denitrification : $\quad CH_2O + 0.8NO_3 \rightarrow 0.4N_2 + 0.4H_2O + HCO_3^- + 0.2H^+$
Iron (3) reduction: $\quad FeOOH + CH_2O \rightarrow Fe^{2+} + HCO_3^-$
Sulfate reduction: $\quad SO_4^{2-} + 2CH_2O \rightarrow H_2S + HCO_3^-$
Pyrite oxidation: $\quad FeS_2 + 3.75O_2 + 3.5H_2O \rightarrow Fe(OH)_3 + 2SO_4^{2-} + 4H^+$

The acidity generated by the dissolution of carbon dioxide in water or the excess protons are commonly expected to bring about calcite dissolution, i.e.,

$$CO_2 + H_2O + CaCO_3 \rightarrow 2HCO_3^- + Ca^{2+}$$

The kinetics of the degradation is also dependent upon the aqueous temperature (Prommer and Stuyfzand 2005; Greskowiak 2006) for oxidation of pyrites and organic matter respectively. Modeling temperature of the influx water and the aquifer is also carried out considering temperature as a dummy species that undergoes sorption following a linear isotherm, as water travels through the aquifer. The reaction networks are described in detail in the PHT3D reaction module PHREEQC-2 database and follow on the lines of (Prommer and Stuyfzand 2005; Greskowiak et al. 2006). It is often seen that a small amount of organic carbon does persist even when oxidizing agents are present. This is because the organic carbon may also consist of a more persistent form. So a basic persistent level of 2E-05 mol/L of organic carbon is modeled to remain in the aquifer.

The RBF and ASR models were first set to run for equilibrium and dissolution reactions with oxidation of organic carbon in a calcite aquifer, typical of the area considered. A second set of simulations was done injecting the water obtained from the carbonate RBF system into an ASR aquifer with arsenopyrites to check for mobility of arsenic. The pyritic aquifer was assumed to contain 0.1% arsenopyrite, and the simulations considered sole oxidation of pyrites by omitting the presence of the DOC in the filtrate water. If DOC was considered, it would be competing with pyrite for oxidation, and reducing the possibility of release minerals. As it is, the water obtained from the RBF would be low on degradable DOC, as most of it is likely to be consumed in the RBF process. The geochemical modeling of this latter scenario is done in a manner similar to that used by Prommer and Stuyfzand (2005) wherein pyrite oxidation is modeled to be depending upon the concentrations of oxygen, nitrate, proton concentrations, mineral surface area and the concentration of pyrites.

3. Results and Discussion

3.1. The RBF Subsystem

The simulation results for investigation into organic carbon redox reactions, calcite dissolution and the general equilibrium reactions show that the organic carbon degradation is the first predominant reaction that takes place. The model allows for the presence of some dissolved organic carbon as refractory and some as readily degradable (Greskowiak et al. 2005). The simulation result of concentrations of different species in the RBF well is shown in Figure 3.4. It should be noted that the pump is stopped during the periods 230–410 days and 590–770 days.

The model showed that in the RBF well, consumption of DO is rapid as it enters the aquifer. The dissolved oxygen introduced with the infiltrating water is consumed in overcoming the initial anoxic situation in the aquifer as well as in oxidizing the DOC coming in through the river bed. The river water contains 3.5E-04 mol/L of DOC which is reduced to less than 0.7E-04 mol/L at the first phase of injection. There are seen initial increases in DOC and the SO_4^{2-} (denoted by S(+6) in the plot) during the early start up time in the plots shown. The small increase in DOC in the plot initially is due to the reason that no DOC was considered present in the aquifer at the start, so its concentration increases to the minimum level assumed to be present in the aquifer. The SO_4^{2-} and Fe(+3) also increase initially as they are also present in higher concentrations in the river than in the aquifer.

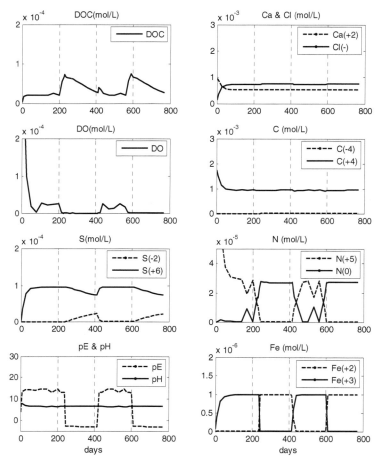

Figure 3.4. Concentration time series data at the RBF pumping well. The pump is turned off during the periods 230–410 and 590–770 days.

At around 200 days, aerobic degradation diminishes and denitrification takes place rapidly as shown by consumption all of the N(+5) and increase N(0) concentrations. It is seen that denitrification is not able to degrade the entire DOC coming in with the river water as the DOC concentration picks up slightly. At the end of the pumping period, at 230 days, N(+5) is exhausted and rapid iron reduction also takes place converting all the Fe(+3) present into Fe(+2). A strongly reducing situation then develops corresponding to higher electron activity, indicated by low *pE*. The degradation of remaining DOC is then gradually degraded to the minimum level due to sulfate reduction, wherein S(+6) in SO_4^{2-} is converted to S(−2). Some SO_4^{2-} is still present till the start of the next cycle showing that sulfate reduction is incomplete and no methane generation takes place.

In the next cycle of pumping from 410 to 590 days, the infiltrating water from the river supplies more DO and DOC so that aerobic reactions start taking place. The pE reverts to its higher value. The concentrations of N(+5) and SO_4^{2-} increase with corresponding decreases in N(0) and S(−2) respectively. At the end of the pumping period, denitrification, iron reduction also take place followed by some sulfate reduction when the pump is stopped. This cycle repeats itself subsequently.

The concentrations of the aqueous species in the pumped RBF filtrate water is shown in Table 3.3 along with that of the Flint River and the final ASR output. These values are within acceptable limits for injection into the ASR aquifer.

TABLE 3.3. Maximum concentrations of various aqueous species in RBF and ASR water.

Aqueous component	Flint River (mol/L)	Upper Floridan Aquifer (mol/L)	RBF filtrate, source for ASR max values (mol/L)	ASR abstraction water max values (mol/L)
DOC	0.00035	0	7.54E-05	6.25E-05
Temperature	9.8–30.6	20.4	20.6	14.2
Dissolved oxygen	4.4E-4 to 6.9E-4	5.44E-04	5.45E-04	5.66E-06
Ca(+2)	5.24E-04	5.65E-04	9.65E-04	5.28E-04
Mg(+2)	5.76E-05	2.17E-05	5.80E-05	8.68E-05
Na(+)	2.94E-04	9.18E-05	2.94E-04	6.66E-04
K(+)	4.60E-05	8.19E-06	4.60E-05	5.40E-05
Fe(+3)	1.38E-06	5.37E-08	9.99E-07	1.63E-06
Fe(+2)	–	–	9.91E-07	9.94E-07
Mn(+3)	7.81E-07	1.82E-08	9.99E-07	9.87E-07
Mn(+2)	–	–	2.38E-08	2.29E-18
Si	1.09E-04	1.33E-04	1.33E-04	1.82E-04
Cl(−)	7.48E-04	1.44E-04	7.47E-04	8.97E-04
C(+4)	5.98E-04	1.76E-03	1.75E-03	9.66E-04
C(−4)	–	–	6.24E-08	2.48E-04
S(+6)	9.68E-05	1.87E-06	9.62E-05	9.61E-05
S(−2)	–	–	2.30E-05	1.07E-04
N(+5)	2.71E-05	2.04E-04	2.03E-04	2.94E-05
N(+3)	2.14E-07	7.14E-08	7.09E-08	–
N(0)	–	–	2.73E-05	2.94E-05
Ammonium(+)	1.61E-06	0	2.00E-06	1.98E-06
pH	7.3	7.9	7.96	7.4

Units: mol/L except temperature (°C) and pH is dimensionless.

3.2. The ASR Subsystem

The RBF filtrate is of adequate quality to be allowed for injecting into the deeper aquifer. The injection periods into the ASR aquifer are 0–230 days and 410–590 days; and the extraction periods are 230–410 days and 590–770 days. The time series plot of the concentrations of various species in the ASR well is shown in Figure 3.5.

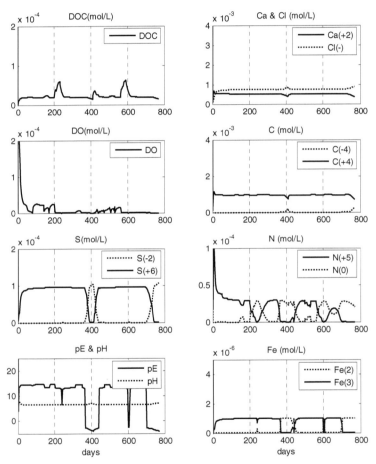

Figure 3.5. Concentration time series data at the ASR well. The pump is injecting RBF water into the deeper aquifer in the periods 0–230 days and 410–590 days, and extracting at other times.

Anoxic conditions are fully developed in the aquifer; some DO seen at the start of the simulations is during the start up simulations. A closer look at the plot shows that complete denitrification occurs at around 230, 410, and 590 days at the

A COMBINED RBF AND ASR SYSTEM FOR PROVIDING DRINKING WATER 45

ASR well, corresponding to times when the aquifer becomes strongly reducing with low pE values. Further degradation of organic carbon is caused by partial iron reduction at 230 and 590 days, where as full iron reduction takes place around 410 days and after 750 days. The S(6) and S(-2) plots also show that reduction of SO_4^{2-} occurs at around 400 days, and 750 days when the pE is inverted to reducing conditions. All the sulfate is consumed briefly at the end of the extraction periods and extreme reducing conditions exist at those short intervals, with production of 2.48e-04 mol/L of C(-4) for a short time. This situation is immediately corrected when RBF water is injected into the ASR well.

One important aspect in limestone aquifers is that calcite dissolution can be triggered by repeated flow reversals and by lowered pH. Degradation of organic carbon is associated with an increase in acidity, which should, to some extent, promote carbonate dissolution. It was observed in the ASR model simulation (results not shown here) that the calcium concentration was slightly raised near the wells and the zone of this elevated concentration would expand outwards as the water was pumped in and decrease when it was extracted. This shows that there is some flux of calcium ions in the near vicinity of the wells and it could help increase porosity. It is understood that this mechanism can help offset the well clogging phenomenon that can occur due to precipitates near the well or deposition of other organic complexes. A more detailed qualitative analysis was not done here but there is considerable literature available on it (Mirecki et al. 1998; Greskowiak 2005; Rinck-Pfieffer et al. 2000; Pyne 2005).

3.3. Pyrite Oxidation and Arsenic Mobility

Results from the set of runs modeling pyrite oxidation with kinetics are presented in Figure 3.6. The simulations assuming the absence of DOC did not yield any significant presence of arsenic in the extracted water.

The figure shows that the aquifer remains anoxic, because the RBF water that is injected has already reduced levels of oxygen, due to oxidation of DOC in the RBF set up. So any remaining oxygen is rapidly used up in pyrite oxidation. The model shows that the rapid consumption of NO_3^- and SO_4^{2-} during the model run up period also occurs and it subsequently attains equilibrium at very low concentrations during the normal operation of the ASR system. It is seen that the pH averages around the near neutral range. The concentrations of As(3) and SO_4^{2-} start rising once the injection stops (230 days) and stop when it starts again (410 days). The peak concentration of As(3) and As(5) observed in the ASR well are 5.27E-11 mol/L and 9.12E-11 mol/L respectively, which are well below the EPA permissible limits in potable water. During pyrite dissolution, ferrous iron released from the pyrite is oxidized to hydrous ferric oxides and this precipitates out. Sorption of arsenic onto these neoformed oxides are known to occur and these further attenuate the arsenic concentration in water. If these conditions are prevalent then any appreciable mobilization of As will probably not take place. Mobilization could be higher if more electron acceptors are available. This would occur if more oxic waters were to be injected into the aquifer with pyrites.

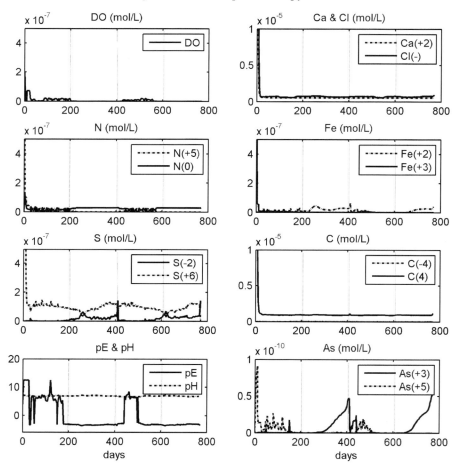

Figure 3.6. Concentration time series plot at the ASR well for the second case when pyrite oxidation is considered. The pump is injecting RBF water into the deeper aquifer in the periods 0–230 days and 410–590 days, and extracting at other times.

4. Conclusions and Future Research Recommendations

The study shows that a coupled RBF and ASR scheme is possible and would help alleviate the water scarcity problems in the drier seasons if there is an excess of water in the wetter seasons and suitable geo-hydrologic conditions exist. It was determined this would be a viable option to store and pump water. It was observed through numerical modeling that the degradation of DOC reduces all of the Fe(+3) and nitrates, and some SO_4^{2-}. In our simplified case, there appeared no significant mobilization of arsenic and the ASR water was found to be of acceptable quality. Hydraulic aspects such as possible storage volume, recovery efficiency and rates of injection, permissible aquifer head build-ups and fluctuations of it also need to be carefully looked at.

Transport and fate of arsenic is an immediate concern for the field of RBF and ASR schemes. A better elucidation backed by benchmark studies is essential to restore confidence in these schemes. Mobilization of arsenic and other heavy trace elements in different scenarios could be modeled more accurately if it is backed by accurate in-situ geochemical data.

The modeling did not look at the aspects of transport and fate of micro-organisms and other chemicals of concern such as disinfection by-products (DBPs) and pharmaceuticals and personal care products which are pressing issues that need to be looked at in the immediate future. The extent of dissolution and potential clogging of the wells also need to be carefully examined to estimate the ease of injection and efficiency of aquifer recovery.

One approach that is deemed worthy of further consideration for bank filtration of surface water prior to ASR recharge, is to apply a well designed RBF system that is capable of strongly reducing pathogenic microbiota, providing natural filtration, and also reducing the Eh of the recharge water so that it is less likely to dissolve metals from the limestone during ASR storage. Recovered water from this type of bank-filtration ASR system, must comply with EPA standards before being distributed for drinking water purposes. It is expected that the water treatment costs would not only be minimized, but it would also provide water during drought seasons.

Acknowledgments The authors would like to thank USGS for their support for this study and the two anonymous reviewers for their help in refining the paper.

References

Arthur JD, Dabous AA, Cowart JB (2002) Mobilization of arsenic and other trace elements during aquifer storage and recovery, Southwest Florida. US Geological Survey Artificial Recharge Workshop Proceedings, Sacramento, CA, pp 2–4

Arthur JD, Dabous AA, Cowart, JB (2005) Water-rock geochemical consideration for aquifer storage and recovery: Florida case studies. In: Tsang CF, Apps JA (eds) Underground Injection Science and Technology, Developments in Water Science, Elsevier. Amsterdam, vol 52, pp 327–339

Dillon P (2005) Future management of aquifer recharge. Hydrogeol J 13(1):313–316

Doussan C, Poitevin G, Ledoux E, Detay M (1997) River bank filtration: Modeling of the changes in water chemistry with emphasis on nitrogen species. J Contam Hydrol 25:129

Eckert P, Appelo CAJ (2002) Hydrogeochemical modeling of enhanced benzene, toluene, ethylbenzene, xylene (BTEX) remediation with nitrate. Water Resour Res 38(8):1130

EPD (2008) Georgia comprehensive state-wide water management plan. Georgia Water Council, GA, USA

Greskowiak J, Prommer H, Massman G, Nutzmann G (2006) Modeling seasonal redox and dynamics and the corresponding fate of the pharmaceutical residue phenazone during artificial recharge of groundwater. Environ Sci Technol 40(21):6615–6621

Greskowiak J, Prommer H, Vanderzalm J et al (2005) Modeling of carbon cycling and biogeochemical changes during injection and recovery of reclaimed water at Bolivar, South Australia. Water Resour Res 41, W10418, doi:10.1029/2005WR004095

Harbaugh AW, Banta ER, Hill MC, McDonald MG (2000) MODFLOW-2000, the US Geological Survey modular ground-water model – User guide to modularization concepts and the Ground-Water Flow Process. US Geological Survey

Hicks DW, Krause RE, Clarke JS (1981) Geohydrology of the Albany Area, Georgia. Prepared in cooperation with the US Geological Survey and the Albany Water, Gas, and Light Commission, Georgia Geologic Survey, Environmental Protection Division, Department of Natural Resources, Atlanta, GA

Hiscock KM, Grischek T (2002) Attenuation of groundwater pollution by bank filtration. J Hydrol 266:139

Hubbs SA (ed) (2006) Riverbank filtration hydrology; impacts on system capacity and water quality. Proc. NATO Advanced Research Workshop on Riverbank Filtration Hydrology, Bratislava, 7–10 Sept. 2004, Springer NATO Science Series IV, Earth and Environ Sciences 60, 344 p

Jones GW, Pichler T (2007) Relationship between pyrite stability and arsenic mobility during aquifer storage and recovery in Southwest Central Florida. Environ Sci Technol 41:723–730

Khan S, Mushtaq S, Hanjra MA, Schaeffer J (2008) Estimating potential costs and gains from an aquifer storage and recovery program in Australia. Agr Water Manage 95(4):477–488

Kuehn W, Muller U (2000) Riverbank filtration. J Am Water Works Assoc 92(12):60–69

Massmann G, Dünnbier U, Heberer T, Taute T (2008) Behaviour and redox sensitivity of pharmaceutical residues during bank filtration – Investigation of residues of phenazone-type analgesics. Chemosphere 71(8):1476–1485

Meyer KU (1999) A numerical model for multicomponent reactive transport in variable saturated porous media. University of Waterloo, Waterloo, Canada

Miettinen IT, Martikainen PJ, Vartiainen T (1994) Humus transformation at the bank filtration water-plant. Water Sci Technol 30(10):179–187

Miller JA (1986) Hydrogeologic Framework of the Floridan Aquifer System in Florida and in parts of Georgia, Alabama, and South Carolina. US Geological Survey Professional Paper 1403-B

Mirecki JE, Campbell BG, Conlon KJ, Petkewich MD (1998) Solute changes during aquifer storager and recovery testing in a limestone/clastic aquifer. Ground Water 36(3):394–403

NRC (2007) Prospects of managed underground storage of recoverable water. Committee on Sustainable Underground Storage of Recoverable Water, Water Science and Technology Board, National Research Council, The National Academies Press, Washington, DC, 256 pp

Parkhurst DL, Appelo CAJ (1999) User's guide to PHREEQC (version 2) – A computer program for speciation, batch-reaction, one-dimensional transport, and inverse geochemical calculations. US Geol Survey, Water-Resour Inv Rep 99-4259, 312 p

Petrunic BM, MacQuarrie KTB, et al (2005) Reductive dissolution of Mn oxides in river-recharged aquifers: A laboratory column study. J Hydrol 301(1–4):163–181

Price RE, Pichler T (2002) Oxidation of Framboidal Pyrite as a Mobilization Mechanism During Aquifer Storage and Recovery in the Upper Floridan Aquifer, Southwest Florida EOS. Trans, Am Geophys Union, 83, 47

Price RE, Pichler T (2005) Abundance and mineralogic association of arsenic in the Suwannee Limestone (Florida): Implications for arsenic release during water-rock interaction. Chem Geol 228:44–56

A COMBINED RBF AND ASR SYSTEM FOR PROVIDING DRINKING WATER 49

Prommer H, Stuyfzand PJ (2005) Identification of temperature-dependent water quality changes during a deep well injection experiment in a pyritic aquifer. Environ Sci Technol 39(7):2200–2209

Prommer H, Barry DA, Zheng C (2003) MODFLOW/MT3DMS-based reactive multicomponent transport modeling. Ground Water 41(2):247–267

Pyne D (2005) Aquifer storage recovery: A guide to groundwater recharge through wells ASR Systems. CRC Press, FL, USA

Ray C, Grischek T, Schubert J et al (2002) A perspective of riverbank filtration. J Am Water Works Assoc 94(4):149

Ray C (2004) Modeling RBF efficacy for migrating chemical shock loads. J Am Water Works Assoc 96(5):114–128

Rinck-Pfeiffer S, Ragusa S, Sztajnbok P, Vandevelde T (2000) Interrelationships between biological, chemical, and physical processes as an analog to clogging in aquifer storage and recovery (ASR) wells. Water Res 34(7):2110–2118

Saaltink MW, Ayora C, Stuyfzand PJ, Timmer H (2003) Analysis of a deep well recharge experiment by calibrating a reactive transport model with field data. J Contam Hydrol 65(0169–7722)

Schubert J (2002) Hydraulic aspects of riverbank filtration—Field studies. J Hydrol 266(3-4):145–161

Smedley PL, Kinniburg GD (2002) A review of the source behavior, and distribution of arsenic in natural waters. Appl Geochem 17:517–668

Stuyfzand PJ (1989) Hydrology and water quality aspects of Rhine bank ground water in The Netherlands. J Hydrol 106:341–363

Stuyfzand PJ (1998) Quality changes upon injection into anoxic aquifers in the Netherlands: Evaluation of 11 experiments. In: Peters JH (ed) Artificial Recharge of Groundwater. Rotterdam, The Netherlands: A A Balkema

Stuyfzand PJ (2010) Hydrogeochemical processes during river bank filtration and artificial recharge of polluted surface waters: zonation, identification and quantification. This volume

Su GW, Jasperse J, Seymour D et al. (2007) Analysis of pumping-induced unsaturated regions beneath a perennial river. Water Resour Res 43, W08421, doi:10.1029/2006WR005389

Tufenkji N, Ryan JN and Elimelech M (2002) The promise of bank filtration. Environ Sci Technol 36:21:423A

Vanderzalm JL, Le Gal La Salle C, Dillon PJ (2006) Fate of organic matter during aquifer storage and recovery (ASR) of reclaimed water in a carbonate aquifer. Appl Geochem 21(7):1204–1215

Wang JZ, Hubbs SA, Song R (2002) Evaluation of riverbank filtration as a drinking water treatment process. American Water Works Association Research Foundation, Report No. 90922

Water Resource Solutions (2006) Flint River Basin near Albany, Georgia aquifer storage and recovery (ASR) feasibility assessment, Water Policy Working Paper #2006-005, The Georgia Water Planning and Policy Center, Albany, GA

Wilson D (2007) The feasibility of using aquifer storage and recovery to manage water supplies in Georgia. In: Rasmussen T (ed) Proceedings of the 2007 Georgia Water Resources Conference University of Georgia, Athens, GA, USA

Zheng C, Wang PP (1999) MT3DMS: A modular three-dimensional multispecies transport model for simulation of advection, dispersion, and chemical reactions of contaminants in groundwater systems, documentation and user's guide. Contract Report SERDP-99-1, US Army Engineer Research and Development Center, Vicksburg, MS

Chapter 4 Behavior of Dissolved Organic Carbon During Bank Filtration Under Extreme Climate Conditions

Dagmar Schoenheinz and Thomas Grischek*

University of Applied Sciences Dresden, Faculty of Civil Engineering & Architecture, Friedrich-List-Platz 1, 01069 Dresdan, Germany. E-mail: schoenheinz@htw-dresden.de, grischek@htw-dresden.de

Abstract Based on laboratory experiments concerning the influence of temperature on dissolved organic carbon (DOC) transport processes during soil passage, the impact of possible hydrological changes on the efficiency of bank filtration and artificial recharge are discussed. Possible climate change as discussed in established climate scenarios and its effect on bank filtration as cost-efficient sustainable technique for drinking water supply is evaluated. In particular, the increased biological activity at higher temperatures in the aquifer improves the cleaning efficiency of the soil passage, and indeed the soil passage itself acts as a buffer. For the case of an increased occurrence of extreme events such as floods and low-flow periods, application strategies for safeguarding the drinking water supply are presented.

Keywords: Water temperature, DOC removal efficiency, changing boundary conditions

1. Introduction

The increased stress on water resources due to the growth in world population, shifting water demand due to changes in life style and changes in water availability due to climate change represent significant challenges for water supply. For more than a century, a significant portion of the drinking water supply in Europe has depended on the abstraction of bank filtrate and artificially recharged groundwater. The use of these techniques is one of the most promising approaches to integrated water resources management in facing the upcoming threat to water supply. However, the continuously successful application of both bank filtration

---

*Thomas Grischek, University of Applied Sciences Dresden, Faculty of Civil Engineering & Architecture, Friedrich-List-Platz 1, 01069 Dresdan, Germany, e-mail: grischek@htw-dresden.de

C. Ray and M. Shamrukh (eds.), *Riverbank Filtration for Water Security in Desert Countries*, DOI 10.1007/978-94-007-0026-0_4, © Springer Science+Business Media B.V. 2011

and artificial groundwater recharge requires an integrated evaluation of sustainability under more extreme hydrological boundary conditions as are expected due to changing climate conditions.

Increased hydrological variability might trigger problems for water resources management and water supply particularly in densely populated areas. Both quantitative effects (e.g., in terms of flood management or less snowmelt runoff and water allocation in low flow periods) and qualitative effects (e.g., regarding increased sediment transport in the running waters, higher concentrations of harmful substances or reduced dilution potential against wastewater discharge in low flow periods) are expected (Kundzewicz et al. 2007, Meyer et al. 1999).

As a result of the discussions about the climatic change and hence possible variation in hydrological boundary conditions, the question of how to ensure water supply in the required quantity and quality is of great concern for water suppliers. In Germany, about 13% of the raw water demand is met through bank filtration and artificial groundwater recharge (DVGW 2008). In particular, cities such as Dresden, Berlin, Düsseldorf, and Bochum are supplied by 20–100% bank filtrate from the Elbe, Spree, Rhine, and Ruhr rivers, respectively. The quantity and quality of bank filtrate are significantly affected by hydrological conditions and hydro-chemical properties of the river water.

2. Climate Projections and Their Relevance to the Abstraction of Bank Filtrate

As a consequence of climate change, changes in the moisture and radiation balances as well as in the average, minimum and maximum temperatures and precipitation amounts are expected. There are still significant uncertainties concerning the global climate projections as well as their regional transformation (Grünewald 2008). Nevertheless, increased global air and water temperatures, greater precipitation intensity, and longer periods of low flows as well as intense rainfall with higher flood risks seem to be predicted with high confidence (Kundzewicz et al. 2007, Merkel et al. 2007). In Germany, the total annual precipitation is expected to remain more or less constant, although the winter months might receive slightly more and summer months slightly less precipitation (Hagemann and Jacob 2007). Though more precipitation during winter will not affect the typical conditions for bank filtration. However, more frequent occurrence and longer duration of low flow periods during summer combined with a higher probability of heavy rain events and floods will be of significant impact. According to Kabat et al. (2002), a paradigm shift is required in water management strategies towards living with floods and droughts. Changes in extreme hydrological events as observed in the Elbe River catchment during the summer flood of 2002 and the subsequent extreme low flow period in 2003, might become more rapid and occur within 1 year's time due to the accelerated hydrological cycle caused by climate change. Thereby, the moderate hydrological conditions in Middle Europe might move towards more semi-arid conditions typical of warmer regions (Kundzewicz et al. 2007). At a large scale, there is evidence of a broadly coherent pattern of change in annual river runoff. Water resources in many semi-arid areas are expected to experience a decrease and drought-affected areas are expected to increase in extent (Kundzewicz et al. 2007). In regions suffering from droughts, a greater incidence of diarrhoeal and other water-related diseases will reflect the deterioration in water quality (Patz 2001). Also, due to sea-level rises, groundwater salinisation will very likely increase. Reduced lowering of the water table due to abstraction of bank filtrate lowers the risk of salt water rise, in contrast to groundwater abstraction alone.

3. Temperature Patterns of Water Resources

Temperature variations in surface waters follow with some delay the fluctuations of air. Depending on the degree of communication with surface water, groundwater temperatures more or less reflect the annual mean air temperature. Non-geothermal surface water temperatures vary from 0°C to 36°C (Table 4.1), and mean values in Germany are between 10°C and 13°C.

TABLE 4.1. Temperature ranges of various surface waters worldwide.

Water	Min	Max	Mean	Reference
Danjiangkou Reservoir, China	8.0	30.6	19.0	Li et al. (2009)
Han River, China	12.2	35.7	21.3	Li and Zhang (2009)
Ganga River, India	13	30	–	Sarin et al. (1989)
Ohio River, Louisville, USA	3.6	29.7	–	Partinoudi et al. (2003)
Missouri, Bolton, USA	1.0	29.0	15.0	Caldwell (2006)
Loire River, France	1.8	28.5	14.2	Grosbois et al. (2000)
Girnock River, Scotland	–	–	7.0	Langan et al. (2001)
Tarn River, UK	0.8	18.6	–	Hannah et al. (2009)
Spring water at Tarn River, UK (groundwater discharge)	3.3	13.8	–	Hannah et al. (2009)
Calabra River ~ groundwater, Southeast Nigeria	–	–	28	Edet and Worden (2009)
Elbe River, Germany	-0.5	12.8	11.6	own data
Rhine River, Karlsruhe, Germany	–	–	12.4	DKR (2000)
Bodensee, Germany	–	–	10.4	KHR (2008)

By passage through bank soils an equilibration of water temperature as well as concentrations of water compounds due to dispersion occurs. As a function of the flow paths, the temperature will assume the background temperature of the ambient groundwater. This is demonstrated by the example of an observation cross-section at a bank filtration site in Torgau, on the Elbe River (Figure 4.1).

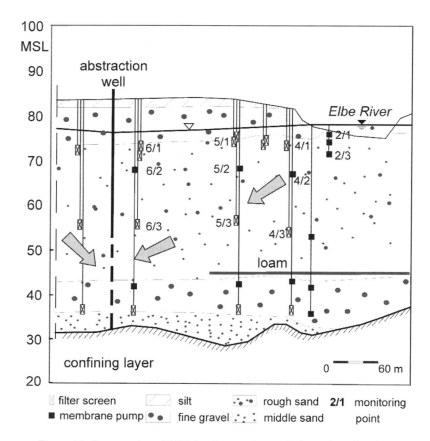

Figure 4.1. Cross-section of RBF site, Torgau. Arrows indicate flow directions.

For the monitoring points MP 4/1, MP 5/1, and MP 6/1, the annual temperature behavior plotted with the water level fluctuations of the river is shown for the period between 1999 and 2002 (Figure 4.2). The flowpath direction is nearly constant during low, medium and high flow. The monitoring wells MP 4/1, 5/1 and 6/1 are the shallow wells in a layer having the highest hydraulic conductivity. The residence time changes according to the river stage and pumping of the production wells. All temperature probes were calibrated to a common reference.

The temperature in the river fluctuates by 22 K. In the upper observation points, this fluctuation is reduced to 13 K in a distance of about 60 m (MP 4/1), to 5 K in a distance of about 120 m (MP 5/1), and to 4 K in about 250 m distance at MP 6/1 (Table 4.2). Since the area between the river and the abstraction well is small and due to the silt layer covering the aquifer, groundwater recharge between river and abstraction well is low. Therefore, the upper observation points are representative for the upper flow path (Figure 4.2).

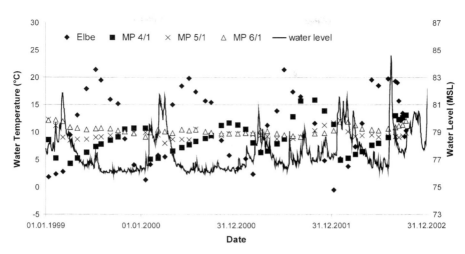

Figure 4.2. Temperature changes in river water and bank filtrate at the RBF site Torgau at the River Elbe.

The average annual air temperature calculated from monthly average values (DWD 2009) at the considered site is about 10°C, which is close to the equilibrated average temperature data at observation point MP 6/1.

TABLE 4.2. Statistical data for River Elbe water level and temperature data along the flow path.

1999–2002	Water level (MSL)	Temperature (°C)			
	River Elbe	River Elbe	MP 4/1	MP 5/1	MP 6/1
Median	76.9	11.6	8.9	9.7	10.3
Maximum	84.6	21.8	17.7	13.0	12.2
Minimum	75.7	–0.5	4.3	8.0	7.9

4. Effects of Low Flow Periods on Bank Filtrate Quality

Temperature increase, higher global radiation and evaporation combined with scarce precipitation in summer result in low flow periods. Low flow periods result in lower discharge rates combined with a greater portion of wastewater in surface water and thus a decreased dilution potential of the water body. In anthropogenically influenced surface waters, increased concentrations are observed of trace organic compounds that are not efficiently retained by wastewater treatment, such as EDTA (Grischek et al. 1997), radio-opaque substances (Claus et al. 2008) and pharmaceuticals and cosmetics (Reemtsma et al. 2006). Higher global radiation, abundant nutrient supply and higher water temperatures result in increased biomass production, which in turn causes a higher suspended matter load. This, together with low flow velocities and reduced shear stress characteristic of low flow periods, leads to more intense clogging at the bottom of the water body (Heeger 1987, Grischek 2003). The infiltration rate and consequently the bank filtration portion in the raw water will thereby be reduced. Oxygen depletion by microbial degradation of organic substances along with lower solubility of oxygen due to higher temperatures may cause a shift from aerobic to anoxic conditions in the bank filtrate, as observed at the Rhine River in the extreme summer of 2003 (Eckert et al. 2008).

5. Effects of Flood Events on River Bank Filtration

Heavy rainfall events that cause floods are usually accompanied by diffuse matter input due to the elution from temporarily flooded areas that undergo different human uses. Such inputs are (i) pesticides, insect repellents, and nutrients from fertilizers, (ii) increased bacteriological loads as consequence of elution from grazing land or from stormwater overflow out of sewer systems, and (iii) higher concentrations of dissolved organic compounds of natural and anthropogenic origin. An instance of increased dissolved organic carbon (DOC) concentrations was observed during the August 2002 flood at the Elbe River in Dresden. While on average the DOC concentration in river water is about 6 mg/L, during the flood values of up to 11 mg/L were measured (Schoenheinz 2004). Another effect of floods is the at least partial removal of the clogging layer due to higher flow velocities and higher shear stress occurring at many sites. Better hydraulic conductivity at the river bottom, and the relatively high water levels, result in higher infiltration rates and an increase of the bank filtration portion at the abstraction wells.

6. Evaluation Based on the Sum Parameter DOC

6.1. Boundary Conditions

Whether the anticipated low flow periods and flood events lead to a worsening of the bank filtrate quality has to be evaluated both site-specifically and as a function of the relevant parameters. For the DOC parameter, it will be shown by example how the projected changes in boundary conditions influence the retention capacity of the aquifer. Considered changes of boundary conditions are:

- temperature,
- changing organic loads as a consequence of changes between flood and drought conditions,
- migration velocity to reflect longer retention times during low flow conditions and shorter retention times during floods.

6.2. Materials and Methods

Bench-scale soil column experiments on the degradation of DOC were performed at three different temperatures, 5°C, 15°C, and 25°C. The stainless steel columns (length 0.5 m, diameter 0.075 m) were filled with pumice stone and operated according to the experimental set-up of either a flow-through regime (Figure 4.3a) or a circulating flow regime (Figure 4.3b).

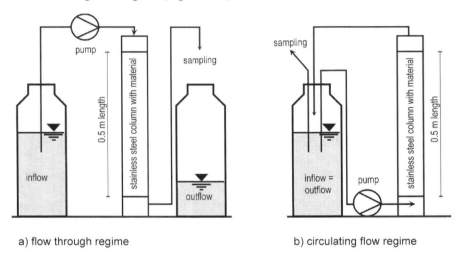

Figure 4.3. Experimental set-up.

Generally, the columns were operated for a period of 4 and 8 weeks respectively. During this period, an active biofilm was established.

Flow-through regime. Downward percolation through the column (Figure 4.3a) represents the soil passage within the first half-meter of river bank infiltration, which is the most active biological zone and experiences the strongest temperature changes as shown in Figure 4.2. The column was operated with the same inflow for cycles of 2–3 weeks. Each period in which the column was operated with the same source water and thus constant boundary conditions with respect to source concentrations and constant flow velocities was called a run. For each run, 3–6 samples of column in- and out-flow were taken. Water samples were filtered through a 0.45 µm cellulose-nitrate filter and analyzed for:

- DOC concentrations with a Carbon Analyzer (model TOC-5000, Shimadzu),
- UVA$_{254}$ (ultraviolet absorption coefficient for 254 nm) with a UV/VIS-spectrometer (model Spekol, Zeiss Jena) using a cell with a 5-cm path length,
- oxygen using a WTW probe Oxi 315i Weilheim, and
- nitrate and sulfate (spot samples only) with an ion chromatograph (model DX 100, Dionex).

Circulating flow regime. Column operation in a circulating flow regime (Figure 4.3b) enables monitoring of the DOC degradation kinetics. Therefore, the same water percolated repeatedly through the column and the inflow containers served temporarily as outflow container. By dropping the column effluent into the container, oxygen insertion is supported. In discrete time intervals as function of the realized contact time, samples of a defined volume (100 mL) were taken out of the container (Figure 4.3b). The contact time was calculated as a function of the real residence time in the column (Schoenheinz et al. 2006). The samples were analyzed for DOC and UVA$_{254}$. The experiments lasted 1–3 months.

As a feeding source, water of the Elbe River was used which is the source water at various bank filtration sites in Saxony, Germany. To operate at different temperatures, the columns where installed in three thermostatic cabins at 5°C, 15°C, and 25°C (Figure 4.4). To ensure reproducibility of the results, the experiments were performed at each temperature with two identically prepared columns, fed by the same source water.

Column material. Pumice stone is supposed to support the formation of biofilm only but allow no sorption of organic material. This was controlled in preliminary tests by analysing DOC concentration changes in river water inhibited by NaN$_3$ with both batch and column experiments. Figure 4.5 displays the DOC concentration results for a column experiment operated in circulating flow. Here, river water with an initial DOC concentration of 5.6 mg/L was applied. The biological activity in the river water was inhibited by 400 mg/L NaN$_3$. Likewise, an experiment with the same water but without the addition of inhibitor was applied to another column. Since under biologically active conditions for the realized contact time of 15 days, a DOC degradation of more than 40% was observed while the DOC concentration in the inhibited water remained stable for both in and outflow.

Thus, during the soil passage observed decreases in DOC concentration can be ascribed to biological degradation and sorption only in the biofilm. The characteristics of pumice stone are summarized in Table 4.3.

Changes in organic load were accomplished through the application of river water of the same origin sampled during average flow conditions as well as during flood events but also by the switch to the application of lake water with generally higher DOC concentrations compared to river water. After the incorporation

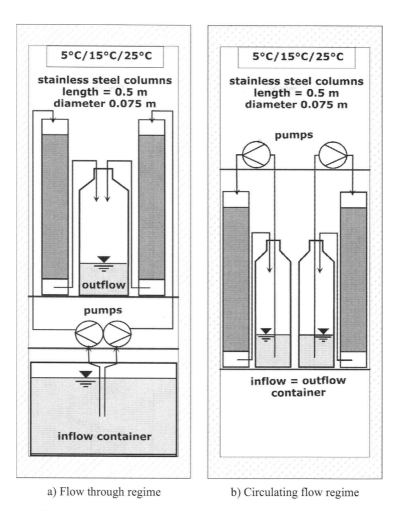

Figure 4.4. Temperature-controlled experimental set-up. The outer hatching indicates the thermostatic cabin.

period, the DOC degradation within the flow-through columns was investigated for 7 different types of water and/or changed migration velocities. For the first 5 runs, river water with a DOC concentration range of 5–11 mg/L was applied. For runs 6 and 7, lake water with another DOC matrix and naturally higher DOC concentration up to almost 12 mg/L was investigated.

For the assessment of different retention times on the biological removal efficiency, the flow velocity was changed by controlling the pumping rate. Initially, the migration velocity was set to about 0.14 m/day for runs 1–3, which is equivalent to a retention time of 3.5 days. Subsequently, the migration velocity was increased to 1.0 m/day, which is equivalent to a contact time of 12 h in the column (runs 4–7).

At the same time, the degradation behavior of organic compounds as function of contact time was investigated in a circulating column regime.

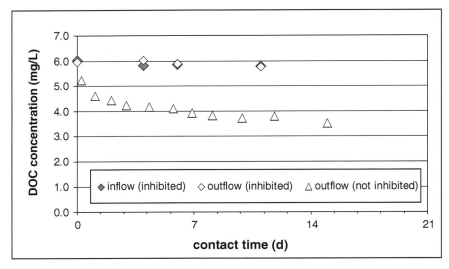

Figure 4.5. Comparison of inhibited and biologically active river water in column experiments.

TABLE 4.3. Characteristics of pumice stone as column material.

Material classification	–	volcanic rock
Organic carbon content	f_{OC} (%)	0.024
Hydraulic conductivity	k_f (m/s)	$4.3 \cdot 10^{-2}$
Uniformity	d_{60}/d_{10} (-)	1.45

6.3. Results and Discussion

Temperature. The comparison of DOC concentrations at the inflow and outflow of the flow-through columns showed the expected elevated level of biological activity and higher removal rates at higher temperatures. Figure 4.6 shows the results for 5°C and 25°C for the discussed runs. Apparently, this higher efficiency of biological degradation is valid for all investigated changes of the boundary conditions.

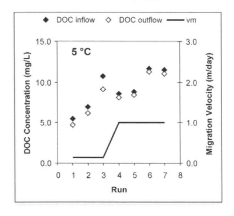 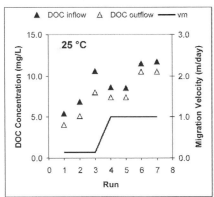

Figure 4.6. Comparison of DOC concentrations in the inflow and outflow of the flow-through columns experiencing different boundary conditions.

Changing organic loads. An increase in the DOC concentration of river water up to 10.7 mg/L as result of a flood event did not significantly change the percentage of the degraded portion during the column passage but resulted in higher DOC concentrations of the outflow (Table 4.4). In contrary, the change in DOC matrix between runs 5 and 6 due to the application of lake water instead of river water reduced the portion of DOC that is degradable within the 0.5 m flow path length in the column at temperatures of 15°C by 2% and at 25°C by 5%. Evidently, there is an adaptation time for the bacteria required to get used to the changed DOC matrix. Anthropogenic vs. natural DOC was not characterized.

TABLE 4.4. Comparison of inflow and outflow concentrations from columns operating in a flow-through regime as a function of temperature and flow velocity; runs 1–5 river water; runs 6 and 7 lake water.

Run	v_m	n	Inflow concentration (mg/L)			n	Outflow concentration (mg/L)			Residual concentration (%)		
–	m/day	–	5°C	15°C	25°C	–	5°C	15°C	25°C	5°C	15°C	25°C
1	0.14	3	5.5	5.4	5.4	6	4.7	4.3	4.1	85	80	75
2	0.14	3	7.0	7.0	6.8	6	6.2	5.6	5.1	89	81	74
3	0.14	3	10.7	10.7	10.6	4	9.1	8.6	8.0	85	80	75
4	1.0	3	8.6	8.6	8.6	6	8.1	7.8	7.4	94	91	86
5	1.0	3	8.8	8.6	8.5	6	8.4	7.9	7.4	96	91	86
6	1.0	3	11.7	11.4	11.5	6	11.3	10.6	10.5	96	93	91
7	1.0	3	11.5	11.7	11.7	10	11.1	10.8	10.5	96	92	90

r, run; v_m, migration velocity; n, number of samples for each temperature

Migration velocity. Furthermore, the influence of longer retention times as a function of migration velocity on the biological degradation can be detected. While the removal rate for river water DOC is about 13% at 5°C and 25% at 25°C for migration velocities of 0.14 m/day (Figure 4.6, runs 1–3), the removal rate decreases to 5% at 5°C and 14% at 25°C for migration velocities of 1 m/day (Figure 4.6, runs 4 and 5).

These results are also confirmed by the comparison with the degradation kinetics derived from column experiments in a circulating flow regime performed simultaneously at room temperature of approximately 20°C (Figure 4.7).

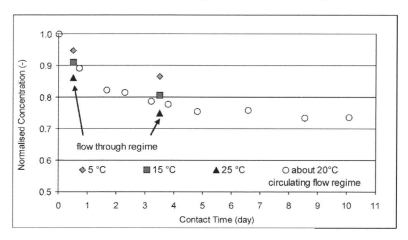

Figure 4.7. Dependency of DOC degradation on the contact time; comparison of column experiments operating in flow-through and circulating flow regimes.

Besides the known effect of increasing DOC removal, Figure 4.7 suggests that there is a final minimum DOC concentration reached after about 7–10 days that cannot be degraded further. This coincides with the widespread opinion that DOC removal at bank filtration sites is mainly limited to the first decimeters of infiltration. However, various field observations indicated that DOC concentrations are often lower compared to the results in the laboratory which suggests a further ongoing degradation along longer flow paths (Gimbel and Mälzer 1987, Mälzer et al. 1992, Drewes and Fox 1999, Sacher et al. 2000, Grischek 2003).

This assumption was investigated by a number of additional column experiments in circulating flow regime with river water, realizing contact times between 14 and 50 days. The normalized results for the DOC degradation changes within those experiments are summarized in Figure 4.8. Apparently, there is a recognizable additional degradation of 10–15% for contact times longer than 14 days.

As shown from Figures 4.7 and 4.8, the longer the contact time, the more efficient is the ongoing DOC removal of degradable compounds, as well. In the above graph, an additional degradation of about 15% was gained between the contact time of about 14 and 50 days. How far this is depending on the type of DOC and soil grain chemistry has not yet been investigated.

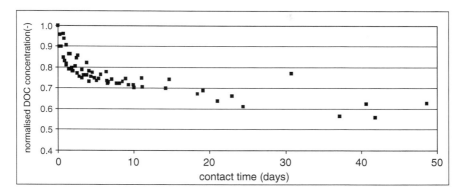

Figure 4.8. Influence of contact time on the DOC removal rate. Columns operating in circulating flow regime.

Both changes in DOC concentration within one source water type as well as changes in the source water specific DOC matrix by the application of lake water instead of river water resulted in higher DOC concentrations over the same travel distance. The lower removal efficiency is supposed to be the consequence of the disturbed biogeochemical equilibrium and can be aggravated by contemporaneous desorption of dissolved organic carbon compounds (e.g., from the clogging layer). This effect was shown for investigations with organic-rich river sediment and in combination with high DOC inflow concentrations of more than 10 mg/L, resulting in anoxic conditions in the laboratory experiments that are again accompanied by slower DOC degradation kinetics (Schoenheinz 2004).

Disturbances of biogeochemical equilibrium are spatially and temporarily limited processes. Generally, the realization of sufficiently large distances between wells and river bank and thus of longer flow paths and retention times can compensate for such disturbances both by adaptation of biological activity to changed conditions, adsorption processes and thus by the re-establishment of equilibrium flow conditions. However, for the occurrence of persistent organic compounds at higher concentrations in the source water, an increased breakthrough risk at wells must be taken into account. During flood events, the hazardous breakthrough of persistent compounds will be increased even more and has to be dealt with in the water treatment process. Similarly, it is expected that breakthroughs of pathogens are more likely during floods (Schijven et al. 2002). However, this is not yet sufficiently proven.

7. Adaptation Measures

Possible changes of boundary conditions for bank filtration sites as a consequence of the anticipated climate change are shown with their effects on bank filtrate water quality and optional adaptive measures in Table 4.5.

TABLE 4.5. Boundary condition changes during bank filtration and relevant adaptive measures.

Boundary condition	Response	Relevance for bank filtrate quality	Adaptive measures
Increased temperature in the aquifer	Increased biological activity	Increased cleaning efficiency	Not required
Lower flow velocities in the clogging zone (low flow)	Release of organics during higher temperatures	Increased concentration of organics	Sufficiently long flow paths
Anoxic conditions due to algae/organic load and low oxygen concentration (low flow)	Slower biological degradation	Reduced cleaning efficiency	Sufficiently long flow paths
High fluctuation in concentrations of inorganic and organic compounds	Disturbance of hydrochemical equilibria	Desorption processes	Sufficiently long flow paths
Increased flow velocities (flood)	Shorter retention times	Breakthrough of pathogens	Safe emergency disinfection
Increased concentrations of persistent, hardly adsorbable compounds (low flow)	No retention or degradation	Breakthrough of compounds	Activated carbon filtration as additional barrier

The most effective tool for neutralizing fluctuations in organic compound concentrations is the optimisation of contact time as a function of the flow path length and the hydraulic gradient between the source and the abstraction well. Sufficiently long contact time to compensate for concentration variations, desorption processes or slower anoxic degradation can be achieved by sufficiently long flow paths, adequate selection of abstraction well locations with the appropriate distance to the bank and site-specific optimisation of abstraction quantities.

To be capable of acting in response to the exposure of a potential breakthrough of persistent organic compounds or pathogens, an event-related monitoring of the flow paths between river and abstraction wells by groundwater observation wells is suggested. Additionally, planning and installation of emergency water treatment measures might be required.

Another hazard for safe water supply can be the temporary deficit of water with respect to both quantity and quality (e.g., due to droughts). To face this potential stress on supply, a multi-component water supply plan including different independent water resources is suggested. Examples can be found in the city of

Dresden, Germany (Fischer et al. 2006) and the city of Haridwar, India (Dash et al. 2010).

The city of Dresden makes an integrated use of surface water from reservoirs in the Ore Mountains, bank filtrate from the Elbe River and, to a low percentage, groundwater. During the Elbe River flood of 2002, the abstraction of bank filtrate was cut down due to technical problems such as well flooding and lack of emergency energy supply. Nevertheless, the increased use of reservoir water maintained a secure water supply for the whole city. The situation changed completely 1 year later during the low flow period of 2003. While the reservoirs were facing problems with algae growth due to the hot, sunny summer, the bank filtrate abstraction provided an important supply.

Technical problems during flood events as mentioned before (e.g., energy supply breakdown, flooding of well chambers) can be avoided by structural preventive measures. Examples were given by the waterworks Fernwasserversorgung Elbaue-Ostharz GmbH in 2002. There, by protecting energy supply units from flooding the abstraction wells could be operated despite flooding of the facility. The bank filtration scheme in Torgau guaranteed drinking supply in adequate quality and quantity during the flood event (Krueger and Nitzsche 2003).

Acknowledgment This study was conducted with financial support from the German Federal Ministry of Education and Research (BMBF) as well as the Saxon State Ministry of Science and Arts. Furthermore, the authors wish to thank the Dresden University of Technology and Institute for Water Chemistry for their support in the data collection.

References

Claus E, Fink G, Krämer T (2008) The water quality of River Elbe during drought periods (in German). Proc. Magdeburger Gewässerschutzseminar October 2008, Magdeburg: 81–83

Caldwell, TG (2006) Presentation of data for factors significant to yield from several riverbank filtration systems in the U.S. and Europe. In: Hubbs SA (ed) Riverbank Filtration Hydrology. NATO Sci Ser IV Earth Environ Sci 60:299–344

Dash RR, Prakash EVPB, Kumar P, Mehrotra I, Sandhu C, Grischek T (2010) River bank filtration in Haridwar, India: Removal of turbidity, organics and bacteria. Hydrogeol J 18(4):973–983. doi:10.1007/s1004001005744

DKR (2000) Rhine water quality report 2000 (in German). Deutsche Kommission zur Reinhaltung des Rheins

Drewes JE, Fox P (1999) Fate of natural organic matter (NOM) during groundwater recharge using reclaimed water. Water Sci Technol 40(9):241–248

DVGW (2008) Sector overview of German water supply and distribution. wvgw Wirtschafts- und Verlagsges. Gas und Wasser mbH, Bonn

DWD (2009) Climate data online (in German). German Weather Service. www.dwd.de. Accessed on 13 October 2009

Eckert P, Lamberts R, Wagner C (2008) The impact of climate change on drinking water supply by riverbank filtration. Water Sci Technol 8(3):319–324

Edet A, Worden RH (2009) Monitoring of the physical parameters and evaluation of the chemical composition of river and groundwater in Calabar (Southeastern Nigeria). Environ Monit Assess 157(1–4):243–258

Fischer T, Day K, Grischek T (2006) Sustainability of riverbank filtration in Dresden, Germany. UNESCO IHP-VI, Ser on Groundwater 13:23–28

Gimbel R, Mälzer H-J (1987) Testfilter experiments to evaluate drinking water relevance of organic compounds in running waters (in German). Vom Wasser 69:139–153

Grischek T (2003) Management of bank filtration sites along the Elbe River. PhD thesis, Faculty of Forestry, Geo and Hydro Sciences, Dresden University of Technology

Grischek T, Neitzel P, Andrusch T (1997) Fate of EDTA during infiltration of Elbe River water and identification of infiltrating river water in the aquifer (in German). Vom Wasser 89:261–282

Grosbois C, Negrel P, Fouillac C, Grimaud D (2000) Dissolved load of the Loire River: chemical and isotopic characterization. Chem Geol 170(3–4):179–201

Grünewald U (2008) Climate change and water resources management. In: Kleeberg H-B (ed.) Klimawandel – Was kann die Wasserwirtschaft tun? Proc Symp Klimawandel 24/25 June 2008 Nürnberg, Forum für Hydrologie und Wasserbewirtschaftung 24(08):5–18

Hagemann S, Jacob D (2007) Results of the climate model REMO for Germany and the Rhine chatchment (in German). IWW-Schriftenreihe 20. Mülheimer Wassertechnisches Seminar 22 November 2007, 46:23–40

Hannah DM, Malcolm IA, Bradley C (2009) Seasonal hyporheic temperature dynamics over riffle bedforms. Hydrol Processes 23(15):2178–2194

Heeger D (1987) Investigation of clogging in rivers (in German). PhD thesis, Faculty of Water Sciences, Dresden University of Technology, Germany

Kabat P, Schulze RE, Hellmuth ME, Veraart JE (2002) Coping with impacts of climate variability and climate change in water management: A scoping paper. DWC-Report no. DWCSSO-01 International Secretariat of the Dialogue on Water and Climate, Wageningen, Netherlands

KHR (2008) Annual report of KHR 2008. International Commission for Hydrology of the Rhine River Catchment (in German). www.chr-khr.org Accessed on 13 October 2009

Krueger M, Nitzsche I (2003) The 100-years flood of the Elbe River in 2002 and its effects on riverbank filtration sites. In: Melin G (ed) Proc 2nd Int Riverbank Filtration Conf, 16–19 September 2003, Cincinnati, USA, 81–85

Kundzewicz ZW, Mata LJ, Arnell N (2007) Freshwater resources and their management. In: Parry ML, Canziani OF, Palutikof JP (eds) Climate Change 2007: Impacts, Adaptation and Vulnerability. Contribution of Working Group II to the Fourth Assessment Report of the Intergovernmental Panel on Climate Change. Cambridge University Press, Cambridge UK

Langan SJ, Johnston UL, Donaghy MJ (2001) Variation in river water temperatures in an upland stream over a 30-years period. Sci Total Environ 265:195–207

Li S, Cheng X, Xu Z, Han H, Zhang Q (2009) Spatial and temporal patterns of the water quality in the Danjiangkou Reservoir, China. Hydrol Sci J 54(1):124–134

Li S, Zhang Q (2009) Geochemistry of the upper Han River basin, China: 2: Seasonal variations in major ion compositions and contribution of precipitation chemistry to the dissolved load. J Hazard Mat 170(2–3):605–611

Mälzer H-J, Gerlach M, Gimbel R (1992) Development of testfilters to simulate shock loads for bank filtration (in German). Vom Wasser 78:343–353

Merkel W, Leuchs W, Oldenkirchen G (2007) Challenges of global climate change for the water supply and distribution in Germany: Experience report, Handlungsfelder und Forschungsbedarf. IWW-Schriftenreihe 20. Mülheimer Wassertechnisches Seminar 22 November 2007, 46:1–16

Meyer JL, Sale MJ, Mulholland PJ, Poff NL (1999) Impacts of climate change on aquatic ecosystem functioning and health. J Am Water Works Assoc 35:1373–1386

Partinoudi V, Collins MR, Margolin AB, Brannaka LK (2003) Assessment of the microbial removal capabilities of riverbank filtration. In: Melin G (ed) Proc 2nd Int Riverbank Filtration Conf. Cincinnati,USA, 129–136

Patz JA (2001) Public health risk assessment linked to climatic and ecological change. Hum Ecol Risk Assess 7(11):1317–1327

Reemtsma T, Weiss S, Müller J (2006) Polar pollutants entry into the water cycle by municipal wastewater: A European perspective. Environ Sci Technol 40:5451–5458

Sacher F, Brauch HJ, Kühn W (2000) Fate studies of hydrophilic organic micro-pollutants in riverbank filtration. In: Jülich W, Schubert J (eds) Proc Int Riverbank Filtration Conf. IAWR:139–148

Sarin MM, Rishnaswami SK, Dilli K (1989) Major ion chemistry of the Ganga-Brahmaputra river system: Weathering processes and fluxes to the Bay of Bengal. Geochim Cosmochim Acta 53(5):997–1009

Schijven J, Berger P, Miettinen I (2002) Removal of pathogens, surrogates, indicators, and toxins using riverbank filtration. In: Ray C, Melin G, Linsky RB (eds) Riverbank Filtration – Improving Source Water Quality. Kluwer Academic Publ, Dordrecht, Boston, London, 73–116

Schoenheinz D (2004) DOC as control parameter for the evaluation and management of aquifers with anthropogenic influenced infiltration. PhD thesis, Faculty of Forestry, Geo and Hydro Sciences, Dresden University of Technology, Germany

Schoenheinz D, Börnick H, Worch E (2006) Temperature effects on organics removal during riverbank filtration. UNESCO IHP-VI, Ser on Groundwater 13:581–590

Chapter 5 Risk Assessment for Chemical Spills in the River Rhine

Paul Eckert*

Stadtwerke Düsseldorf AG, Höherweg 100, 40233 Düsseldorf, Germany.
E-mail: peckert@swd-ag.de

Abstract Drinking water production based on riverbank filtration has a long tradition along the River Rhine. The river water quality plays a crucial role in ensuring the safe supply of drinking water. Despite the successful restoration of the Rhine, the risk of future chemical spills still remains. A prediction of the contaminant breakthrough at the reach of the river used for riverbank filtration can be performed using a computer model. This knowledge enables the waterworks to prevent the infiltration of polluted river water into the aquifer using an adapted approach to well management. Nevertheless, preventing water pollution must remain the goal of the environmental authorities and local industry.

Keywords: Risk assessment, chemical spills, drinking water supply

1. Introduction

Riverbank filtration (RBF) is a well proven treatment that is part of a multi-barrier concept in drinking water supply (Grischek et al. 2002). In the Düsseldorf area, for instance, bank filtrate has been discharged from wells close to the banks of the Rhine since 1870 (Eckert and Irmscher 2006). RBF is a process whereby surface water is subjected to subsurface flow prior to extraction from vertical or horizontal wells. From a water resources perspective, RBF is characterised by an improvement in water quality, which in general is based on a combination of sorptive-filtration and biodegradation (Kühn and Müller 2000, Ray et al. 2002).

Until the middle of the last century the water quality of the Rhine was so good that the bank filtrate could be used as drinking water. Only biodegradable organic carbon, turbidity and microorganisms had to be removed or significantly reduced during bank filtration. In the 20th century, population growth and increasing industrialization along the Rhine resulted in extensive water pollution, necessitating technical water treatment in order to produce safe drinking water. Friege (2001) reports about incentives beginning in the 1970s, which led to a significant

---

* Paul Eckert, Stadtwerke Düsseldorf AG, Höherweg 100, 40233 Düsseldorf, Germany, e-mail: peckert@swd-ag.de

C. Ray and M. Shamrukh (eds.), *Riverbank Filtration for Water Security in Desert Countries*, 69
DOI 10.1007/978-94-007-0026-0_5, © Springer Science+Business Media B.V. 2011

improvement in river water quality. These incentives included the effective combination of numerous regulations, and monitoring programs. Today, while waterworks along the Rhine are no longer threatened by a polluted river, the risk of chemical spills still remains. This paper presents measures to ensure the safe supply of drinking water in case of contamination events along the Rhine.

2. Site Description

The River Rhine, which is 1,320 km long and has a catchment area of 185,000 km², is the third largest river and the largest source of drinking water in Europe. The city of Düsseldorf is situated in the North-West of Germany, in the lower Rhine valley (Figure 5.1). The mean discharge from the Rhine in the Düsseldorf area is 2,200 m³/s of which the waterworks use less than 2 m³/s. During times of flood, the discharge increases to 10,000 m³/s. The width and dynamics of the Rhine, which has its source in the Alps, allow the sustainable application of RBF for drinking water supply.

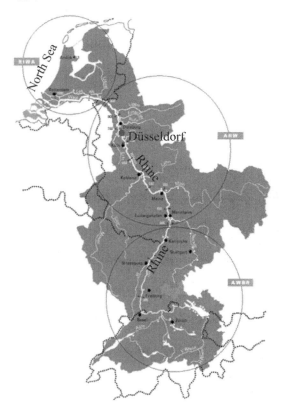

Figure 5.1. The Rhine catchment and the location of Düsseldorf.

The raw water, containing between 50 and 90% bank filtrate, is pumped from a quaternary aquifer. At present three waterworks supply 600,000 inhabitants with 50 million m^3 of treated bank filtrate annually, meeting demand of up to 200,000 m^3/day.

The vertical wells and the horizontal collector wells are situated between 50 m and 300 m from the river bank. Figure 5.2 shows a line of vertical wells at Flehe waterworks that have been in operation since 1870. The raw water is discharged using a siphon system. Depending on the hydraulic situation, the residence time of the bank filtrate in the aquifer varies between 1 week and several months (Schubert 2002a, Eckert et al. 2005). In general, the flow path and the retention time are long enough for the purification processes to work, resulting in the complete degradation of biodegradable organic carbon and the effective elimination of pathogenic bacteria, viruses and protists (Schubert 2002b).

Figure 5.2. Vertical wells at Flehe waterworks and the River Rhine.

3. Development of River Water Quality at the Rhine

Levels of contamination in the Rhine increased rapidly after World War II. In the 1950s and 1960s, sewage systems in the cities destroyed during the war were built prior to waste water purification plants. As a result, a lot of untreated sewage was discharged into the river leading to increasing pollution levels (Friege 2001). For a long time, local industries were successful in opposing pressure from the public and local government to construct wastewater treatment facilities.

In the 1950s the pollution of the Rhine reached such a high level that the RBF purification processes were no longer able to ensure good quality drinking water. The taste and odor of the bank filtrate became so bad that the waterworks were forced to develop and apply sophisticated new treatment steps. In addition to the application of technical treatment methods, the waterworks reinforced their efforts to achieve better river water quality by forming a common organization. The International Association of Waterworks in the Rhine Catchment Area (IAWR) was founded in 1970 in Düsseldorf, Germany. Its goal was to demand measures for water protection. In 1973, the IAWR published its first "Memorandum" on raw water quality that served as a "yardstick" for local government bodies and for the public debate on Rhine water quality. Together with other stakeholders, such as environmental groups, the IAWR promoted a public discussion on water protection. At the end of this process the federal government issued its first program for environmental protection, which included measures to ensure that river water quality would attain a high standard within 20 years. Target values for surface water were, for example, 8 mg/L for oxygen, 0.2 mg/L for ammonia and 0.1 µg/L for pesticides. State authorities started to control effluents thoroughly and levied a charge for certain pollution parameters. These measures forced local industries and communities to meet high purification standards in a very short time (Friege 2001). For example, in the state of North Rhine-Westphalia with 17 million inhabitants, the amount of effluents treated biologically increased from less than 22% in 1965 to 90% in 1985.

As a consequence of this collaboration between the waterworks and the government, the role of the chemical industry evolved from that of an opponent to that of a key partner, which now publishes its efforts and successes in reducing industrial effluents. Many actions were necessary to reduce nutrients and pollutants. The numerous measures taken to reduce nutrients and pollutants were consistent with the best available technology in wastewater treatment and production along the Rhine. Consequently, river water quality has improved significantly since the mid-1970s with the return of salmon to the river in 2000.

The historical development of water pollution of the Rhine can be illustrated by the concentration-time plot of oxygen (Figure 5.3). The oxygen concentration in the Rhine decreased continuously until the beginning of the 1970s. One of the many negative consequences of this decrease was the occurrence of manganese in the anaerobic well water, which increased the cost of treatment. Then, as a consequence of the restoration efforts, the oxygen concentrations returned to saturation level at the beginning of the 1990s. The higher oxidation capacity, combined with the lower oxygen demand of the infiltrating river water, led to more efficient natural attenuation processes within the aquifer. This, in turn, enabled the waterworks to reduce their treatment expenses (Eckert and Irmscher 2006). However, the occurrence of chemical pollutants in the river water, like pesticides and pharmaceuticals, remained an issue. (Verstraeten et al. 2002).

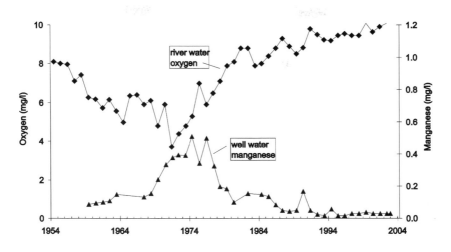

Figure 5.3. Rhine oxygen concentrations and well water manganese.

In addition to the issue of reliable water quality, periodic contamination caused by chemical spills has implications for the supply of drinking water using RBF. Ship accidents and problems with industrial wastewater treatment are the primary causes of periodic river pollution. For example, the accidental release of the insecticide Endosulfan by the chemical industry in 1969 resulted in a huge spike in the death rate of fish in the river. This was one of the most serious pollutant accidents. However, together with the aforementioned efforts for greater water protection in the 1970s the risk of chemical spills was reduced during subsequent years.

Nevertheless, in 1986, the so called SANDOZ accident caused the death of large numbers of eel in the Rhine. The pollution was the result of a warehouse fire at the chemical production or plant SANDOZ, Switzerland, where some 1,200 tons of deadly agricultural chemicals were stored. Firemen attempting to put out the blaze accidentally washed some 20 tons of highly toxic pesticides into the river, where they soon formed a 50 km-long trail that moved downstream at 3 km/h. One of the measures taken was to close the drinking water wells along a 900 km stretch of the Rhine and as far as the Netherlands. As a consequence, the International Commission for the Protection of the Rhine (ICPR) formulated recommendations for the "Prevention of accidents and security of industrial plants."

4. Risk Assessment

4.1. Rhine Alarm Model

Water protection is a high priority for waterworks in the Rhine valley. Ship accidents and problems with industrial wastewater treatment are still the primary reasons for periodic river pollution. When source water becomes extremely polluted, effective forecasting is essential for water supply companies in order to be able to take the necessary preventive measures in time. Along the Rhine, water quality data is provided by the ICPR.

If, despite all preventive measures, an accident occurs and large amounts of hazardous substances flow into the Rhine, the water suppliers depend on quick and accurate information about the chemical composition, the pollution source, and the estimated concentration and breakthrough at the reach close to the supply wells. This data is particularly needed to undertake a risk assessment of the drinking water supply.

The ICPR monitoring stations, as well as those in border countries where the Rhine flows, constantly check the chemical composition of the river water using, inter alia, biological tests. Based on this data and the information provided by the polluter, the international Warning and Alarm Plan (WAP) comes into effect providing reliable information about the accident. The WAP data enables water supply companies to forecast pollutant breakthrough at the relevant reach of the river using the Rhine alarm model (Mazijk et al. 2000).

The River Rhine Alarm Model was developed by the "International Commission for the Hydrology of the River Rhine" (CHR) and the ICPR. For this kind of predictive model, a great deal of effort and money must be spent on calibration using extensive in-situ tracer measurements (van Mazijk 2002). In the case of the River Rhine Alarm Model, which uses a two-dimensional analytical approximation for the travel time and concentration curve, a dispersion coefficient and a lag coefficient have to be calibrated. The alarm model covers the river from Lake Constance to the North Sea, including the Aar, Neckar, Main, and Moselle tributaries. The model calculations take into account the location and conditions of the initial pollution, decomposition and drift capacity of the harmful substances released, discharges and/or water levels, geometry and dispersion. The calibration was performed using tracer tests.

For calculation purposes, the following input data are required: location, time and duration of the discharge, and the amount and biodegradable fractions of the harmful substances discharged. Information on whether the discharge is a floating harmful substance (such as oil) or a substance that mixes with the river water is also required. Finally, data on the discharge of the Rhine must be entered since the discharge largely determining the speed at which the pollution is carried downstream. The model then calculates how the pollution is transported downstream.

Figure 5.4 shows a series of calculated breakthrough curves for the Rhine at Düsseldorf calculated from different locations where the pollution occurred. The calculation was performed based on a virtual pollutant of 1,000 kg over a period of 24 h during average river water level. The peak concentration and the arrival time varied depending on the location of the accident. For example, an accident in Leverkusen would result in a peak concentration of almost 5µg/L after 2 days while the same accident in Basel produces a peak concentration of just over 3 µg/L after 8 days.

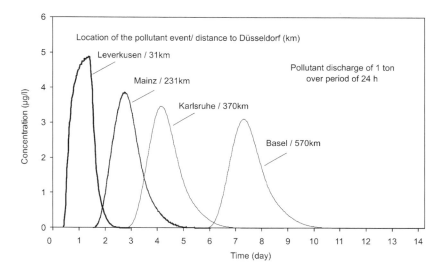

Figure 5.4. Calculated breakthrough curves at the Rhine near Düsseldorf.

4.2. Well Management

Based on the expected river water concentration an estimate is needed if the maximum thresholds for drinking water are exceeded. In some cases the amount of pollution is not very significant and the attenuation process in the river results in a low concentration, so that no action is necessary. Alternatively, if the expected concentration exceeds acceptable limits for drinking water and a significant decrease through attenuation processes within the aquifer can't be accurately predicted, it becomes necessary for the water supplier to react. Even if technical treatment steps are taken in order to ensure a decrease in the level of pollution, well protection remains the main priority.

The drinking water supplier may react to a pollution accident by an adapted well management to prevent contaminated river water from reaching the drinking water wells. The presence of an extremely toxic and concentrated substance should result in the switching off of all wells while the pollutant passes the adjacent reach of the river. Depending on the length of time the river is polluted, this may negatively impact the water supply. Since pollution breakthrough often lasts for several days (see Figure 5.4) thereby exceeding the drinking water storage capacity of approximately 12 h, a concept was developed to safeguard well operation during this period.

An appropriate and adequate approach involves switching off only half of the wells shortly before the pollutant plume arrives. As shown in Figure 5.5, the reach where the wells are switched off becomes effluent and infiltration only occurs close to the wells that are still in operation. Since the pollutant breakthrough usually only lasts hours or in some cases days, the pollutant can only infiltrate a small section of the aquifer. In the case of Düsseldorf, the infiltration velocity is about 2 m/day and the wells are more than 50 m away from the river bank.

Following the breakthrough of the pollutant plume, the wells that were in operation before must be switched off and the water supply is then drawn from the other wells (Figure 5.5). Without the drawdown the system becomes quickly effluent so that the contaminated groundwater flows back towards the river.

This approach depends on a reliable prediction of the pollutant breakthrough by the Rhine alarm model. Usually the pollutant plume is approaching so quickly that chemical sampling and analytical results are not available in time.

4.3. The Chloracetophenon Case in 2003

In 2003, contaminated wastewater containing 0.7 tons of chloracetophenone was released untreated into the river Main, a tributary of the Rhine. The treatment process of a local industry had failed, but the problem was recognized very quickly and the environmental authorities were informed within a short time of the accident. The polluter provided information about the duration of the discharge and the mass of the contaminant. This data made it possible to develop a reliable prediction of the pollution breakthrough in the Düsseldorf area (Figure 5.6).

The arrival of the pollutant plume was expected on 30 January 2001, about 60 h after the accident occurred 260 km upstream. During the breakthrough half of the production wells were switched off per the well management concept described above. After the plume had left Düsseldorf on 2 February, the wells that were switched off initially were switched back on. Simultaneously, the other wells were shut down for a 1-week period to ensure that the contaminants flowed out of the aquifer. Measured river water concentrations of chloracetophenone confirmed the model results (see squares in Figure 5.6) and showed that the emergency well management measures were applied at the correct time.

RISK ASSESSMENT FOR CHEMICAL SPILLS IN THE RIVER RHINE

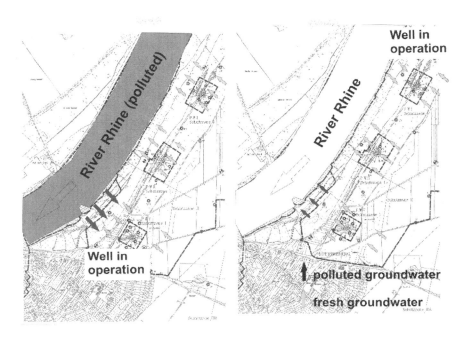

Figure 5.5. Well Management in case of a chemical spill in the River Rhine (left: period of river pollution; right: period after the pollutant breakthrough).

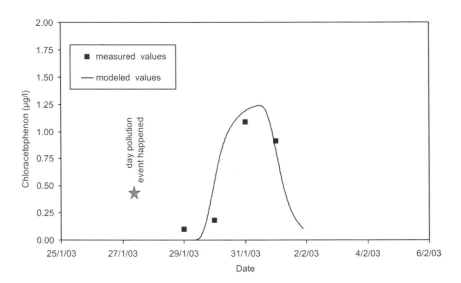

Figure 5.6. Measured and modeled chloracetophenon data.

5. Conclusions

In the Rhine valley RBF has proved a reliable method for ensuring the supply of safe drinking water for more than 130 years though sophisticated technical treatment methods were necessary to overcome the massive river pollution levels that existed in the middle of the last century. The commitment of key stakeholders to water protection was a particularly important step towards re-establishing a healthy ecosystem. Nowadays, the risk of water pollution by chemical spills remains relatively low.

In case of accidental spills, due to the intensive monitoring by the environmental authorities and the good collaboration with the industrial concerns along the Rhine, the waterworks are generally given sufficient advance warning to react. Water supply companies and water boards use the Rhine alarm model to accurately predict pollutant breakthrough and implement preventive measures. The model results enable the water suppliers to decide on appropriate steps for well management or drinking water treatment. Monitoring and modelling based on data provided by/included in the international Warning and Alarm Plan help the waterworks ensure the provision of safe drinking water even in critical situations.

References

Eckert P, Rohns HP, Irmscher R (2005) Dynamic processes during bank filtration and their impact on raw water quality, In: Luck F et al. (ed) Recharge systems for protecting and enhancing groundwater resources, UNESCO-IHP-VI, Ser on Groundwater No. (Proc. Int. Conf. Aquifer Recharge, held 10–16 June 2005, Berlin, Germany), 17–22

Eckert P, Irmscher R (2006) Over 130 years of experience with Riverbank Filtration in Düsseldorf, Germany. J Water Supply Res T 55:283–291

Friege H (2001) Incentives for the improvement of the quality of river water. In: Proc. of the International Riverbank Filtration Conference, Düsseldorf, Germany, 2–4 November 2000, IAWR – Rheinthemen 4, Amsterdam, 13–29

Grischek T, Schoenheinz D, Worch E, Hiscock KM (2002) Bank filtration in Europe – An overview of aquifer conditions and hydraulic controls. In: Management of aquifer recharge for sustainability (Dillon, P. ed.), Swets&Zeitlinger, Balkema, Lisse, 485–488

Kühn W, Müller U (2000) Riverbank filtration. J Am Water Works Assoc 92(12):60–69

Mazijk A van, Gils JAG van, Weitbrecht V (2000) Analyse und Evaluierung der 2-D-Module zur Berechnung des Stofftransportes in der Window-Version des RHEINALARMMODELLS in Theorie und Praxis, CHR-Report no. II-16. http://www.chr-khr.org

Mazijk A van (2002). Modelling the effects of groyne fields on the transport of dissolved matter within the rhine alarm-model. J Hydrol 264:213–229

Ray C, Melin G, Linsky RB (eds) (2002) Riverbank filtration: Improving source water quality. Kluwer Academic Publishers, Dordrecht

Schubert J (2002a) Hydraulic aspects of riverbank filtration–field studies. J Hydrol 266:145–161

Schubert J (2002b) Water-Quality improvements with riverbank filtration at Düsseldorf Waterworks in Germany. In: Ray C, Melin G, Linsky RB (eds) Riverbank filtration: Improving source water quality. Kluwer Academic Publishers, Dordrecht, 267–277

Verstraeten I, Heberer T, Scheytt T (2002) Occurrence, characteristics, transport and fate of pesticides, pharmaceuticals, industrial products, and personal care products at riverbank filtration sites. In: Ray C, Melin G, Linsky RB (eds) Riverbank filtration: Improving source water quality. Kluwer Academic Publishers, Dordrecht, 175–227

Chapter 6 Fluorescent Microspheres as Surrogates in Evaluating the Efficacy of Riverbank Filtration for Removing *Cryptosporidium parvum* Oocysts and Other Pathogens

Ronald Harvey[1]*, David Metge[1], Rodney Sheets[2], and Jay Jasperse[3]

[1] U.S. Geological Survey, 3215 Marine Street, Suite E-127, Boulder, CO 80305 USA.
E-mail: rwharvey@usgs.gov, dwmetge@usgs.gov

[2] U.S. Geological Survey Eastern Region Science Office, 6480 Doubletree Avenue,
Columbus, OH 43229 USA. E-mail: rasheets@usgs.gov

[3] Sonoma County Water Agency, 404 Aviation Boulevard, Santa Rosa, CA 95403 USA.
E-mail: jay.jasperse@scwa.ca.gov

Abstract A major benefit of riverbank filtration (RBF) is that it provides a relatively effective means for pathogen removal. There is a need to conduct more injection-and-recovery transport studies at operating RBF sites in order to properly assess the combined effects of the site heterogeneities and ambient physicochemical conditions, which are difficult to replicate in the lab. For field transport studies involving pathogens, there is considerable interest in using fluorescent carboxylated microspheres (FCM) as surrogates, because they are chemically inert, negatively charged, easy to detect, available in a wide variety of sizes, and have been found to be nonhazardous in tracer applications. Although there have been a number of in-situ studies comparing the subsurface transport behaviors of FCM to those of bacteria and viruses, much less is known about their suitability for investigations of protozoa. Oocysts of the intestinal protozoan pathogen *Cryptosporidium* spp. are of particular concern for many RBF operations because of their ubiquity and persistence in rivers and high resistance to chlorine disinfection. Although microspheres often have proven to be less-than-ideal analogs for capturing the abiotic transport behavior of viruses and bacteria, there is encouraging recent evidence regarding use of FCM as surrogates for *C. parvum* oocysts. This chapter discusses the potential of fluorescent microspheres as safe and easy-to-detect surrogates for evaluating the efficacy of RBF operations for removing pathogens, particularly *Cryptosporidium*, from source waters at different points along the flow path.

Keywords: Riverbank filtration, bank filtration, fluorescent microspheres, pathogens, *Cryptosporidium*, oocysts,

---

* Ronald Harvey, U.S. Geological Survey, 3215 Marine Street, Suite E-127, Boulder, CO 80305 USA,
e-mail: rwharvey@usgs.gov, dwmetge@usgs.gov

C. Ray and M. Shamrukh (eds.), *Riverbank Filtration for Water Security in Desert Countries*,
DOI 10.1007/978-94-007-0026-0_6, © Springer Science+Business Media B.V. 2011

1. Introduction

Satisfying the increasing demand for safe drinking water is one of the major environmental/health challenges that humans face. Surface waters, particularly rivers in close proximity to population centers, are typically contaminated with pathogens from treated and un-treated domestic wastewater discharges, as well as runoff that might originate from animal feeding operations and agricultural fields where manure has been applied as fertilizers. Increasingly, river-bank filtration (RBF), an old technology first utilized two centuries ago by the Glasgow Waterworks Company (UK), is being utilized as a cost-effective means of providing pre-treatment and, in some cases, complete treatment of surface water (Ray et al. 2002). One of the major benefits of RBF is it provides a relatively effective means for removal of pathogens as contaminated river water is drawn into wells completed in aquifer materials adjacent to the river (Partinoudi and Collins 2007). In some cases, the microbial quality of water extracted from municipal wells at bank filtration sites exceeds that of water filtration plants (Dash et al. 2008, Gollnitz et al. 2003) and, in at least one case, even that of the local aquifer (Shamrukh and Abdel-Wahab 2008). Also, the passage of river water through the subsurface during RBF has been shown to compensate for peak and shock loads (Malzer et al. 2002). However, the interplay of site-specific parameters, e.g., retention time, pore water velocity, river and groundwater chemistries, and characteristics of the aquifer sediments make pathogen removal difficult to predict based upon colloid filtration theory (Tufenkji et al. 2002, Gupta et al. 2009). Consequently, the efficacy of each RBF site for removing pathogens must be evaluated separately (Kuehn and Mueller 2000) and, in the United States, regulators are reluctant to assign RBF "treatment credits" without clear evidence showing removal of indicators and pathogens.

Although much has been learned about the transport of viruses, bacteria, and protists through geologic media from column studies (Harvey et al. 2007), it has been shown that lab-scale studies can underestimate considerably, sometimes by many orders of magnitude, transport potential of pathogens in aquifers at scales relevant to RBF (Harvey et al. 2008). Also, it is difficult to accurately replicate in the lab, the various physicochemical conditions and physical heterogeneities that come into play when surface water is drawn through an aquifer to water-supply wells. Consequently, there is a need to conduct more injection-and-recovery transport studies at operating RBF sites in order to properly assess the combined effects of the site heterogeneities and ambient physicochemical conditions. For field transport studies involving pathogens, there is considerable interest in using fluorescent carboxylated microspheres (FCM) as surrogates, because they are chemically inert, negatively charged, easy to detect, available in a wide variety of sizes, and have been found to be nonhazardous in tracer applications (Behrens at al. 2001).

Of particular concern for many RBF operations are oocysts of the intestinal protozoan pathogen *Cryptosporidium* spp. because of their ubiquity in many surface water environments (LeChevallier at al. 1991), persistence in aquatic environments for long periods of time (Pokorny et al. 2002), low infective dose rate (Okhuysen at al. 1999), and high resistance to chlorine disinfection (Carpenter et al. 1999). Failures of granular media filtration to remove *Cryptosporidium* from source water during water treatment have resulted in serious outbreaks of cryptosporidiosis; the most publicized being the massive 1993 outbreak in Milwaukee that resulted in over 400,000 infections (Mac Kenzie et al. 1994). In contrast, definitive data are lacking that link Cryptosporidiosis to RBF operations. However, the potential for oocyst transport from rivers to RBF water-supply wells remains a concern, in part because outbreaks of cryptosporidiosis from consumption of improperly treated, contaminated groundwater are known to occur (e.g., Willocks et al. 1998). Because of the episodic nature of infectious doses and low infective does rate characterizing *Cryptosporidium* contamination events, early detection can be challenging. The use of microbial surrogates is thought to be important in evaluating the efficacy of RBF sites for removing *Cryptosporidium* (Tufenkji et al. 2002). However, the lack of correlation between the microbial surrogates and pathogenic protists in RBF-produced water (e.g., Gollnitz et al. 2003) suggest that reliable microbial surrogates have yet to be identified.

Fluorescent microspheres may be particularly useful surrogates for use in in-situ transport studies designed to assess the degree of removal of oocysts and other pathogens at RBF sites because they are available in sizes and buoyant densities that are similar to those of oocysts and because their aspect ratios compare favorably to the near-spherical oocysts (Harvey et al. 2008). Microspheres have been used as safe surrogates for oocysts in a variety of bench-scale water-treatment studies, involving water-treatment filters (Brown and Emelko 2009, Dai and Hozalski 2003, Emelko and Huck 2004), carbon-block filtration (Lau et al. 2005), biofilm removal (Stott and Tanner 2005), ozonation (Tang et al. 2005), and sandy media (Tufenkji et al. 2004). However, much less is known about their suitability as surrogates in field-scale studies designed to assess the vulnerability of water-supply wells at RBF sites. This chapter discusses the potential of fluorescent microspheres as safe and easy-to-detect surrogates for evaluating the efficacy of RBF operations for removing pathogens, particularly *Cryptosporidium*, from source waters at different points along the flow path.

2. Comparisons of Microsphere and Microbial Transport Behaviors in the Subsurface

FCM have been used as surrogates in groundwater injection-and-recovery studies since the mid 1980s in order to gain information about the abiotic aspects of subsurface microbial transport behavior in a variety of aquifers (Harvey and Harms 2002). In most of these studies, the microspheres are added to the aquifer concomitantly with a conservative solute tracer, typically a halide salt or non-reacting dye. By comparing the breakthrough curves of the conservative tracer and the microspheres, information is gained about the role of the "particulate" and "reactive" natures of the microbial-sized microspheres in determining its transport behavior in various geohydrologic settings. However, in order to relate how transport of microspheres and pathogens might compare at the field scale, it is important to conduct side-by-side transport comparisons of the microorganism and the surrogate microsphere using the site-specific conditions and geologic media that characterize the field site. For pathogens, such comparisons must be done in the lab for safety and permitting reasons. However, the most meaningful transport comparisons are accomplished in the field, because it is often exceedingly difficult to replicate the field geochemical/physical heterogeneities at the bench scale.

Table 6.1 lists groundwater tracer studies where FCM were injected into aquifers along with viruses, bacteria, or protozoa and comparisons made between their respective transport behaviors. The first such test conducted in 1986 involved both forced- and natural-gradient conditions (Harvey et al. 1989). In that test, transport behaviors of a variety of sizes of microspheres were compared to those of the unattached indigenous bacterial communities that had been recovered from a sandy aquifer (Cape Cod, Massachusetts, USA), concentrated, stained with the DNA-specific fluorochrome DAPI (4',6-diamidino-2-phenylindole) and returned to the aquifer with the microspheres and conservative tracers (bromide and chloride). Also in the 1980s, the transport of FCM and the bacterium *Escherichia coli* were assessed in a fractured granite aquifer within the Canadian Shield (Ontario, Canada) (Champ and Schroeter 1988). A more recent subsurface transport study involving the concomitant addition of bacteria and microspheres to a fractured-granite near Mirror Lake, New Hampshire, USA used several morphologically and physicochemically different bacterial strains (*Pseudomonas stuzeri*, *Microbacterium* sp., and Staphylococcus sp.) (Becker et al. 2003). Also, microspheres and the bacterium *Ralstonia eutropha* were added to unsaturated epikarst/karst limestone at Gännsbrunnen, Switzerland (Sinreich et al. 2009). For the five studies listed in Table 6.1, retardation of the peak concentrations of FCM (relative to a conservative tracer) was reasonably close (within half a log unit) to those of the bacteria. However, immobilization of the microspheres was generally much greater, in some cases by as much as 2 log units (Champ and Schroeter 1988), than that experienced by bacteria traveling through the same flow paths. Consequently, FCM may generally be expected to under-predict bacterial transport potential in a variety of hydrologic settings, including granular systems that are similar in basic structure to some of the alluvial aquifers where RBF operations are sited. However, a greater attenuation of the microspheres may not always be the case, because of the species-to-species differences in surface properties. For example, a greater attenuation of the bacterium than the FCM was observed in the Mirror Lake, NH study, but only for the *Microbacterium* strain (a Gram positive rod) (Becker et al. 2003).

TABLE 6.1. Field injection-and-recovery studies comparing subsurface transport behaviors of carboxylated microspheres with those of microorganisms.

Field test site	Media	Microsphere diameter, μm	Microbe (size, μm)	Distance (m)	Relative velocity*	Relative attenuation**	Reference
Viruses							
Cape Cod, MA (USA)	well sorted sand	0.7	PRD1, phage (0.06 diameter)	12	-	---	(Bales et al. 1995)
Borden, Ont. (Canada)	well sorted sand	0.01	PRD1, phage (0.06 diameter)	2	+	---	(Bales et al. 1995)
			M1, phage (0.025 × 0.11)	2	+/-	---	(Bales et al. 1995)
Jura Mtns. (Switz.)	karst limestone	1.0	H40, marine phage (85nm)	1250	-	+	(Auckenthaler et al. 2002)
Bacteria							(Harvey et al. 1989)
Cape Cod, MA (USA)	well sorted sand	0.2, 0.7, 1.2	aquifer community (0.2–1.4)	1.7	+/-	++	(Harvey et al. 1993)
Cape Cod, MA (USA)	well sorted sand	0.7	aquifer community (0.2–1.6)	6	+/-	++	(Sinreich et al. 2009)
Gännsbrunnen (Switz.)	limestone (epikarst)	1.0	*Ralstonia eutropha* (0.5 × 1.8)	10	+/-	+	(Becker et al. 2003)
White Mtns, NH (USA)	fractured granite	1.0	*Pseudomonas stuzeri* (1.5–2.2)	36	+/-	+/-	(Becker et al. 2003)
			e.g., Microbacterium sp. (1.4–1.9)	36	+/-	+	(Becker et al. 2003)
			Staphylococcus sp. (0.5–0.8)	36	+/-	-	(Champ and Schroeter
Chalk R Lab (Canada)	fractured granite	2.0	*Escherichia coli* (unspecified)	13	+/-	+	1988)
Protists							
Cape Cod, MA (USA)	well sorted sand	2.0	*Spumella guttula Kent* (2–3)	1	+/-	+/-	(Harvey et al. 1995)

* "Relative velocity" was estimated based upon the arrival times of the peak concentrations relative at the downgradient (sampling) well. "+" means that the microspheres traveled >1.5 times faster than the test microorganism. "+/-" means that the microspheres and viruses traveled at velocities that differed less than a factor of 1.5. "-" means that the microspheres traveled at a velocity that was at least 1.5 times slower than that of the test microorganism. For the H4/H40 phage test (Jura Mtns, Switzerland), relative velocity is based upon first detection. Size of H40 phage from (Flynn et al. 2004).

** "Relative attenuation" was estimated from the differences between the microspheres and microorganisms in their respective fractional recoveries. "++" means that the fractional loss (immobilization) of microspheres was 1.5–2.4 log units higher than that of the test microbe. "+" means that the fractional loss of microspheres was 0.5–1.4 log units higher than that of the test microbe, "+/-" means that there was less than 0.5 log units of difference between fractional loss of microspheres and test microbe, "-" means that the fractional loss of microspheres was 0.5–1.4 log units lower than that of the test microbe, "--" means that the fractional loss (immobilization) of microspheres was 1.5–2.4 log units lower than that of the test microbe, "---" means that the fractional loss (immobilization) of microspheres was at least 2.5 log units lower than that of the test microbe. Relative attenuation for the Borden site study was estimated from the changes in peak concentrations as the viruses and microspheres were advected from well ML4-4 to ML5-4 .

Injection tests involving both microspheres and viruses (Table 6.1) indicate that it would be difficult to make any generalizations about the suitability of microspheres as surrogates for viruses in field injection-and-recovery experiments. However, judging from results of tracer tests involving sandy aquifers at sites in Cape Cod, MA USA (Bales et al. 1995) and at Borden, Ontario Canada (Bales et al. 1997), it is likely that FCM would over-predict the rate of virus transport by a factor of up to $\sim10^3$. Microspheres traveled significantly faster than the phage at the latter site, but considerably slower than the phage at the former site. For a 1250 m-long colloid transport study involving karst limestone (Switzerland), 1 µm microspheres traveled ~7 times faster than the H40 bacteriophage based upon time of first arrival (Auckenthaler et al. 2002). However, the more than ten-fold difference in size of the two colloids would be expected to have at least some affect on time of first detection, judging from the results of other tests performed in karst limestone (e.g., Harvey et al. 2008).

Very little information is available on how the subsurface transport behaviors of FCM compare to those of protozoa. Although there were substantial differences in the transport behaviors of 1 µm FCM and similar sized groundwater bacteria in a well-sorted, sandy glacial outwash aquifer in Cape Cod, Massachusetts (Harvey and Garabedian 1991), larger (2 and 3 µm) FCM reasonably captured the transport behavior of the 2–3 µm groundwater protist (protozoa), *Spumella guttula* Kent in a subsequent study at the same site (Harvey et al. 1995). In particular, the breakthrough of microspheres and protozoa were reasonably similar in terms of both immobilization and retardation (Table 6.1). Also, the FCM did a reasonable job of replicating the multi-peaked pattern of the protozoan breakthrough curves at the sampling wells downgradient from point of injection. Consequently, it was suggested in the latter study that FCM may be useful as abiotic analogs for *C. parvum* oocysts in subsequent field transport tests.

3. Microspheres for Assessing Vulnerability of RBF Wells to *Cryptosporidium* Contamination

3.1. Comparison of Oocyst and Microsphere Properties

A comparison of physicochemical characteristics (buoyant densities, aspect ratios, average diameters, zeta potentials) of *C. parvum* oocysts and oocyst-sized FCM that would affect subsurface transport behavior is detailed in Harvey et al. (2008). In that study, aspect ratios of oocysts were generally slightly greater than those of the microspheres, although both were nearly spherical. The oocysts exhibited a range of buoyant densities (1.03–1.07 g/cm^3) that bracketed that of the microspheres (~1.05 g/cm^3). However, under ionic strength (10^{-2}) and near-neutral to slightly alkaline conditions, FCM exhibited substantially more negative zeta potentials than those measured for the C. parvum oocysts. In a preliminary study involving static minicolumns, artificial groundwater, and fragments of crushed limestone core, both oocysts and 4.9 μm microspheres exhibited a tendency for increasing sorption on limestone surfaces in response to increasing dissolved calcium (Osborn et al. 2000).

Unfortunately, a dearth of information exists about surface properties for *C. parvum* oocysts found in natural waters, because of the difficulties in concentrating a large enough number to perform surface measurements. However, the range of surface properties reported for *C. parvum* oocysts suggests that it is probably incorrect to assume that oocysts in aquifers will always be highly charged. A more cautious approach is suggested that assumes some oocysts in natural waters may be characterized by low charge, at least until there is more definitive evidence to the contrary. Although FCM proved useful surrogates in the 2004 Northwest Well Field vulnerability test (Miami, FL), it is recommended that a mixture of different types and sizes of microspheres representing a wider variety of surface charge having car-boxylated, neutral, and carbonyl surfaces be employed in future well field vulnerability assessments regarding this pathogen (Harvey et al. 1989).

3.2. Bench-Scale Comparisons of Microsphere and Oocyst Transport

Several recent laboratory studies have compared the attachment and (or) transport behaviors of oocyst-sized FCM and *C. parvum* oocysts in the presence of granular porous media (Dai and Hozalski 2003, Emelko et al. 2003, Tufenkji et al. 2004). Emelko et al. (2003) reported similar rates of removal for formalin-inactivated oocysts and 4.7 μm FCM in a pilot-scale dual-composition media (anthracite and sand) filter. Similarly, Tufenkji et al. (2004) noted similar transport behaviors (dimensionless concentration histories in the eluent and fractional removals) for heat-inactivated oocysts and 4.1 μm FCM in clean quartz sand (1 mM ionic strength and pH 5.6–5.8). In contrast, other microsphere-oocyst comparisons suggest that their attachment and transport behaviors in porous media can also differ substantially. Bradford and Bettahar (2005) reported differences in breakthrough and tailing behaviors for transport of oocysts (3–6 μm) and microspheres (3.2 μm) being advected through a sand column. In another study (Brush et al., 1998), 3–4 fold greater attachment was observed for 4.5 μm carboxylated polystyrene microspheres onto polystyrene surfaces at 50 mM ionic strength as compared with DIS-purified oocysts.

The effectiveness of FCM as surrogates for *C. parvum* oocysts in in-situ subsurface transport studies depends upon how well the microspheres represent their surface properties for a particular set of chemical and physical conditions. In several studies, zeta potentials (ζ) of the FCM were more negative than those of oocysts under neutral to slightly alkaline conditions (Bradford and Bettahar 2005, Dai and Hozalski 2003, Harvey et al. 2008). Although some reports suggest a very weak surface charge at circumneutral pH (e.g., Brush et al. 1998, Butkus et al. 2003), others (Considine et al. 2000, Hsu and Huang 2002, Ongerth and Pecoraro 1996) suggest that oocysts may carry a more substantive negative charge. Reported differences in oocyst ζ where similar physicochemical conditions were used are probably due to a variety of factors, including source (Butkus et al. 2003), age, exposure to antibiotics, and method of purification (Brush et al. 1998).

Given the variability of surface characteristics of oocysts and differing geologic media from site to site, surrogate microspheres can potentially over-predict and under-predict transport of *C. parvum* oocysts. For a transport study involving intact core samples of lime-stone characterized largely by matrix porosity, 4–6 μm oocysts were transported, respectively, ~4 and ~6 times more readily than 3 and 5 μm FCM (10 mM ionic strength, pH 8) (Harvey et al. 2008). In contrast, recovery of 3 μm FCM in static columns packed with sediments from a RBF site (Russian River, CA) was substantively greater than oocysts of similar size (Metge et al. 2010). Unfortunately, there is little information about surface properties for oocysts found in natural waters, because of the difficulties involved in recovering and concentrating a sufficient number on which to perform surface measurements. However, available evidence suggests that FCM are better surrogates for some *C. parvum* oocysts than others. It is also likely that the suitability of FCM as transport surrogates for *Cryptosporidium* oocysts will vary from species to species.

It is also evident that within the typical *C. parvum* oocyst size range, colloidal diameter can substantively affect their rate of attenuation within geologic media. Figure 6.1 depicts the breakthrough curves for pulse injections for 2.5–6 μm oocysts and a monodisperse suspension of microspheres being advected through 10 cm of repacked aquifer sediments recovered from the Greater Miami River (Ohio, USA) RBF site (Sheets et al. 1989). Average size of the oocysts decreased ~2 fold with time and distance in the column. This suggests that the larger sized oocysts are subject to a higher rate of immobilization. Similarly, 3 μm FCM being advected through in-tact lime-stone cores were attenuated to a lesser degree than were 5 μm FCM (Harvey et al. 2008). The differences in attenuation in the latter study appeared to be due largely, but not entirely, to predicted differences in settling rates.

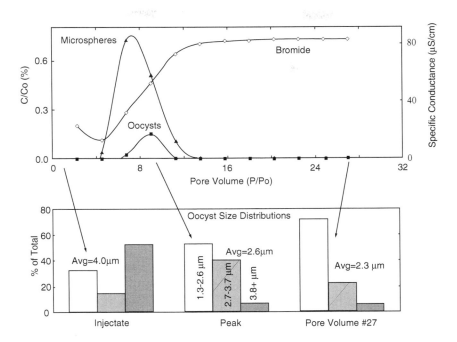

Figure 6.1. Dimensionless concentrations of carboxylated microspheres (1 μm) and 2.5–7 μm *C. parvum* oocysts in the eluent of a column of aquifer sediment recovered from the Greater Miami River RBF site located in southwestern Ohio, USA. Breakthrough of the constant injection of the conservative tracer (bromide) was measured by specific conductance. The bar graphs indicate the changes in the size classes and average diameter for the oocysts as they were advected through the column.

3.3. Field Transport Studies Using Oocyst-Sized Microspheres

Several injection-and-recovery tests have been conducted in which the subsurface transport of protozoan-sized microspheres have been assessed. In 1998, the transport potential of 5 and 15 μm microspheres were assessed in a forced-gradient injection and recovery study involving a granular aquifer in Idaho USA (Petrich et al. 1998). More recently, oocyst-sized FCM were used as transport surrogates to assess the vulnerability of two karst-limestone aquifers to *Cryptosporidium* contamination. In 2005, 5 μm microspheres were added to a cave stream in southern Germany, some of which were recovered 2.5 km away (Goppert and Goldscheider 2008). Although greatly attenuated within the aquifer, the appearance of microspheres at a spring 2.5 km downgradient from point of injection demonstrated that oocyst-sized colloids were capable of traveling quickly over considerable distances in karst limestone settings and could be detected in very low numbers.

In a forced-gradient, convergent study conducted in 2004, the transport of 2, 3, and 5 μm FCM were assessed relative to that of a conservative solute (SF6) from a borehole placed into the karst limestone aquifer to a water-supply well 97 m away (Harvey et al. 2008). The Miami, FL (USA) study established the utility of a polydispersed suspension of fluorescence microspheres in tracer application in the vicinity of water-supply wells. The early appearance of the center of mass of the three size classes of microspheres relative to the conservative tracer underscored the need to use colloidal tracers in studies assessing the vulnerability of wells to oocyst contamination. Although not a RBF site, the water-supply well drew water from an aquifer that was hydraulically connected to surface water and in the vicinity of borrow pit lakes with similar depths as the well's production zone. Oocyst-sized microspheres could be quantified with reasonable accuracy at a distance of 97 m downgradient using either epifluorescence microscopy or by the less labor-intensive flow cytometry. For the latter method, microspheres in well water samples had to be concentrated by filtration. Lastly, 2, 3, and 5 μm FCM were employed as safe surrogates in order to gather in-formation about the transport potential of oocysts in the shallow bottom sediments 25 m above a lateral collector (Russian River RBF site, Sonoma County, CA, USA) (Metge et al. 2007). The latter study pointed out the effect of site-to-site variability in transport parameters; the smallest microspheres (2 μm diameter) were transported preferentially at a location characterized by relatively lower vertical hydraulic conductivity (and slower vertical flow rate) site, but that the intermediate size microspheres (3 μm diameter) were transported preferentially at a nearby site characterized by higher vertical hydraulic conductivity and faster vertical flow rates.

The experimental design of injection-and-recovery studies involving the use of protozoan-sized microspheres in a given water-supply aquifer depends, in large part, upon their transport potential in a given system. Table 6.2 illustrates the variations in removal efficiency of 2 μm FCM during field studies involving three different types of aquifers. Removal (expressed per meter of travel through the aquifer) were 5–6 log units greater in the poorly sorted, Fe-rich sediments beneath the Russian River (CA) relative to sediments in a glacial outwash aquifer (Cape Cod, MA), in spite of the coarser grains at the former site. These differences reflect both differences in structure (sorting) and differences in extractable (grain-surface) iron. As expected, transport of the same size microspheres through the karst limestone characterized by preferential flow was much greater (by ~77 log units) than in the sandy, outwash aquifer.

TABLE 6.2. Role of media characteristics in the removal efficiency of 2 μm carboxylated microspheres in field injection and recovery studies.

Field test site	Depth (m)	Flow (m/d)	Media characteristics			Log_{10} removal	Reference
			Description	Grain size	Extractable Fe		
Russian R. bank filtration site (Sonoma County, CA, USA)	0–0.5	0.15–0.38	poorly sorted granular	2.74 mm (D_{50}) 0.50 mm (D_{10})	20–34 mg/g*	6.6/m	Metge et al. 2007
USGS Cape Cod test site (Falmouth, MA, USA)	9.3	0.5	well-sorted sand	0.59 mm (Avg)	0.2–0.3 mg/g**	1.0/m	Harvey et al. 1995
Northwest Well Field (Miami, FL, USA)	11–13	variable	karst limestone	n/a	ND	0.013/m	Harvey et al. 2008

n/a = not applicable, ND = not determined
*Metge 2010
**Ryan et al. 1999 and Scholl et al. 1992

4. Limitations of Microspheres as Surrogates

Although commercially available fluorescent microspheres are easy to detect, available in a variety of sizes (tens of nm to tens of μm), and chemically stable over long periods of time, there are two primary drawbacks that could limit their use as pathogen surrogates in certain field applications. The first is that their surface characteristics and morphology may not match up well with the microorganisms of interest, which can lead to differences in retardation, immobilization, and apparent dispersion. For example, it was observed in forced- and natural-gradient tracer tests performed in a sandy aquifer, that the transport behavior and collision efficiencies of native groundwater bacteria were not well represented by FCM of similar size (Harvey and Garabedian 1991, Harvey et al. 1989). However, it was also observed that a polydispersed suspension (2–5 μm) of FCM could be very useful as a surrogate of *C. parvum* oocysts, providing that differences in surface characteristics are taken into account (Harvey et al. 2008).

One potential solution for improving the utility of microspheres as surrogates in field applications would be to use microspheres whose surface chemistry more closely match the microorganism of interest. In addition to FCM discussed in this report, there are other types of commercially available microspheres that may be more appropriately suited for a particular application. It was observed that different types of microspheres having carboxylated, carbonyl, and neutral surfaces and injected concomitantly into a sandy aquifer were subject to different degrees of retardation and immobilization (Harvey et al. 1989). In a tracer test that involved addition of both amidine (positively charged) and polystyrene (negatively charged) microspheres to a fractured granite aquifer, the microspheres with carboxylated surfaces were transported much more readily that than the microspheres with positively charged surfaces (Becker et al. 2003). The surfaces of at least some microspheres can be chemically modified relatively easily. For example Pang et al. (2009), chemically modified the surfaces of 20 nm FCM with proteins (bovine milk α_s casein) such that the surface electrical properties were fairly similar to those of the MS2 bacteriophage (virus).

A second drawback for the use of microspheres for RBF field application involves cost. Using a large enough quantity of microspheres that ensures accurate and quantifiable breakthrough at wells downgradient can be very expensive. This is particularly true for the larger (protozoa-sized) microspheres, because the cost (per microsphere) is related to their diameters. The microspheres are manufactured to exact tolerances for such use as instrument calibrations and biomedical applications, both of which require much smaller quantities than a field-scale injection-and-recovery test. However, finding and using microbial surrogates in RBF operations can be problematic often because of permitting and safety issues or because good microbial surrogates do not always exist. Also, some surrogates require analysis within 24-h, which, in many cases, could limit the scope of field transport studies (Goppert and Goldscheider 2008).

5. Conclusions

The suitability of fluorescent, microbial sized microspheres as field surrogates for pathogens at RBF sites continues to be explored. Although microspheres often have proven to be less-than-ideal analogs for capturing the abiotic transport behavior of viruses and bacteria, encouraging evidence exists regarding use of FCM as surrogates for *C. parvum* oocysts. Suspensions of variably-sized FCM that bracket the size range of oocysts typically found in natural waters can be useful as surrogates in field-scale injection-and-recovery tests at operating RBF sites (Metge et al. 2007) or other situations where water-supply wells are drawing water, at least in part, from nearby sources of surface water (Harvey et al. 2008). There are commercially available microspheres that have roughly the same aspect ratios and buoyant densities as oocysts. However, the use of microspheres is limited because of the high costs and discrepancies between their surface characteristics. Following a recent test involving the transport of oocyst-sized microspheres through a drinking-water karst limestone aquifer, it was recommended that a mixture of different types of microspheres having a wider variety of surface characteristics be used in future field-scale studies designed to investigate the transport potential of pathogens. Modification of microsphere surfaces with proteins (Pang et al. 2009) may facilitate our ability in the future to construct mixtures of microspheres that would bracket not only the size of pathogen of interest, but also its surface charge. Field-scale studies by Passmore et al. (2010) found that microspheres with similar surface properties and size as microorganisms of interest can be useful surrogates to trace transport pathways of in the subsurface.

Acknowledgments The authors gratefully acknowledge assistance and funding from the Sonoma County Water Agency, the USGS Ohio Water Science Center, the USGS Massachusetts Science Center, and from the USGS Toxics program.

References

Auckenthaler A, Raso G, Huggenberger P (2002) Particle transport in a karst aquifer: natural and artificial tracer experiments with bacteria, bacteriophages and microspheres. Water Science and Technology 46: 131–138

Bales RC, Li SM, Maguire KM, Yahya MT, Gerba CP, Harvey RW (1995) Virus and bacteria transport in a sandy aquifer, Cape-Cod, MA. Ground Water 33: 653–661

Bales RC, Li SM, Yeh TCJ, Lenczewski ME, Gerba CP (1997) Bacteriophage and microsphere transport in saturated porous media: forced-gradient experiment at Borden, Ontario. Water Resources Research 33: 639–648

Becker MW, Metge DW, Collins SA, Shapiro AM, Harvey RW (2003) Bacterial transport experiments in fractured crystalline bedrock. Ground Water 41: 682–689

Behrens H, Beims U, Dieter H, Dietze G, Eikmann T, Grummt T, Hanisch H, Henseling H, Käß W, Kerndorff H, Leibundgut C, Müller-Wegener U, Rönnefahrt I, Scharenberg B, Schleyer R, Schloz W, Tilkes F (2001) Toxicological and ecotoxicological assessment of water tracers. Hydrogeology Journal 9: 321–325

Bradford SA, Bettahar M (2005) Straining, attachment, and detachment of cryptosporidium oocysts in saturated porous media. Journal of Environmental Quality 34: 469–478

Brown TJ, Emelko MB (2009) Chitosan and metal salt coagulant impacts on *Cryptosporidium* and microsphere removal by filtration. Water Research 43: 331–338

Brush CF, Walter MF, Anguish LJ, Ghiorse WC (1998) Influence of pretreatment and experimental conditions on electrophoretic mobility and hydrophobicity of *Cryptosporidium parvum* oocysts. Applied and Environmental Microbiology 64: 4439–4445

Butkus MA, Bays JT, Labare MP (2003) Influence of surface characteristics on the stability of *Cryptosporidium parvum* oocysts. Applied and Environmental Microbiology 69: 3819–3825

Carpenter C, Fayer R, Trout J, Beach MJ (1999) Chlorine disinfection of recreational water for *Cryptosporidium parvum*. Emerging Infectious Diseases 5: 579–584

Champ DR, Schroeter J (1988) Bacterial Transport in Fractured Rock – a Field-Scale Tracer Test at the Chalk River Nuclear Laboratories. Water Science and Technology 20: 81–87

Considine RF, Dixon DR, Drummond CJ (2000) Laterally-resolved force microscopy of biological microspheres-oocysts of *Cryptosporidium parvum*. Langmuir 16: 1323–1330

Dai XJ, Hozalski RM (2003) Evaluation of microspheres as surrogates for *Cryptosporidium parvum* oocysts in filtration experiments. Environmental Science & Technology 37: 1037–1042

Dash RR, Mehrotra I, Kumar P, Grischek T (2008) Lake bank filtration at Nainital, India: water-quality evaluation. Hydrogeology Journal 16: 1089–1099

Emelko MB, Huck PM (2004) Microspheres as surrogates for *Cryptosporidium* filtration. Journal American Water Works Association 96: 94–105

Emelko MB, Huck PM, Douglas IP (2003) *Cryptosporidium* and microsphere removal during late in-cycle filtration. Journal American Water Works Association 95: 173–182

Flynn R, Hunkeler D, Guerin D, Burn C, Rossi C, Aragno M (2004) Geochemical influences on H40/1 bacteriophage inactivation in glaciofluvial sands. Environmental Geology 45: 504–517

Gollnitz WD, Clancy JL, Whitteberry BL, Vogt JA (2003) RBF as a microbial treatment process. Journal American Water Works Association 95: 56–66

Goppert N, Goldscheider N (2008) Solute and colloid transport in karst conduits under low- and high-flow conditions. Ground Water 46: 61–68

Gupta V, Johnson WP, Shafiean P, Rhu H, Alum A, Abbaxzadegan M, Hubbs SA, Rauch-Williams T (2009) Riverbank filtration: Comparison of pilot scale transport with theory. Environ Sci Technol 43: 669–676

Harvey RW, Garabedian SP (1991) Use of colloid filtration theory In modeling movement of bacteria through a contaminated sandy aquifer. Environmental Science & Technology 25: 178–185

Harvey RW, George LH, Smith RL, Leblanc DR (1989) Transport of microspheres and indigenous bacteria through a sandy aquifer – Results of natural-gradient and forced-gradient tracer experiments. Environmental Science & Technology 23: 51–56

Harvey RW, Harms H (2002) Tracers in groundwater: use of microorganisms and micro-spheres, p. 3194–3202. In G. Bitton (ed.), Encylopedia of Environmental Microbiology, vol. 6. John Wiley & Sons, Inc., New York

Harvey RW, Harms H, Landkamer L (2007) Transport of microorganisms in the terrestrial subsurface: In situ and laboratory methods., p. 872–897. In C. J. Hurst, R. L. Craw-ford, J. L. Garland, D. A. Lipson, A. L. Mills, and L. D. Stetzenbach (ed.), Manual of Environmental Microbiology, 3rd ed. ASM Press, Washington

Harvey RW, Kinner NE, Bunn A, Macdonald D, Metge D (1995) Transport behavior of groundwater protozoa and protozoan-sized microspheres in sandy aquifer sediments. Applied and Environmental Microbiology 61: 209–217

Harvey RW, Kinner NE, Macdonald D, Metge DW, Bunn A (1993) Role of physical heterogeneity in the interpretation of small-scale laboratory and field observations of bacteria, microbial-sized microsphere, and bromide transport through aquifer sediments. Water Resources Research 29: 2713–2721

Harvey RW, Metge DW, Shapiro AM, Renken RA, Osborn CL, Ryan JN, Cunningham KJ, Landkamer L (2008) Pathogen and chemical transport in the karst limestone of the Biscayne aquifer: 3. Use of microspheres to estimate the transport potential of *Cryptosporidium parvum* oocysts. Water Resources Research 44, W08431, doi:10.1029/2007WR006060.

Hsu BM, Huang CP (2002) Influence of ionic strength and pH on hydrophobicity and zeta potential of Giardia and Cryptosporidium. Colloids and Surfaces a-Physicochemical and Engineering Aspects 201: 201–206

Kuehn W, Mueller U (2000) Riverbank filtration – An overview. Journal American Water Works Association 92: 60–69

Lau BLT, Harrington GW, Anderson MA, Tejedor I (2005) Physicochemical aspects of *Cryptosporidium* surrogate removal in carbon block filtration. Journal American Water Works Association 97: 92–101

LeChevallier, MW Norton WD, Lee RG (1991) Occurrence of *Giardia* and *Cryptosporidium* spp. in surface water supplies. Applied and Environmental Microbiology 57: 2610–2616

Mac Kenzie WR, Hoxie NJ, Proctor ME, Gradus MS, Blair KA, Peterson DE, Kazmierczak JJ, Addiss DG, Fox KR, Rose JB (1994) A massive outbreak in Milwaukee of *Cryptosporidium* infection transmitted through the public water supply. The New England Journal of Medicine 331: 161–167

Malzer HJ, Schubert J, Gimbel R, Ray C (2002) Effectiveness of riverbank filtraiton sites to mitigate shock loads, p. 229–259. In Ray C, Melin G, Linsky RB (ed.), Riverbank Filtration, vol. 43. Springer

Metge DW, Harvey RW, Aiken GW, Anders R, Lincoln G, Jasperse J (2010) Influence of organic carbon loading, sediment associated metal oxide content and sediment grain size distributions upon *Cryptosporidium parvum* removal during riverbank filtration operations. Water Research 44: 1126–1137

Metge DW, Harvey RW, Anders R, Rosenberry DO, Seymour D, Jasperse J (2007) Use of carboxylated microspheres to assess transport potential of *Cryptosporidium parvum* oocysts at the Russian River water supply facility. Geomicrobiology Journal 24: 231–245

Okhuysen PC, Chappell CL, Crabb JH, Sterling CR, DuPont HL (1999) Virulence of three distinct *Cryptosporidium parvum* isolates for healthy adults. The Journal of Infectious Diseases 180: 1275–1281

Ongerth JE, Pecoraro JP (1996) Electrophoretic mobility of *Cryptosporidium* oocysts and *Giardia* cysts. Journal of Environmental Engineering-ASCE 122: 228–231

Osborn AM, Pickup RW, Saunders JR (2000) Development and application of molecular tools in the study of Inc N-related plasmids from lake water sediments. Fems Microbiology Letters 186: 203–208

Pang L, Nowostawska U, Ryan JN, Williamson WM, Walshe G, Hunter KA (2009) Modifying the surface charge of pathogen-sized microspheres for studying pathogen transport in groundwater. Journal of Environmental Quality 38: 2210–2217

Partinoudi V, Collins MR (2007) Assessing RBF reduction/removal mechanisms for microbial and organic DBP precursors. Journal American Water Works Association 99: 61–71

Petrich CR, Stormo KE, Ralston DR, Crawford RL (1998) Encapsulated cell bioremediation: Evaluation on the basis of particle tracer tests. Ground Water 36: 771–778

Passmore JM, Rudolph, DL, Mesquita, MF, Cey EE, Emelko MB (2010) The utility of microspheres as surrogates for the transport of *E. Coli* RS2g in partially saturated agricultural soil. Water Research 44: 1235–1245

Pokorny NJ, Weir SC, Carreno RA, Trevors JT, Lee H (2002) Influence of temperature on *Cryptosporidium parvum* oocyst infectivity in river water samples as detected by tissue culture assay. The Journal of Parasitology 88: 641–643

Ray C, Melin G, Linsky RB (ed) (2002) Riverbank filtration: improving source-water quality. Kluwer Academic Publishers, Dordrecht

Ryan JN, Elimelech M, Ard RA, Harvey RW, Johnson PR (1999) Bacteriophage PRD1 and silica colloid transport and recovery in an iron oxide-coated sand aquifer. Environmental Science & Technology 33: 63–73

Scholl MA, Harvey RW (1992) Laboratory investigations on the role of sediment surface and groundwater chemistry in transport of bacteria through a contaminated sandy aquifer. Environmental Science & Technology 26: 1410–1417

Shamrukh, M., Abdel-Wahab, A. (2008) Riverbank filtration for sustainable water supply: application to a large-scale facility on the Nile River. Clean Technologies and Environmental Policy 10: 351–358

Sheets RA, Bair ES, Rowe GL (1989) Use of 3H/3He ages to evaluate and improve groundwater flow models in a complex buried-valley aquifer. Water Resources Research 34: 1077–1089

Sinreich M, Flynn R, Zopfi J (2009) Use of particulate surrogates for assessing microbial mobility in subsurface ecosystems. Hydrogeology Journal 17: 49–59

Stott R, Tanner CC (2005) Influence of biofilm on removal of surrogate faecal microbes in a constructed wetland and maturation pond. Water Science and Technology 51: 315–322

Tang G, Adu-Sarkodie K, Kim D, Kim JH, Teefy S, Shukairy HM, Marinas BJ (2005) Modeling *Cryptosporidium parvum* oocyst inactivation and bromate formation in a full-scale ozone contactor. Environmental Science & Technology 39: 9343–9350

Tufenkji N, Miller GF, Ryan JN, Harvey RW, Elimelech M (2004) Transport of *Cryptosporidium* oocysts in porous media: Role of straining and physicochemical filtration. Environmental Science & Technology 38: 5932–5938

Tufenkji N, Ryan JN, Elimelech M (2002) The promise of bank filtration. Environmental Science & Technology 36: 422a–428a

Willocks L, Crampin A, Milne L, Seng C, Susman M, Gair R, Moulsdale M, Shafi S, Wall R, Wiggins R, Lightfoot N (1998) A large outbreak of Cryptosporidiosis associated with a public water supply from a deep chalk borehole. Communicable Disease and Public Health 1: 239–243

Chapter 7 Hydrogeochemical Processes During Riverbank Filtration and Artificial Recharge of Polluted Surface Waters: Zonation, Identification, and Quantification

Pieter J. Stuyfzand*

KWR Watercycle Research Institute, PO Box 1072, 3430 BB Nieuwegein, Netherlands, Phone: +31610945021.

VU University Amsterdam, Dept. Hydrology & Geo-Environmental Sciences, FALW, Boelelaan 1085, 1081 HV Amsterdam, Netherlands

Abstract The hydrogeochemical processes acting during riverbank filtration (RBF) and artificial recharge using basins (BAR), aquifer storage recovery (ASR) and deep well injection with remote recovery (ASTR) are reviewed, and quantified by the chemical mass balance approach, also referred to as inverse modeling. The various systems are classified and described, together with the various hydrogeochemical zones and processes within them. The zones discerned include the surface water compartment, the recharge proximal, distant aquifer and discharge proximal zone in general, and a typical redox zonation for RBF, BAR, ASR and ASTR in particular. Mass balances are drawn up by the EXCEL spreadsheet code REACTIONS+ version 6, which is a significant upgrade of an earlier version and explained in detail. The model only requires data entry for the chemical composition of the input (infiltration water), the output (sample of infiltrate in the aquifer, taken from an observation or pumping well), and admixed local groundwater (if relevant), and may necessitate a change of default settings for specific reactions. The output consists of a concise list of the hydrogeochemical reactions in terms of losses and gains (in mmol/L) by the following reactions: oxidation of NH_4^+ and DOC in the input; oxidation of organic material, pyrite or desorbing Fe^{2+} and NH_4^+ from the aquifer; dissolution of various carbonate minerals, silica, gypsum or halite; reductive dissolution of ferrihydrite; and exchange of cations and anions. Examples of application include five different RBF samples with differing water sediment interaction. They illustrate the importance of the redox environment and additional gas inputs and gas outputs.

---

* Pieter J. Stuyfzand, KWR Watercycle Research Institute, PO Box 1072, 3430 BB Nieuwegein, Netherlands, Phone: +31610945021. VU University Amsterdam, Dept. Hydrology & Geo-Environmental Sciences, FALW, Boelelaan 1085, 1081 HV Amsterdam, Netherlands, e-mail: pieter.stuyfzand@kwrwater.nl

C. Ray and M. Shamrukh (eds.), *Riverbank Filtration for Water Security in Desert Countries*, DOI 10.1007/978-94-007-0026-0_7, © Springer Science+Business Media B.V. 2011

Keywords: Riverbank filtration, artificial recharge, ASR, redox, inverse modeling, mass balance, water quality

1. Introduction

River bank filtration (RBF) and artificial recharge (AR) are expected to boom in the near future, because they can partly solve or mitigate the mondial water crisis which is deemed to worsen due to climate change and growing water demands (Gale 2005, Dillon 2005). Important advantages of these managed aquifer recharge (MAR) techniques consist of the transformation of unreliable, often polluted surface water into hygienically safe groundwater of much better quality, and the subterranean storage which protects the water against evaporation losses, algae blooms, atmospheric fallout of pollutants, and earthquake hazards (Huisman and Olsthoorn 1983, Sontheimer 1991, Ray 2002, Pyne 2005). Disadvantages may consist of cumbersome clogging phenomena (Hubbs 2006), water losses due to mixing with brackish groundwater, and natural reactions with the porous medium. The latter may raise the concentrations of for instance Fe, As, F, Mn, NH_4, Ca, DOC, ^{222}Rn, and ^{226}Ra (Stuyfzand 1989, 1998b; Dillon and Toze 2005) and thus necessitate a post-treatment.

In this contribution the focus is on the zonation, identification and quantification of the hydrogeochemical processes that act on infiltrating polluted surface water and normally improve its quality. Without this information no reactive transport model can be built, which is the next step and only slightly addressed here.

2. RBF and AR Types

The various types of managed aquifer recharge systems have been classified as indicated in Figure 7.1. The focus here is mainly on induced recharge by river bank filtration (RBF), surface spreading by recharge basins (BAR), and subsurface recharge by ASTR and ASR. These MAR techniques, schematized in Figure 7.2, are among the most frequently used for supply of drinking, agricultural and industrial water, and certainly have been studied in most detail.

Some important characteristics of RBF, BAR, ASTR and ASR systems as applied in the Netherlands are listed in Table 7.1. Together they contribute for ca. 20% to the national drinking water production of 1,250 Mm^3/year, in the Netherlands.

RBF systems can be classified into 6 types from mountainous areas downstream towards their estuary or delta, as indicated in Table 7.2. The classification is based on important changes in river type, river bed material, flow velocity, infiltration regime and driving force, the presence of an unsaturated zone and aquitard, and the predominant redox environment (explained further in Section 3). Not included factors are the fraction of admixed groundwater of local origin and the type of aquifer recharged.

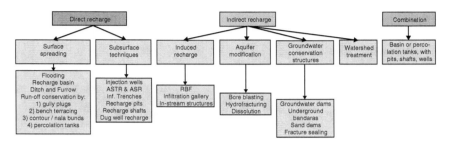

Figure 7.1. Classification of managed aquifer recharge systems (based on a scheme by Chadha 2002).

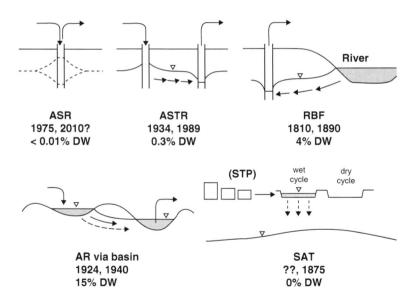

Figure 7.2. The most common MAR types (modified after Dillon 2005), in the Netherlands with the indicated year of first application (left: small scale; right: for centralized drinking water supply), and the contribution to the national drinking water supply (% DW). ASR = Aquifer Storage Recovery; ASTR = Aquifer Storage Transfer Recovery; RBF = River Bank Filtration; AR = Artificial Recharge; SAT = Soil Aquifer Treatment; STP = Sewage Treatment Plant.

100   P.J. STUYFZAND

TABLE 7.1. Characteristics of RBF and AR systems in the Netherlands (from Stuyfzand 1999).

Process / charact	River Bank Filtration (RBF)	Artificial Recharge		
		Basins (BAR)	Wells (ASTR)	Well (ASR)
WHOLE SYSTEM (anno 2005)				
Start of (surface) water infiltration (AD)	1810[S], 1890[L]	1924[S], 1940[L]	1970[P], 1989[L]	1975[S], 2005[P]
Number of abandoned sites in Neths (2005)	17	1	0	3
Number of active sites in Netherlands (2005)	27	11	3	1
Size well field (range in Mm^3/y)	1 - 11	1 - 57	1 - 5	0.1 - 0.3
Possibility of 2 intake points	-	+	+	+
Relation river discharge - recharge rate	++	-	-	-
Selective intake	-	++	+++	+++
Pre-treatment (near the intake)	-	++[A]	+++[A]	+++[B]
Transport-chlorination	-	++	+	-
Post-treatment	+++[C]	++[D]	+[D]	-/+[B]
THE INFLUENT				
Source [##]	R	R,M,Y	R,M,Y	DW
Mean quality [$$]	SI_c>0, SOx	-0.6<SI_c<0.2, Ox	-0.6<SI_c<0.2, Ox	-0.6<SI_c<0.2, Ox
INFILTRATING MEDIUM (river, basin, well)				
Drying up	-/+	+	-	-
Clogging	-	+	++	+
Sludge removal	-	++/+	-/+	-/+
Quality changes	+	++	-	-
AQUIFER PASSAGE				
Passage of Holocene aquitards	20-100%	5-30%	0%	0%
Permeability aquifer (K_h, in m/d)	50	12	25	25
Redox environment	(Deep) Anoxic	(Sub)- & Anoxic	(Sub)- & Anoxic	(Sub)- & Anoxic
Modal travel time (d)	1000	70	120	1-120
Modal distance travelled (m)	500	70	100	1-300
Modal flow velocity groundwater (m/d)	0.5	1	0.5-1	0.5-2
Variations in detention time (y)	2-100	0.1-30	0.3-5	0.2-1
Attenuation of quality fluctuations	+++	++	++	+
RECOVERY				
Admixing native groundwater	5-80%	5-20%	5-20%	<1%
Clogging type of wells	Particles-out, Fe(OH)3, Bio	Fe(OH)3, Bio	Fe(OH)3, Bio	Particles-in, Bio

- = not the case; + = of minor importance c.q. low; ++ = of great importance c.q. high; +++ = of paramount importance c.q. very high.
A = coagulation, pH-correction, rapid sand filtration; B = aeration, rapid sand filtration; C = mostly: aeration, rapid sand filtration,
aeration, filtration over activated carbon; D = often: hardness correction, aeration, rapid sand filtration, filtration over activated carbon.
##: DW = Drinking Water; R = Rhine River; M = Meuse River; Y = Lake IJsselmeer;
$$: SIc = calcite saturation index; Ox = oxic; SOx = (sub)oxic.

TABLE 7.2. Classification of River Bed Filtration (RBF) into six types, from the headwaters in the mountains to the river mouth in sea.

RBF Type (downstream direction)	1	2	3	4	5	6
Site characteristics	Mountains	Hills	Fluvial plain	Fluvial plain	Estuary / Delta	
	Alluvial fan	Valley	Upper	Lower	River	Lake
River type	Braided	Meandering			Anastomosing	-
River / Lake bed	Gravel		Sand	Sand	Sand	Silt / clay
Flow velocity	Extreme	High	Moderate	Slow	Very slow	
Flow drive behind RBF	G	P	P	D + P	D + P	
Infiltration regime	Periodical	Reversing	Constant	Constant	Constant	
River bed contact with aquifer	Direct		Direct	Direct	Intercalated aquitard	
Unsaturated zone below river bed	Yes		Rarely	No	No	
Redox environment	(sub)oxic		mixed	anoxic	deep anoxic	
PS Well Fields Neth's, active	0	1	2	15	3	2
PS Well Fields Neth's, abandoned	0	0	2	6	9	0
Example Netherlands	none	Roosteren	Remmerden	Opperduit	Ridderkerk	Holl Diep

D = Drainage of adjacent land (polder)   G = Gravity   P = Pumping

3. Compartments and Processes

RBF and BAR systems are typically composed of the four compartments depicted in Figure 7.3. ASR and ASTR lack the surface water compartment. In the following these compartments are briefly discussed together with the most important physico-chemical processes. A compact overview of the most important processes acting in the various compartments, the most affected water quality parameters and the factors that influence these processes is presented in Table 7.3.

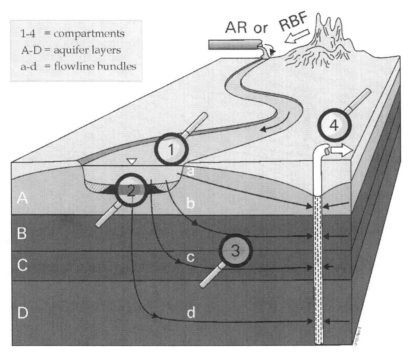

Figure 7.3. The four compartments of RBF and BAR systems, with schematization of the aquifer system, flow, interaction with underwater muds and recovery (after Stuyfzand 2002b).

P.J. STUYFZAND

TABLE 7.3. The most important physicochemical processes acting during RBF and AR, with specification of the most influenced water quality parameters and factors influencing these processes (after Stuyfzand and Lüers 2000).

Processes	Water quality parameters	Factors
INFILTRATION MEANS		
volatilization & gas exchange	VOX, TA, ATT	t, temp, D, X
atmospheric deposition	PAH, Pb, V, PO4, NH4	t, D, Loc
sedimentation/precipitation	SS, OM, NU, TEs, OMPs, Ca, HCO3	t, D, X, input (ZB, NU, pH, SI)
neoformation DOM	DOM, TOC, NU	t, temp, D, LP
degradation (incl.photolysis)	AOX, TA, ATT	t, temp, D, input (NU), LP
reinfection	B&V	t, D, X, loc
dispersion/mixing	ATT	t, D, X, config
AQUIFER PASSAGE		
filtration	SS, OMPs, B&V, ATT	t, Acc, v
oxidation & reduction	O2, NO3, SO4, NH4, DOM, TEs, TA, ATT	t, temp, X, pH, RB, v
	Fe, Mn, OMPs	
SORPTION:		
net adsorption	RS, TEs, OMPs, B&V, PO4, TA, temp	RAP, input
net desorption	Ca, SiO2, DOM	RAP, orig
continuous exchange@	temp, OMPs, SiO2, F, RZ, ATT	RAP
DISSOLUTION (no redox reactions)		
carbonates	pH, Ca, HCO3, Sr, PO4	temp, SI, RAP, DOC, i
phosphates	PO4, Ca, Fe, F	t, X, SI, RAP, i
silicates	SiO2, Al, Ca, Na, K	t, X, temp, SI, RAP, v
sulfates	Ba	t, X, SI, RAP, DOC, i, H2, CH4
PRECIPITATION (no redox reactions)		
carbonates	pH, Ca, HCO3, Sr, Mg, PO4, Ni	temp, SI, DOC, i, Mg, PO4
hydroxides	Fe, Al, PO4, TEs	DOC, pH, O2, NO3
phosphates	PO4, Ca, Fe, F	t, DOC
silicates	SiO2, (Al, Ca, Na, K)	t, temp, SI, DOC, v
sulfides	Fe, H2S, TEs	temp, SI, v
sulfates	Ba	SI, DOC, i
decay/degradation	RAD, B&V, OMPs, ATT	t, R, temp, NU
dispersion	ATT	t, X
RECOVERY		
mixing	ATT	config
admixing authochtonous water	all, ATT, NU, As, HCO3	config
precipitation	Fe(OH)3, TEs, PO4	config, pH, temp
as infiltration means (above)#	nearly all	t, D, X, loc, config

= if recollection open to atmosphere, then same processes as in infiltration basins; @ = by seasonal fluctuations.

WATER QUALITY PARAMETERS		FACTORS	
ATT =	ATTenuation quality fluctuations	Config =	Configuration system
AOX =	non-volatile Active carbon adsorbable Organohalogens	D =	water depth recharge basin
B&V =	Bacteria & Viruses	i =	ionic stength
DOM =	Dissolved Organic Matter	Input =	quality infiltration water
NU =	Nutriens, ie. NH4, PO4, K, SiO2	Loc =	Local conditions (eolian input, birds)
OM =	Organic Material	LP =	Light Penetration
OMPs =	Organic MicroPollutants	ORC =	Oxidation-Reduction capacity of water
RAD =	RADio-active substances like ^{137}Cs, ^{90}Sr, ^{106}Ru and ^{131}I	Orig =	original groundwater composition
RS =	River Salts (Na, K, Ca, Mg)	R =	Retardation factor of compound
SS =	Suspended Solids	RAP =	Reactive Aquifer solid Phases
TEs =	Trace Elements	SI =	Saturation Index mineral
TA =	TAste causing substances	t =	travel/detention time
temp =	temperature	v =	flow velocity of water
VOX =	Volatile Organohalogens and other volatile OMPs	X =	distance travelled.

3.1. The Surface Water Compartment

The most important processes in the surface water compartment (river or recharge basin) are sedimentation, evaporation, nutrient uptake by biota, precipitation of $CaCO_3$ (by CO_2 uptake, evaporation, heating), volatilization, photolysis, (bio)degradation, some attenuation by mixing, and the admixing of rain water (+ dry deposition) and, only in case of RBF, also of effluent discharges or tributary rivers.

This compartment is quite heterogeneous because of differences in depth and morphology, flow velocity, vegetation cover (frequently reeds), transit time and exposition to sunlight and wind. Distinct patterns develop in infiltration basins, especially because of preferential sedimentation of fines in the deeper parts (Figure 7.4). In fluvial compartments with RBF, fines settle preferentially where river flow is significantly reduced, like behind dams, in cut-off meander loops, and in between groins that protect the river banks against erosion. In the deeper parts of rivers, contrary to recharge basins, normally little underwater muds will deposit, also because shipping may concentrate there. The formation of underwater muds is extremely important as explained below.

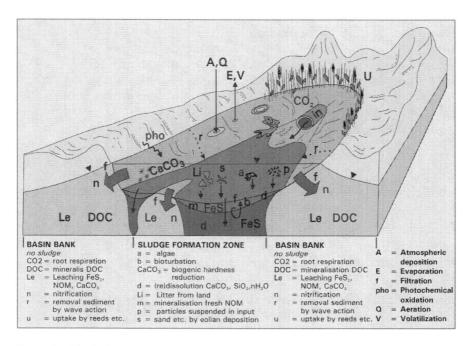

Figure 7.4. Physical and chemical processes in an infiltration basin, showing heterogeneity due to the formation of underwater muds (after Stuyfzand 2002b).

3.2. The Recharge Proximal Aquifer Zone

The recharge proximal aquifer zone is defined here as the first few (ca. 5) meters of aquifer surrounding the recharge facility (river, basin or well), where the bulk of the relatively steady reactions (like those in a slow sand filter) takes place. It is characterized by a high biological activity, a high filtration efficiency, relatively short detention times (<5 d), and an exponential decline of reaction rates with distance. In RBF systems the recharge proximal aquifer zone may overlap with the socalled hyporheic zone, defined as the zone beneath and adjacent to a river or stream where groundwater and surface water mix.

At the water sediment interface a clogging layer may form by filtration of suspended fines, microbiological slimes and $CaCO_3$ precipitates. This layer, which in basins and rivers may form underwater muds, also concentrates or scavenges hydrophobic, silt-bound pollutants like many heavy metals, PAHs, PCBs etc. A clogging layer not only determines infiltration intensities and flow patterns, but also the redox environment which is decisive for the behavior of many pollutants (Section 4).

The most important processes (Figure 7.4) include: cake and deep bed filtration of fines, additional O_2 inputs where an unsaturated zone is present, additional CO_2 inputs where root respiration is important, mineralization of accumulated organic matter, dissolution of $CaCO_3$ (from the clogging layer and aquifer, if present and the infiltrate undersaturated), dissolution of diatoms from the clogging layer, precipitation of sulphides (especially in and below the clogging layer), oxidation of NH_4 (nitrification) and DOC where sufficient O_2 remains (in areas without clogging layer). The highest elimination rate for pathogenic microorganisms, NH_4 and DOC is normally achieved within the recharge proximal zone (Medema and Stuyfzand 2002, Stuyfzand 1989, Schoenheinz 2004).

The highest accumulation rate of pollutants takes place in the recharge proximal zone, most of which concentrates in the first few centimeters (Stuyfzand et al. 2007a). Aquifer leaching will be noticed first within this compartment, but within the first few (deci)meters this leaching may be partly compensated for by the accumulation of deep bed filtrated particles deriving from the surface water.

3.3. The Distant Aquifer Zone

The distant aquifer zone is situated in between the recharge proximal zone and the discharge proximal zone (the recovery system). Biological activity is relatively low, detention times are high (65–1,000 d; Table 7.1) and reaction rates remain rather constant. The most important processes consist of the displacement of native groundwater, cation exchange, oxidation of pyrite and organic matter, dissolution of carbonate minerals (if present), reductive dissolution of $Fe(OH)_3$ and MnO_2, sorption of trace elements, Organic MicroPollutants (OMPs) and pathogens, radioactive decay, dying-off of pathogens and (bio)degradation of OMPs. Sorption prolongs the subterranean detention time of OMPs, radionuclides and pathogens, and thereby contributes to a much higher elimination rate.

After a long time of MAR operation reactive aquifer solid phases, like carbonates that prevent aquifer acidification and pyrite and soil organic matter that protect against aquifer oxidation, may become leached. The high infiltration intensities and resulting aquifer flushing rates may accelerate leaching by a factor of 1,000 as compared to natural groundwater systems (Stuyfzand 2002a, Stuyfzand et al. 2007a). The leaching of soil organic material and a pH decrease by carbonate leaching may lead to a significant reduction of the sorption capacity of the aquifer for heavy metals and OMPs (Stuyfzand 2002a).

3.4. The Recovery System and the Discharge Proximal Zone

The discharge proximal zone is close to the recovery system, surrounding the wells, drains or canals in a zone of various meters wide. Typically, 2 additional processes may show up here: the removal of fine particles from the aquifer by high flow velocities (Hofmann 1998), and well clogging just outside of the borehole wall by fine particles (Van Beek et al. 2009) or on the well screen by precipitates of iron (hydr)oxides (Houben and Treskatis 2007). The latter form of well clogging arises when infiltrate containing oxygen (but no Fe^{2+}) is mixed with anoxic water ($O_2 = NO_3 = 0$) containing Fe^{2+}.

When the recovery is composed of open canals or ditches, then all processes described in Section 3.1 will operate again. Reinfection is an important disadvantage of an open recollection system.

4. Redox Zonation

4.1. Definition and Measurement

A redox environment (zone) is defined as a confined space with a specific range of the oxidation reduction potential (ORP; degree of aerobicity), and with a specific set of dissolved constituents with characteristic concentration levels, i.e. ions and gasses that remain stable only within those ORP ranges. There are 3 reasons to prefer a subdivision into redox zones on the basis of all redox sensitive main components of water, above direct measurement of the redox potential (Stuyfzand 1993, Chapelle 2001):

a)   direct measurement of the redox potential using electrodes is difficult and yields unreliable results, whereas the redox zone can be determined in an unambiguous way from all redox sensitive main components of water;

b) the analytical results of the redox potential do not correspond with the thermodynamic expectations. The theoretical redox potentials (essential input for chemical computer models) can be approximated by selected redox couples for each redox zone; and
c) redox zones constitute a readily mappable unit.

The redox environments occurring in nature are depicted in Figure 7.5. The shown redox sequence is also observed in the case of aquifer passage during RBF or AR, where water is closed from the atmosphere once infiltrated. Criteria for determining the redox environment are given in Table 7.4. The determination has been included in HYCA, a computer program for the storage, retrieval, elaboration and interpretation of water quality data (Mendizabal and Stuyfzand 2009, www.hyca.nl).

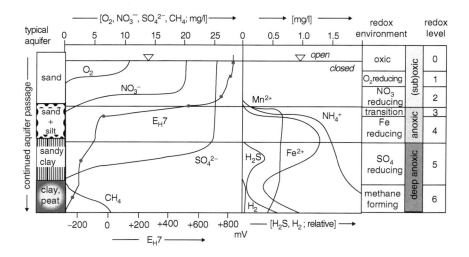

Figure 7.5. The ideal redox sequence for a system closed from the atmosphere like groundwater, where no mixing occurs and first all oxygen, then all nitrate and subsequently all sulphate is used to oxidize dead organic material, and finally methane appears (modified after Stuyfzand 1993). Also indicated are the theoretical redox potentials at pH7 (E_H7), the concentration logs for iron, manganese, hydrogen, hydrogen sulphide and ammonium, and a subdivision into redox environments. Criteria for subdivision are presented in Table 7.4.

TABLE 7.4. Practical criteria for determining the redox environment (slightly modified after Stuyfzand 1993; concentrations in mg/L).

Level	Environment	Criteria [mg/L]						
		O_2	NO_3^-	Mn^{2+}	Fe^{2+}	SO_4^{2-}	H_2S #	CH_4
0	Oxic	$O_2 \geq 0.9\ (O_2)_{SAT}$		< 0.1	< 0.1	$\geq 0.9\ (SO_4)_0$	no	< 0.1
1	O2-reducing (Penoxic)	$1 \leq O_2 < 0.9\ (O_2)_{SAT}$		< 0.1	< 0.1	$\geq 0.9\ (SO_4)_0$	no	< 0.1
2	NO_3-reducing (Suboxic)	< 1	≥ 1	< 0.1	< 0.1	$\geq 0.9\ (SO_4)_0$	no	< 0.1
3	Transition (Mn-reducing)	< 0.5	< 1	≥ 0.1	< 0.1	$\geq 0.9\ (SO_4)_0$	no	< 0.1
4	Iron reducing	< 0.5	< 0.5		≥ 0.1	$\geq 0.9\ (SO_4)_0$	no	< 0.1
5	Sulphate reducing	< 0.5	< 0.5			A	yes	< 1
6	Methanogenic	< 0.5	< 0.5			B		≥ 1

Redox clusters:		O_2	NO_3^-	Mn^{2+}	Fe^{2+}	SO_4^{2-}	H_2S #	CH_4
0-2	(sub)oxic	≥ 1 or	≥ 1	< 0.1	< 0.1		no	< 0.1
3-4	anoxic	< 0.5	< 0.5		≥ 0.1	$\geq 0.9\ (SO_4)_0$	no	< 0.25
5-6	deep anoxic	< 0.5	< 0.5			C	yes or	≥ 0.5
M	mixed @	≥ 1 or	≥ 1	≥ 0.15 or ≥ 0.15			or	> 0.1

#: yes/no = yes/no clear H_2S-smell in field $\qquad$ $(SO_4)_0$ = original SO_4 concentration in mg/L.
A: if $Cl \leq 300$ mg/L then $0.1(SO_4)_0 < SO4 < 0.9(SO_4)_0$; if $Cl > 300$ mg/L then $0.5(SO_4)_0 < SO4 < 0.9(SO_4)_0$
B: if $Cl \leq 300$ mg/L then $SO_4 < 0.1(SO_4)_0$ or $SO_4 < 1$ mg/L ; if $Cl > 300$ mg/L then $SO_4 < 0.5(SO_4)_0$
C: $SO_4 < 0.9(SO_4)_0$ or, if $Cl < 300$ mg/L , $SO_4 < 1$ mg/L
$(O_2)_{SAT}$ = $14.594 - 0.4\ t + 0.0085\ t^2 - 97\ 10^{-6}\ t^3 - 10^{-5}\ (16.35 + 0.008\ t^2 - 5.32/t)\ Cl$, with t = temp. in °C, Cl in mg/L.
@: sample composed of a mix of water from different redox environments

The term 'penoxic' is a contraction of pene oxic, with pene meaning 'nearly' in Greek (compare peninsula, peneplain). The sulphate reducing environment may be hard to determine when H_2S measurements or organoleptic observations are lacking (which is often the case). In RBF and AR systems, the original SO_4 concentration can mostly be easily established via a positive regression with the Cl concentration in the source water.

The redox environment is the chemical master-variable for RBF and AR systems (Stuyfzand 1998), because it dictates to a high degree the mobility, dissolution, breakdown and toxicity of many inorganic and especially organic compounds in or in contact with the water phase. Variations in the other master variables like pH, temperature and ionic strength are normally smaller with less impact in MAR systems. In many cases (where some data are lacking or maps become too complicated) the redox clustering according to Table 7.4 proves an excellent simplification.

The redox environment in an aquifer is a measure of the capacity of the aquifer to reduce undesired oxidants like NO_3, SO_4, (per)chlorate and Cl_2, and to eliminate specific pollutants through by-products of reduction (like H_2S which triggers the precipitation of many metalsulfides), reductive dehalogenation (chlorinated hydrocarbons) or specific oxidation or reduction processes (many organic micro-pollutants including pesticides and pharmaceuticals).

4.2. Typical Zonation Patterns

RBF

The observed spatial distribution pattern of redox zones in a typical RBF system in the lower Rhine fluvial plain (type 4 according to Table 7.2) is shown in Figure 7.6. The river water infiltrating via the sandy river bed incised in Holocene clay and peat deposits is initially (sub)oxic but soon becomes anoxic when contacting the Pleistocene sandy aquifer (Figure 7.6). This is caused by (1) the relatively high chemical oxygen demand of the infiltration water itself, due to raised concentrations of DOC and NH_4 and (2) the relatively high reduction capacity of the sediments contacted.

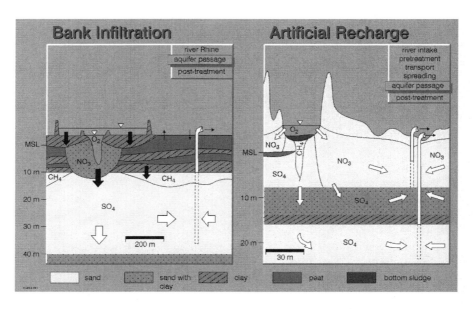

Figure 7.6. Schematized cross section over a representative Rhine bank infiltration system and an artificial recharge system in the Netherlands, showing the position of aquitards, groundwater flow and the following redox environments (for definition see Table 7.4): O_2 = oxic; NO_3 = penoxic and suboxic; SO_4 = anoxic; CH_4 = deep anoxic. After Stuyfzand (1998a).

Due to quality improvements of the Rhine River roughly since 1975 and continued leaching of the aquifer, the suboxic infiltrate is slowly advancing towards the well fields in this region. Once arrived this flow branch may provoke chemical well clogging, which has not been observed to date.

The river water that infiltrates via the endiked fluvial plain during floods, needs to pass clay and peat beds and thereby becomes deep anoxic, which does not change once entering the Pleistocene aquifer.

As shown in Table 7.2, upstream the redox zonation normally becomes less anoxic, but still showing an inversion in the aquifer (on top more reduced than below).

Basin AR

The observed spatial distribution pattern of redox zones in a typical basin artificial recharge (BAR) system in the coastal dunes of the western Netherlands is shown in Figure 7.6. The situation differs from the one in RBF systems, by (a) its more oxic character thanks to pretreatment, a selective intake which may exclude the most polluted surface water, and the very low reduction capacity of eolian sands; and (b) a normal redox stratification, with suboxic infiltrate on top of (deep) anoxic infiltrate. Here the more oxic water infiltrates via the fixed basin banks, where much less (if any) clogging layer develops. Water infiltrating via the deeper parts of the basin has a much higher chance of contacting underwater muds and also Holocene marine clay or silty sand layers with a high reduction capacity.

Most recovery systems pump both the shallow (sub)oxic infiltrate and the deeper flows of mostly anoxic (rarely deep anoxic) infiltrate, and therefore are extremely susceptible to chemical clogging.

ASTR

The observed spatial distribution pattern of redox zones in a typical deep well injection system with a remote recovery well (ASTR) is shown in Figure 7.7. This pattern has been observed after 2 years of continued infiltration (and remote recovery) in an originally deep anoxic aquifer. In the well proximal zone oxic to (sub)oxic conditions prevail, whereas in the distant aquifer zone and discharge proximal zone redox conditions are anoxic due to redox buffering by oxidation of pyrite and SOM (Soil Organic Matter).

With time the (sub)oxic zone is observed to slowly expand at the expense of the anoxic zone, while the infiltrate continues to replace the native groundwater in the aquitards and remote areas. A typical redox and mineral leaching sequence with time is shown in Figure 7.8, as based on many ASTR pilots (Stuyfzand 1998b). The oxidation of pyrite and SOM is a slow process, which contributes to the slowly decreasing O_2 and NO_3 concentrations with distance from the injection well. Their complete removal may take 10–100 days of travel time in a pyritiferous sandy aquifer without earlier leaching (Stuyfzand 1998b). The leaching rate of pyrite and SOM increases significantly with temperature, which has been successfully modeled by Prommer and Stuyfzand (2005).

ASR

The observed spatial distribution pattern of redox zones in a typical Aquifer Storage Recovery (ASR) system is shown in Figure 7.9. Oxic to (sub)oxic surface water (often drinking water) is injected during the low demand, high availability period, stored and recovered via the same well during peak demand.

This normally happens in annual cycles, during which the (sub)oxic zone gradually expands within the ASR bubble. Analogous to ASTR, pyrite and SOM (and occasionally also siderite) may strongly delay this expansion (Stuyfzand et al. 2006b, Pyne 2005). This can create problems with the quality of the recovered water, when the dissolution of Fe^{2+}, Mn^{2+}, NH_4^+, and As (from pyrite) necessitates an undesired post-treatment.

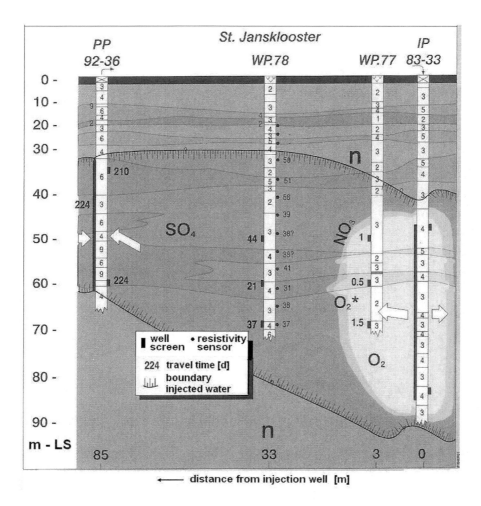

Figure 7.7. Cross section over the ASTR pilot at St. Jansklooster (Eastern Netherlands), showing the hydrogeological schematization, spatial distribution of the water injected and the redox zonation within, after 2 years of continued injection (after Stuyfzand and De Ruiter [1999]). Redox environments (for definition see Table 7.4): O_2 = oxic; O_2^* = penoxic; NO_3 = suboxic; SO_4 = anoxic. n = native groundwater (deep anoxic). Numbers within soil column indicate median grain size (1 = 100, 9 = 900 μm); numbers next to well screens and resistivity sensors indicate travel time (d).

HYDROGEOCHEMICAL PROCESSES DURING RIVERBANK FILTRATION 111

Phase	in quality evolution upon intrusion	Pore Flushes	Water	FeS2	SOM	CaCO3
0	Native groundwater	0 - 0.5	deep an			
1	Displacement native water	0.5 - 1.5				
2	Ion exchange	1.5 - 7	anoxic			
3	Redox reactions at max	7 - 20				
4	NO3- breaking through	20 - 60	suboxic			
5	O2 breaking through	60 - 400				
6	Pyrite leached, calcite still present	400 - 4000	(sub)oxic			
7	Calcite leached (>4000), SOM present	4000-10000				
8	SOM partly leached, redox inactive, still sorbing but less	>10000				

distance from injection well → time

Figure 7.8. Generalized evolution of hydrogeochemical processes in an ASTR system, with continued injection of oxic infiltration water into a deep anoxic, pyritiferous, calcareous aquifer (after Stuyfzand 2001). Deep an = deep anoxic; SOM = Soil Organic Matter; Pore flush ≈ bed volume.

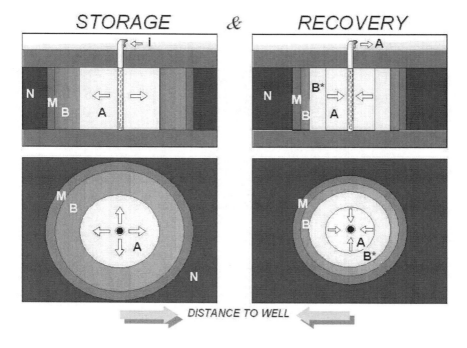

Figure 7.9. Redox zonation around an ASR well in an originally deep anoxic aquifer during storage and recovery, in cross section and planar view (after Stuyfzand 2001). i = oxic infiltration water; A = (sub)oxic infiltrate; B = anoxic infiltrate in Buffer zone; B* = anoxic infiltrate from buffer zone but strongly depleted in Fe^{2+}, Mn^{2+}, NH_4^+ and PO_4^{3-} by sorption to newly formed ferrihydrite; M = mixture of B and N; N = deep anoxic native groundwater.

Two interesting redox phenomena were noticed (Stuyfzand et al. 2006b). The first is the development of an anoxic well proximal zone during storage, which is explained by the dying-off of a microbiological community forming under (sub)oxic conditions during injection. The second is the occurrence of the so-called subterranean iron removal process during the recovery phase, when anoxic infiltrate from the buffer zone passes the (sub)oxic aquifer zone where Fe^{2+} is sorbed to newly formed ferrihydrite.

4.3. Redox Dependent Removal Efficiencies

For many micropollutants the specific 'redox barrier' could be established in field studies (Stuyfzand 1998, Stuyfzand et al. 2007b), the redox environment where these contaminants degrade or precipitate (Figure 7.10). For instance the deep anoxic environment (Table 7.4) poses a reliable redox barrier in aquifers to amongst others: NO_3, trichloroethylene, chloroform, atrazine, diurone, carbendazim and many heavy metals (which (co)precipitate as sulphide minerals). Of course this is only achieved, if the residence time in this redox zone suffices, and if the reducing phases are not leached at a too high rate.

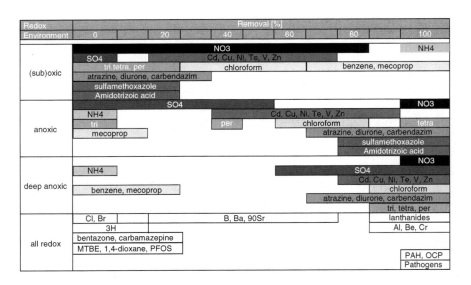

Figure 7.10. Approximate removal efficiencies for selected pollutants in surface water during aquifer passage in RBF or AR systems as a function of the redox environment. Based on numerous field observations in systems operating for many years or decennia (see text). Conditions: sufficient pore flushes to exclude retardation by sorption, low pollution levels (<0.1 mg/l) and pH = 7–8.

It should be noted that the redox zones in Figure 7.10 may perform in a different way, when the OMPs reach high concentration levels or when pH strongly deviates from 7 to 8.

The investigated field systems that form the basis of Figure 7.10, comprise a high number and big variety of RBF, BAR and ASTR systems in the Netherlands. On these sites, with dedicated transects of monitor wells, the effects of sorption could be excluded (by either long term monitoring up to 10 years or by a very hydrophylic pollutant behavior), the redox environment could be precisely assessed with the scheme in Table 7.4, and effects of (ad)mixing could be ignored thanks to using monitor wells with short well screens (0.1–1 m).

5. Identification and Quantification of Reactions by Mass Balancing

5.1. Introduction

The purpose is to identify and quantify the nature and extent of all hydrogeochemical reactions acting on infiltrating surface water, by calculating the mass transfer needed for changing the measured average input (of river water or pretreated surface water) into the output as measured somewhere downgradient in the recharged aquifer system via an observation or pumping well.

The mass balance approach is also referred to as 'inverse modeling' (Plummer 1984, Domenico and Schwartz 1998). Chemical mass balances are drawn up by using a set of linear reaction equations and summing up all resulting mass transfers between an arbitrary starting and ending point in a flow system. Only the main constituents of water are addressed here, thus excluding trace elements and OMPs.

5.2. REACTIONS+6

Mass balances are drawn up here with the EXCEL spreadsheet code REACTIONS+ version 6 (an update of REACTIONS+ version 5 presented in Stuyfzand et al. [2006a]), in 4 consecutive steps (Figure 7.11) as discussed below. The reason not to use more advanced codes like BALANCE (Parkhurst et al. 1982) or PHREEQ-C2 (Parkhurst and Appelo 1999), is that using spreadsheets yields a more direct, simple and transparent route with more degrees of freedom (but perhaps more uncertainties). At this stage our approach has not been compared yet with the above mentioned codes.

Step 1: Sample Selection and Standard Precalculations

The first step consists of the following:

(a) selection of the input and output water composition, by taking preferably the average composition of both the infiltration water and the infiltrate. The latter as monitored in an observation well or recovery system, if possible with a time shift corresponding to the average travel time in the aquifer. These data need to be entered in the spreadsheet as indicated in Figure 7.12;

(b) automatic conversion of the units of chemical analysis of the groundwater sample from mg/L to mmol/L;

(c) automatic distribution of the difference between the sum of cations and anions in a proportional way over all ions (thus tackling the ionic imbalance of water analyses);

(d) automatic calculation of CO_2, CO_3 and saturation index for calcite (SI_C) for input and output (and admixed groundwater if applying), from temperature, pH, HCO_3 and all dissolved major ions. SI_C is very well approximated by the non-iterative solution offered by Feij and Smeenk (1981), provided the water be fresh (SEC < 3,000 µS/cm);

(e) automatic calculation of pH in the output, from the calculated CO_2 and HCO_3 concentrations; and

(f) addition of a 'convective transport' contribution to even out the chloride difference between the input and output, or 'unmix' the output when composed of a mixture of infiltrate and native groundwater. The choice needs to be entered in the spreadsheet (cell O44 in Figure 7.12).

The results of all calculations are shown in Figure 7.12.

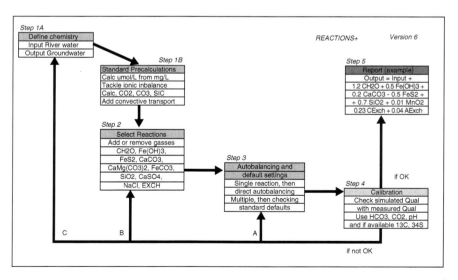

Figure 7.11. Flow chart of REACTIONS+6, showing the various steps from data entry to a reported mass balance for the infiltration water at some point in the aquifer or recovery.

Figure 7.12 — Layout of the REACTIONS+6 model (EXCEL spreadsheet)

		Reactions	O2	NO3	SO4	CO2	CH4	NH4	Fe	Mn	DOC	Cl	HCO3	PO4	Na	K	Ca	Mg	SiO2	Temp	pH	SEC	Sum-A	Sum-C	SI-C
				Oxidants		Acid		Reductants				Other main constituents & parameters										Calculated parameters			
Input		Input (mg/L)	9.1	16.1	65	1.9	0	0.19	0.01	0.11	2.9	146	162	0.31	80.5	6.4	75	12.9	4.7	10.5	8.2	766	8.40	8.48	0.58
		uncorrected (mmol/L)	0.284	0.26	0.68	0.043	0	0.011	0.000	0.002	0.24	4.12	2.65	0.003	3.502	0.16	1.87	0.53	0.078	10.5	8.2	766	8.40	8.48	0.58
		corrected (mmol/L)	0.284	0.26	0.68	0.043	0	0.01	0.000	0.002	0.24	4.14	2.67	0.003	3.48	0.16	1.86	0.53	0.078	10.5	8.2	770	8.44	8.44	0.58
Redox		Unsat zone + Convection			-0.11	0						-0.98										-.119	-1.19	-1.22	
		Nitrification	-0.021	0.01		0.021		-0.01					-0.02												
		DOC oxidation	-0.10	0.00		0.096					-0.10		0.00												
		O2 oxidizing FeS2		0	0.00	0							0												
		NO3 oxidizing FeS2		0	0.00	0			0.00				0												
		O2 oxidizing FeCO3				0																			
		O2-reduction by SOM	-0.167			0.17							0.27												
		NO3-reduction by SOM		-0.27		0.07		0.00					0.02	0.000												
		MnO2-reduction by SOM				-0.013		0.01		0.009			0.45	0.000												
		Fe(OH)3 reduction by SOM				-0.396		0.10	0.23		0.20		0.77	0.001												
		SO4-reduction (FeS2) by SOM			-0.38	0.57		0.10	-0.191				0.77	0.006												
		CH4-formation by SOM				0.01	0.01	0.001			0.002			0.000												
Dissolution		CaCO3-dissolution				-0.37							0.73				0.37								
		CaMg(CO3)2-dissolution				0							0				0	0							
		FeCO3 dissolution				0			0	0			0				0								
		CaSO4.2H2O-dissolution			0												0								
		NaCl-dissolution										0			0										
		SiO2 dissolution																	0.37						
Ex		Cation exchange						0.05			0.15				-0.07	0.02	0.06	-0.06							
		Anion Exchange											-0.147	0.049											
		Fe2+ oxidation	0			0							0												
		Mn2+ oxidation	0			0							0												
Mix		Calculated Sum [mmol/L]	0.00	0.00	0.19	0.20	0.01	0.16	0.04	0.01	0.50	3.16	4.74	0.06	2.52	0.16	2.16	0.44	0.45	10.50	7.79	726	8.46	8.14	0.49
Output		Unmixed output [mmol/L]	0.00	0.00	0.19	0.41	0.01	0.16	0.04	0.01	0.50	3.16	4.42	0.06	2.52	0.16	2.16	0.44	0.45	10.50	7.45	705	8.14	8.14	0.12
		Corrected admix [mmol/L]	0		0	0.07	0.013	0.11	0.072	0.07	0.17	3.61	5.83	0.01	0.62	0.04	2.52	0.25	0.70	10.0	7.20	523	6.48	6.48	0.07
		Admixed groundwater [mg/L]	0	0	0	43.2	0.2	2	4	0.4	2	22	360	1	14	1.5	100	6	42	10.0	7.20	529	6.55	6.40	0.07
		Corrected output [mmol/L]	0	0	0.19	0.41	0.006	0.16	0.035	0.011	0.50	3.16	4.42	0.059	2.52	0.16	2.16	0.44	0.45	10.5	7.45	705	8.14	8.14	0.12
		Measured Output [mg/L]			18	17.5	0.1	3	2	0.6	6	110	265	5.51	59	6.3	88	11	27	10.5	7.45	693	7.99	8.28	0.12
		Cl regression input [mmol/L]			0.109										0.908	0.03	0.129	0.025							

Reactions Inf = H2O +
1.25 CH2O +
0.23 Fe(OH)3 +
-0.19 FeS2 +
0 FeCO3 +
0.37 SiO2 +

+ Admix = 0 %
0.37 CaCO3 +
0 CaMg(CO3)2 +
0.01 MnO2 +
0.13 C-Exch +
0.10 A-Exch +

1.00 = fraction Inf in mix
0 = Fe2+ oxid
-0.11 = NH4 + DOC
0 = CaSO4 +
0 = NaCl

FILL IN ON ROW:
17.72 = Input
21.00 = measured output
0.4 = admixed groundwater
FILL IN HERE:
0 = yes (1) or no (0) admixing

DO NOT FILL IN:
10.40 = reaction controled or calculated
-7.72 = autobalanced
0 = default setting
0.04 = unmixed output to be compared
25.18 = calibration term

Figure 7.12. Layout of the REACTIONS+6 model, set in EXCEL spreadsheet, with the Hollandsch Diep type A case as example (Table 7.6). For further explanations see 'Instructions'.

Anions and cations are corrected for ionic imbalance by taking:

$$A_{CORR} = A (\Sigma A + \Sigma C) / (2 \Sigma A) \qquad (7.1)$$

$$C_{CORR} = C (\Sigma A + \Sigma C) / (2 \Sigma C) \qquad (7.2)$$

with:
A_{CORR}, C_{CORR} = corrected concentration of resp. Anion A and Cation C [mmol/L];
ΣA, ΣC = sum of resp. all Anions and Cations [meq/L].

The saturation index for calcite (SI_C) is standard defined as follows:

$$SI_C = \log ([Ca^{2+}] [CO_3^{2-}] / K_C) \qquad (7.3)$$

with:
[X] = activity of X [mol/kg water], and K_C = solubility constant for calcite [-].

The CO_2 and CO_3 concentrations in mg/L and SI_C are calculated as follows, with HCO_3 and Ca data entry in mg/L:

$$CO_2 = 44010 (10^{-pH} \gamma_1 HCO_3/61020) / K_1 \qquad (7.4)$$

$$CO_3 = 60020*(K_2 HCO_3/61020 / 10^{-pH}) / \gamma_1^3 \qquad (7.5)$$

$$SI_C = pH + \log [(Ca/40080 \ HCO_3/61020 \ K_2) / K_C] - 2.5 \sqrt{i} / (1 + \sqrt{i}) \qquad (7.6)$$

where:
γ_Z = activity coefficient for ions with charge Z [-]; i = ionic strength [mol/L]; K_1 = first dissociation constant of H_2CO_3 [-]; K_2 = second dissociation constant of H_2CO_3 [-]. γ_Z and i are calculated as follows:

$$\gamma_Z = 10^{\wedge}(-0.5 \ Z^2(\sqrt{i}/[1+\sqrt{i}] - 0.3 \ i)) \qquad (7.7)$$

$$i = 0.5 \ \Sigma \ m_i \ Z_i^2 \qquad (7.8)$$

with: m_i = molality of ion i [mol/L]; Z_i = charge of ion i [-].

For the constants K_C, K_1 and K_2 the temperature dependent values given by Plummer and Busenberg (1982) have been taken.

The pH of the output is calculated from the CO_2 and HCO_3 concentrations [mg/L] resulting from the mass balance:

$$pH = -\log [K_1 CO_2/44010) / (\gamma_1 HCO_3/61020)] \qquad (7.9)$$

The 'convective transport' contribution requires an important decision by the user (cell O44 in Figure 7.12). Any Cl difference between input and output indicates that either (i) fluctuations occur in the input that are not exactly matched in the output, by not accounting well enough for the travel time; or (ii) analytical errors are involved; or (iii) native groundwater is admixed in the output.

The first case is relatively easy to account for, by just adding or subtracting the Cl difference to the input and do the same for those water quality parameters that show a clear correlation with Cl in the input, however by utilizing their regression equation with Cl (see Figure 7.12).

The third case is dealt with by determining the fraction of the RBF or AR infiltrate in the mixture (α_{INF}) on the basis of chloride as a tracer, and subsequently by unmixing the ouput so as to obtain the quality of the infiltrated surface water in the aquifer alone (X_{INF}). This requires of course that the quality of the native groundwater (X_N) be known (or can be estimated) and be entered into REACTIONS+6 as well.

$$\alpha_{INF} = (Cl_M - Cl_{INF}) / (Cl_{INF} - Cl_N) \tag{7.10}$$

$$X_{INF} = \{X_M - X_N (1 - \alpha_{INF})\} / \alpha_{INF} \tag{7.11}$$

with:
Cl_M, X_M = measured concentration of resp. Cl and X in the mixed output [mg/L]; Cl_{INF}, X_{INF} = average concentration of resp. Cl and X in the input [mg/L]; Cl_N, X_N = observed average concentration of resp. Cl and X in the admixed native groundwater [mg/L].

Step 2: The Most Plausible Reactions and their Selection
The most important hydrogeochemical reactions included in REACTIONS+6 are listed in Table 7.5. Their selection is based on extensive geochemical and hydrochemical monitoring of various Dutch aquifer systems. Several reactions were excluded from Table 7.5 as they are considered negligible in MAR systems, like the dissolution of (aluminum) silicate minerals (except for quartz or opal) and the dissolution of minerals by strong acids (only CO_2 being active). The reason to exclude these reactions is that most infiltration waters are well buffered, because these reactions already happened during their genesis.

Soil Organic Material (SOM) is represented in Table 7.6 as CH_2O. However, in the mass balance calculations SOM is addressed as $CH_2O(NH_3)_X(H_3PO_4)_Y$ with default values x = 0.151, y = 0.0094 in accordance with Redfield (Froelich et al. 1979). This means that SOM oxidation not only yields $H_2O + CO_2$ (or HCO_3) but also NH_4 and PO_4. This is, however, neglected when SOM is oxidized in (sub)oxic environment (by O_2 or NO_3), because the NH_4 is nitrified and PO_4 has a high chance to sorb to iron (hydr)oxides. In deep anoxic environment also DOC is mobilized, the amount of which is coupled to the amount of NH_4 produced during reactions 11 + 12 in Table 7.5 ($\Delta DOC = 2 \, \Delta NH_4$ in mmol/L).

TABLE 7.5. List of the most relevant hydrogeochemical reactions for infiltrating surface water in either an RBF or AR system.

	Proces	Reaction equation	No.
Initial	Unsat zone + Convection	+ O2 + CO2 - CH4, (Cl-out - Cl-in)	1
	Nitrification	2O2 + NH4 + 2HCO3 --> NO3 + 2 CO2 + 3H2O	2
	DOC oxidation	0.5O2 + 0.4NO3 + CH2O-DOC --> 0.6CO2 + 0.4HCO3 + 0.2N2 + 0.8H2O	3
Redox	O2 oxidizing FeS2	3.75O2 + FeS2 + 4HCO3 --> Fe(OH)3 + 2SO4 +4CO2 + 0.5H2O	4
	NO3 oxidizing FeS2	2.8NO3 + FeS2 + 0.8CO2 + 0.4H2O --> Fe + 2SO4 + 1.4N2 + 0.8HCO3	5
	O2 oxidizing FeCO3	O2 + 4FeCO3 + 6H2O --> 4Fe(OH)3 + 4CO2	6
	O2-reduction	O2 + CH2O --> CO2 + H2O	7
	NO3-reduction	4NO3 + 5CH2O --> 2N2 + CO2 + 4HCO3 + 3H2O	8
	MnO2-reduction	MnO2 + 0.5CH2O + 1.5CO2 + 0.5H2O --> Mn + 2HCO3	9
	Fe(OH)3 reduction	Fe(OH)3 + 0.25CH2O + 1.75CO2 --> Fe + 2HCO3 + 0.75H2O	10
	SO4-reduction (FeS2)	2SO4 + 3.5CH2O + Fe --> FeS2 + 2HCO3 + 1.5CO2 + 2.5H2O	11
	CH4-formation	CO2 + 2CH2O --> CH4 + 2CO2	12
Dissolution	CaCO3-dissolution	CaCO3 + CO2 +H2O <--> Ca + 2HCO3	13
	CaMg(CO3)2-dissolution	CaMg(CO3)2 + 2CO2 + 2H2O <--> Ca + Mg + 4HCO3	14
	FeCO3-dissolution	Fe(1-x)MnxCO3 + CO2 + H2O <--> (1-x)Fe + xMn + 2HCO3	15
	CaSO4.2H2O-dissolution	CaSO4.2H2O --> Ca + SO4	16
	NaCl-dissolution	NaCl <--> Na + Cl	17
	SiO2 dissolution	SiO2 + 2H2O <--> H4SiO4	18
Ex	Cation exchange	aNa + bK+ cMg + [dCa]-EXCH <--> dCa + [aNa,bK,cMg]-EXCH	19
	Anion Exchange	3 HCO3 + [PO4]-EXCH <--> PO4 + [3HCO3]-EXCH	20
Mix	(sub)oxic with (deep)	O2 + 4Fe^{2+} + 8HCO3 + 2H2O --> 4Fe(OH)3 + 8CO2	21
	anoxic	0.5 O2 + Mn^{2+} + HCO3 --> MnO2 + 2CO2	22

TABLE 7.6. The total hydrogeochemical reaction scheme for river bank filtrate as sampled in a monitor well on 5 locations in the Netherlands. Indicated are: site characteristics, gas additions, the geochemical reactions (+ = gain in water phase, by oxidation, dissolution or desorption; - = loss from water phase by precipitation or adsorption), and model validation by comparison of measured with calculated CO_2 and HCO_3.

Location	unit	Roosteren	Opperduit	Hollandsch Diep		
Well code		WP.41-f2	292-b	Type A	Type B	Type C
RBF Type	-	2	4	5	6	6
Redox	-	(sub)oxic	anoxic	deep anoxic	very deep anoxic	
Unsat. Zone	-	yes	no	no	no	no
Mud layer [m]	m	0	0	0.2-0.5	1-2.5	3.5
Extra O2	mmol/L	0.17	0	0	0	0
Extra CO2	mmol/L	0.65	0	0	0	0
Extra CH4	mmol/L	0	0	0	9.50	26.90
CH2O	mmol/L	0.14	0.61	1.25	22.88	58.74
FeS(2)	mmol/L	0.04	-0.04	-0.19	-0.31	-0.50
CaCO3	mmol/L	0.22	0.02	0.37	5.40	13.69
Fe(OH)3	mmol/L	-0.04	0.07	0.23	1.02	1.17
Cation Exchange	mmol/L	0.05	0.39	0.13	3.07	13.04
Anion Exchange	mmol/L	0.01	0.00	0.10	0.22	0.56
SiO2-minerals	mmol/L	0.04	0.05	0.37	0.22	0.19
NH4+DOC oxid	mmol/L	-0.10	-0.14	-0.11	-0.11	-0.11
MnO2	mmol/L	-0.001	0.015	0.009	0.053	0.007
CO2-measured	mmol/L	0.93	0.22	0.41	4.58	14.25
CO2-calculated	mmol/L	0.93	0.38	0.20	4.58	14.25
HCO3-measured	mmol/L	2.76	3.24	4.42	18.25	38.29
HCO3-calculated	mmol/L	2.76	3.26	4.74	17.43	36.14

Additional O_2 or CO_2 inputs may be required, of atmospheric O_2 in the unsaturated zone, and of CO_2 from plant root respiration in either an unsaturated or saturated zone. Also, CH_4 or CO_2 may be lost from the system through upward migrating gas bubbles. These reactions normally become evident when the calculated CO_2, HCO_3 or SO_4 output is too low.

The oxidation of DOC and NH_4 in the input is addressed as follows (based on expert rules). First NH_4 is oxidized by O_2 (reaction 2 in Table 7.5), until – if needed – 50% of all available O_2 has been consumed by this nitrification. This means that a maximum of $O_2/4$ mmol/L can be used for nitrification.

The theoretical maximum of DOC oxidation (ΔDOC_{MAX} in mmol/L) in closed systems (below the groundwater table) is roughly approximated by a function of the DOC input (DOC_{IN} in mmol/L), to be fine-tuned for specific cases:

$$\Delta DOC_{MAX} = -0.17 \, DOC_{IN}^2 + 0.55 \, DOC_{IN} - 0.027 \qquad (7.12)$$

This relation is based on the assumption that a significant portion of DOC cannot be oxidized and that the maximum amount of available oxidants (notably O_2) is limited below the groundwater table. Indeed, insufficient $O_2 + NO_3$ poses an even more severe limit to DOC oxidation. This leads to the following decision:

$$\text{if } \Delta DOC_{MAX} < O_2 - 2 \, \Delta NH_4 + NO_3 + \Delta NH_4 \text{ then:}$$

$$\Delta DOC = \Delta DOC_{MAX} \qquad (7.13A)$$

else:

$$\Delta DOC = O_2 - 2 \, \Delta NH_4 + (NO_3 + \Delta NH_4)/0.8 \qquad (7.13B)$$

where (all in mmol/L): $O_2 = O_2$ in input + extra atmospheric O_2; $NO_3 = NO_3$ in input + convective NO_3; $\Delta NH_4 =$ nitrified NH_4.

The sequence of oxidant consumption by DOC is O_2 first until finished, then NO_3. Pyrite oxidation or formation is automatically diagnosed as follows, with default setting $(SO_4)_{GYPSUM} = 0$:

$$\text{pyrite oxidation if } (SO_4)_{OUT} > (SO_4)_{IN} + (SO_4)_{CT} + (SO_4)_{GYPSUM} \qquad (7.14A)$$

$$\text{pyrite formation if } (SO_4)_{OUT} < (SO_4)_{IN} + (SO_4)_{CT} + (SO_4)_{GYPSUM} \qquad (7.14B)$$

where $(SO_4)_{OUT}$, $(SO_4)_{IN}$, $(SO_4)_{CT}$, $(SO_4)_{GYPSUM} =$ the SO_4 concentration [mmol/L] in resp. the output, input, convective transport and from gypsum dissolution.

In most cases the dissolution of gypsum can indeed be ignored. If pyrite oxidation has been identified, then the consumption sequence of O_2 and/or NO_3 is defined as follows: $\Delta(O_2)_D$ first, then $\Delta(NO_3)_D$ (if still needed) and then $\Delta(O2)_{CT}$ (if still needed), with $\Delta(O_2)_D =$ all O_2 remaining after NH_4 and DOC oxidation;

$\Delta(NO_3)_D$ = NO_3 remaining after NH_4 and DOC oxidation; $\Delta(O2)_{CT}$ = extra atmospheric O_2 from unsaturated zone, to be entered in cell D10 of Figure 7.12.

Although recovery systems may rapidly clog by chemical precipitates, the extent of the responsible redox reactions (numbers 21–22 in Table 7.5) is rarely noticed in the output because of the high fluxes (Houben and Treskatis 2007). Reactions 21 and 22 are quantitatively more important during the injection phase of ASTR and ASR systems, when the (sub)oxic infiltration water oxidizes desorbing Fe^{2+} and Mn^{2+}. Reaction 21 can also be used to change Fe^{2+} that dissolved from pyrite or siderite, into $Fe(OH)_3$.

Step 3: Autobalancing and Default Settings

For each water sample the previous steps are automatically taken care of, as is the 3rd step. In the spreadsheet all reactions are automatically quantified (autobalanced) by directly taking the difference between the in- and output. This works out fine for those reactions where at least one dissolved component is only dictated by that one reaction, like for K and SiO_2 (reactions 18–19 in Table 7.5), and where a fixed reaction sequence or amount is defined. The latter holds for NH_4 and DOC oxidation (Eq. 7.13) and for pyrite oxidation by O_2 and NO_3 (Eq. 7.14).

Default settings are used with subsequent calibration in step 4, where more than one reaction is involved (each with an unknown impact), and where no fixed reaction sequence has been defined. This is the case for many parameters. An example is SO_4, which can be involved in pyrite oxidation or formation on the one hand, and gypsum dissolution on the other. Frequently the dissolution of gypsum can be ignored (default setting 0). The example in Figure 7.12 offers further explanations.

Step 4: Calibration

The best mass balance, with the best distribution over the various reactions, is obtained when the result, the simulated water quality after interaction with the aquifer, closely approximates, in this case, the observed water quality as sampled from the monitoring well or recovery system.

The only final calibration terms in the mass balance as presented, are CO_2 and HCO_3 (and pH which is, however, calculated from the CO_2 / HCO_3 ratio), because CO_2 and HCO_3 are not autobalanced in step 3 but result from adding up all losses and gains.

Losses or gains of important reactive gasses like O_2, CO_2 and CH_4 may lead to inability to obtain good calibration results, if neglected.

If the simulated transformation of water prior to infiltration into the observed groundwater succeeds well, we may have confidence in the mass balance and we have quantified the relative contribution of each mixing end-member and each reaction. Such a simulation approach is, however, never unequivocal and may require further independent evidence from isotopes (like ^{13}C and [34]S), geochemical inspection and laboratory experiments.

Instructions

The following instructions refer to REACTIONS+6, which has been displayed in full in Figure 7.12. The quality of the input needs to be entered in mg/L on row 7, where cells are colored in yellow. On row 38 the linear regression constant a of the relation between X and Cl (X = a Cl + b; mmol/L basis) needs to entered, if the chloride difference between the input and output is due to salinity fluctuations.

The measured quality of the output needs to be entered in mg/L on row 37, where cells are colored in turquoise. The measured quality of the admixed groundwater, if relevant, needs to be entered in mg/L on row 35, where cells are colored in orange.

If admixing is relevant, then change 0 into 1, in blue cell O44. This will automatically fix the Cl autobalance cell M10 at 0, and thereby change the contents of row 33 from the contents of row 36 into the unmixed output as calculated with Eq. 7.11 using the contents of row 34. Default settings may need to be adjusted, depending on the system. All other cells should be left untouched.

5.3. Examples of Application

Mass balances were drawn up with the new REACTIONS+6 model, for the same 5 observation wells (in different hydrochemical RBF environments) as presented by Stuyfzand et al. (2006a). The present results in Table 7.6 slightly deviate from the earlier ones, because of several improvements and simplifications of the model. As an example the whole EXCEL sheet (incl. all inputs and results) is presented in Figure 7.12 for the Hollandsch Diep RBF estuary case (type A), where a mixture of Rhine River water (78%) and Meuse River water (22%) is infiltrating.

A concise form of presenting the results for all 5 observation wells, is to separately express all dissolved/precipitated minerals, all oxidized SOM, all exchanged cations and anions, and the sum of NH_4 and DOC oxidized, in mmol/L. The results of this (shown in Table 7.6), allow to compare the extent of the principal reactions identified. It can be concluded from Table 7.6, that the principal hydro-geochemical reactions are, in general decreasing order: (a) oxidation of SOM from the aquifer; (b) calcite dissolution; (c) cation exchange; (d) reductive dissolution of iron hydroxides; (e) pyrite formation; (f) dissolution of SiO_2 containing phases; (g) anion exchange, (h) the oxidation of imported NH_4 + DOC, and (i) the reductive dissolution of Mn-phases.

The downstream change of RBF types 2–6 is clearly reflected in the reaction schemes given in Table 7.6. There is an increasing trend for all reactions, except for the dissolution of SiO_2 containing phases, the oxidation of imported NH_4 + DOC, and the reductive dissolution of Mn-phases.

The mass balance approach also reveals where specific processes have to be taken into account, which are otherwise easily overlooked (Stuyfzand et al. 2006a). At the Roosteren site this was the additional input of 0.17 mmol/L O_2 and

0.65 mmol/L CO_2 in the unsaturated zone (Table 7.6), because there was otherwise a clear deficit in O_2 to sustain SOM and pyrite oxidation, and in measured CO_2.

At the Hollandsch Diep site, additional CO_2 inputs by methanogenesis were needed for types B and C (Table 7.6). The extra CH_4 input of resp. 9.5 and 26.9 mmol/L yields an equal amount of CO_2 (by reaction 12 in Table 7.5), without which the very high NH_4, DOC, Ca and HCO_3 concentrations cannot be explained. This extreme, additional methane input is explained in 2 ways. First, the permanent upward escape of gas bubbles through the calcareous underwater mud layer probably resulted in a progressive reaction with $CaCO_3$ (CH_4 escaping to the atmosphere, CO_2 being partly consumed by reaction). Second, the measured CH_4 (and to a lesser extent CO_2) concentrations were probably too low due to the escape of pressurized gas during sampling and/or analysis.

6. Reactive Transport Modeling

There are a growing number of non-reactive and reactive transport models that can be used to simulate and predict the water quality changes in MAR systems. The most used, versatile, multicomponent reactive transport models are PHREEQC-2 (Parkhurst and Appelo 1999), for 1D simulations, and PHT3D (Prommer et al. 2003), for 3D simulations. PHT3D couples the transport simulator MT3DMS (Zheng and Wang 1999) with the geochemical model PHREEQC-2.

PHREEQC-2 has been used to model amongst others, within the context of MAR, the hydrogeochemical processes during subterranean iron removal (Appelo et al., 1999) and ASR (Gaus 2001). PHT3D has been applied, again within the context of MAR, to for instance ASR (Greskowiak et al. 2005), ASTR (Prommer and Stuyfzand 2005), BAR (Greskowiak et al. 2006) and RBF (Ray and Prommer 2005).

An alternative way of multicomponent, reactive transport modeling is via an expert model like EASY-LEACHER (EL), which is set in EXCEL spreadsheet, and as such relatively simple to operate and yielding output within seconds (Stuyfzand 1998c, 2001). It has been applied to RBF, BAR and ASTR, and a separate version exists for ASR. Some details are given in Figures 7.13–7.14.

The use of EL allows for a rapid quantification of the behavior of main constituents and pollutants in basins and aquifers, and thus of their potential impact on drinking water quality (or on the required post-treatment) and the environment, notably the aquifer. Perhaps unique is the capability to also calculate the accumulation rate and chemical composition of clogging underwater muds in recharge basins (Figure 7.14).

HYDROGEOCHEMICAL PROCESSES DURING RIVERBANK FILTRATION 123

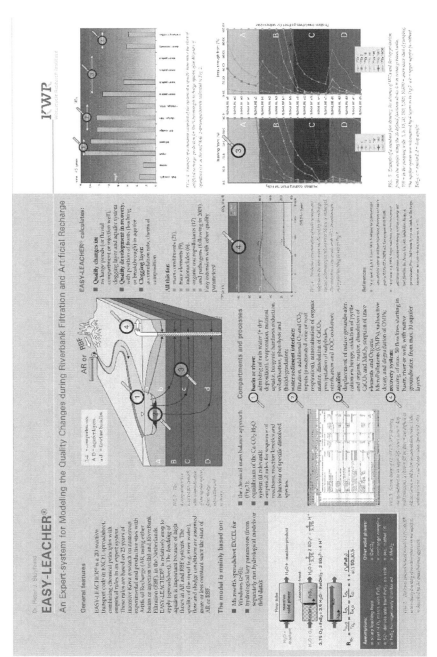

Figure 7.13. General information regarding Easy-Leacher (from Stuyfzand 2001).

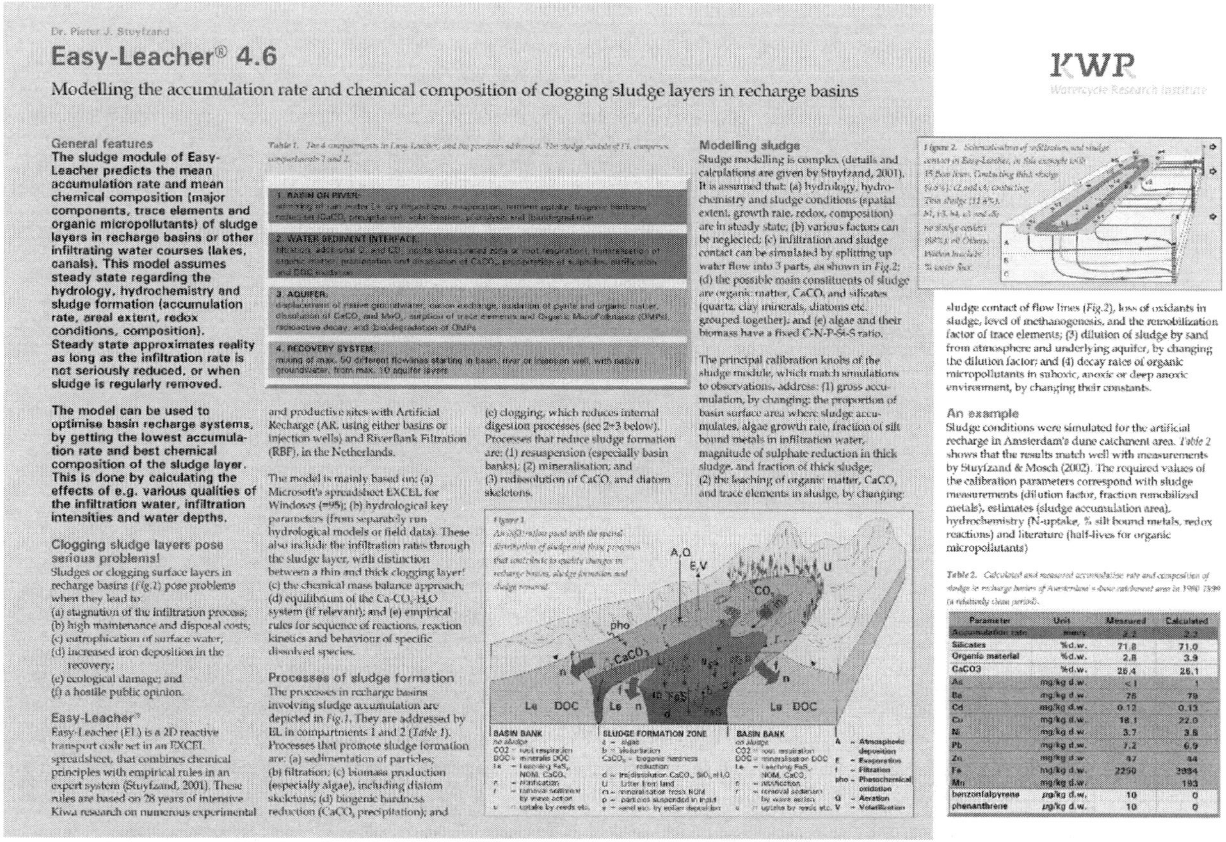

Figure 7.14. Modeling underwater muds in infiltrating water courses with Easy-Leacher (from Stuyfzand 2001).

The pollutants addressed include organic micropollutants like chlorinated hydrocarbons, pesticides and pharmaceuticals, heavy metals, radionuclides and pathogenic micro-organisms. In addition, also the load of acidifying (NH_4, DOC, CO_2) or oxidizing substances (O_2, NO_3, SO_4) is evaluated by calculating the resulting rate of leaching of valuable aquifer constituents like acid buffering calcium carbonate and redox buffering pyrite and Soil Organic Matter. The effects of this leaching on the redox zonation and thereby on the behavior of pollutants are incorporated in the model.

7. Concluding Remarks

Knowledge about the hydrogeochemical processes is crucial especially for (i) optimizing the chemical performance of existing and future RBF and AR systems, (ii) optimizing their water quality monitoring system, and (iii) making predictions about the behavior of exogenic pollutants and endogenic solutes dissolving from the aquifer, which may need to be addressed in a post-treatment.

Basic is the understanding that RBF and AR systems are composed of various compartments (surface water, basin proximal, distant aquifer and discharge proximal zone) and redox environments within them; each compartment and zone with specific removal efficiencies and mobilization potentials, as shown and discussed in this contribution. Further research should focus on the time dependent position of the boundaries of the recharge proximal zone, deep bed filtration of suspended fines, and both seasonal and annual fluctuations of the redox environments and their removal efficiencies for emerging pollutants.

Hydrogeochemical processes can be identified and quantified by a simple mass balance programme, REACTIONS+ version 6, as presented here. This programme is set in EXCEL spreadsheet, and can be downloaded free of charge from www.kwrwater.nl (look for 'publications', and there for 'software'). Any suggestions for improvement are welcomed.

References

Appelo CAJ, Drijver B, Hekkenberg R, de Jonge M (1999) Modeling in situ iron removal from ground water. Groundwater 37:11–817

Chadha DK (2002) State of art of artificial recharge applied on village level schemes in India. In: 'Management of aquifer recharge and subsurface storage; making better use of our largest reservoir.' Netherlands National Committee IAH, Publ. No. 4:19–24

Chapelle FH (2001) Groundwater microbiology and geochemistry. Wiley, New York, 477 p

Dillon PJ (2005) Future management of aquifer recharge. Hydrogeol J 13(1):313–316

Dillon P, Toze S (eds) (2005) Water Quality Improvements During Aquifer Storage and Recovery. American Water Works Assoc. Research Foundation Report 91056F, 286 p + 2 CDs

Domenico PA, Schwartz FW (1998) Physical and chemical hydrogeology. Wiley, New York

Feij LAC, Smeenk JGMM, (1981) Determination of the calcite saturation index of water. H2O 14(6):131–136 (in Dutch)

Froelich PN, Klinkhammer GP, Bender ML et al (1979) Early oxidation of organic matter in pelagic sediments of the eastern equatorial Atlantic: suboxic diagenesis. Geochim Cosmochim Acta 43:1075–1090

Gale I (ed) (2005) Strategies for Managed Aquifer Recharge (MAR) in semi-arid areas. UNESCO IHP, www.iah.org/recharge, 34 p

Gaus I (2001) Physical and geochemical modelling (SWIFT-PHREEQC) of British aquifers for Aquifer Storage and Recovery purposes. Part 2: Geochemical modelling. Brit Geol Surv Report

Greskowiak J, Prommer H, Vanderzalm J et al (2005) Modeling of carbon cycling and bio-geochemical changes during injection and recovery of reclaimed water at Bolivar, South Australia. Water Resour Res 41:W10418

Greskowiak J, Prommer H, Massmann G, Nützmann G (2006) Modelling the seasonally changing fate of the pharmaceutical residue phenazone during artificial recharge of groundwater. Environ Sci Technol 40:6615–6621

Hofmann T (1998) Colloidal and suspended particles: origin, transport and relevance of mobile solids with regard to artificial groundwater recharge. Dortmunder Beiträge zur Wasserforschung, Veröffentlichungen des Instituts für Wasserforschung GmbH Dortmund under der Dortmunder Energie- und Wasserversorgung GmbH, Nr 56, Dortmund (in German)

Houben G, Treskatis C (2007) Water well rehabilitation and reconstruction. McGraw-Hill, New York

Hubbs SA (2006) Riverbank filtration hydrology; impacts on stream capacity and water quality. NATO Science Series IV, Earth and Environmental Sciences, Vol. 60, Springer, 344 p

Huisman L, Olsthoorn TN (1983) Artificial groundwater recharge. Pitman Adv. Publ. Program, London, 320 p

Medema GJ, Stuyfzand PJ (2002) Removal of micro-organisms upon basin recharge, deep well injection and river bank filtration in the Netherlands. In: Dillon PJ (ed) Management of aquifer recharge for sustainability, Proc. 4th Internat. Symp. on Artificial Recharge, Adelaide, Australia, 22–26 Sept 2002, Balkema, 125–131

Mendizabal I, Stuyfzand PJ (2009) HYCA, an analytical database for water quality data; version 3.1. KWR Watercycle Research Institute, www.hyca.nl

Parkhurst DL, Appelo CAJ (1999) PHREEQC-2, a computer program for speciation, reaction-path, 1D-transport and inverse geochemical calculations. http://water.usgs.gov.software/

Parkhurst DL, Plummer LN, Thorstenson DC (1982) BALANCE – a computer program for calculating mass transfer for geochemical reactions in ground water. US Geol Surv, Water Resour Invest 82–14, 29 p

Plummer LN (1984) Geochemical modeling: a comparison of forward and inverse methods. In: Hitchon B, Wallick EJ (eds) First Canadian/American Conference on Hydrogeology, Dublin, Ohio, National Water Well Assoc., 149–177

Plummer LN, Busenberg E (1982) The solubilities of calcite, aragonite and vaterite in CO2-H2O solutions between 0 and 90°C and an evaluation of the aqueous model for the system CaC03-CO2-H2O. Geochim Cosmochim Acta 46:1011–1040

Prommer H, Barry DA, Zheng C (2003) PHT3D – A MODFLOW/MT3DMS based reactive multi-component transport model. Ground Water 42(2):247–257

Prommer H, Stuyfzand PJ (2005) Identification of temperature-dependent water quality changes during a deep well injection experiment in a pyritic aquifer. Environ Sci Technol 39:2200–2209

Pyne RDG (2005) Aquifer storage recovery: a guide to groundwater recharge through wells. 2nd Edition. ASR Press, Gainesville

Ray C (ed) (2002) Riverbank filtration: understanding contaminant biogeochemistry and pathogen removal. Proc. NATO Workshop Tihany, Hungary 5–8 Sept 2001, NATO Science Series IV Earth and Environmental Sciences 14, Kluwer Acad Publ., 253 p

Ray C, Prommer H (2005) Clogging-induced flow and chemical transport simulation in riverbank filtration systems. In: Hubbs SA (ed) Riverbank filtration hydrology; impacts on system capacity and water quality. Proc. NATO Advanced Research Workshop on Effect of Riverbed Clogging on Water Quality and System Capacity, held in Bratislava, Slovak Republic, 7–10 Sept 2004. NATO Science Series: IV: Earth and Environmental Sciences, Vol. 60

Schoenheinz D (2004) DOC als Leitparameter, zur Bewertung und Bewirtschaftung von Grundwasserleitern mit anthropogen beeinflusster Infiltration. PhD TU Dresden, 117 p

Sontheimer H (1991) Trinkwasser aus dem Rhein? Academia Verlag, Sankt Augustin

Stuyfzand PJ (1989) Quality changes of river Rhine and Meuse water upon basin recharge in The Netherlands' coastal dunes: 30 years of experience. In: Johnson AI, Finlayson (DJ (eds) Artificial recharge of groundwater, Proc. Int. Symp. Anaheim USA, 21–28 Aug 1988, Am Soc Civil Eng, New York, 233–245

Stuyfzand PJ (1993) Hydrochemistry and hydrology of the coastal dune area of the Western Netherlands. PhD Thesis Vrije Univ. Amsterdam, published by KIWA, ISBN 90-74741-01-0, 366 p

Stuyfzand PJ (1998a) Fate of pollutants during artificial recharge and bank filtration in the Netherlands. In: Peters JH (ed) Artificial recharge of groundwater, Proc. 3rd Intern. Symp. on Artificial Recharge, Amsterdam the Netherlands, Balkema, 119–125

Stuyfzand PJ (1998b) Quality changes upon injection into anoxic aquifers in the Netherlands: evaluation of 11 experiments. In: Peters JH (ed) Artificial recharge of groundwater, Proc. 3rd Intern. Symp. on Artificial Recharge, Amsterdam the Netherlands, Balkema, 283–291

Stuyfzand PJ (1998c) Simple models for reactive transport of pollutants and main constituents during artificial recharge and bank filtration. In: Peters JH (ed) Artificial recharge of groundwater, Proc. 3rd Intern. Symp. on Artificial Recharge, Amsterdam the Netherlands, Balkema, 427–434

Stuyfzand PJ (2001) Modelling the quality changes upon artificial recharge and bank infiltration: principles and user's guide of EASY-LEACHER 4.6. Kiwa-report SWI 99.199, 2nd ed, 150 p

Stuyfzand PJ (2002a) Quantifying the environmental impact and sustainability of artificial recharge systems. In Dillon PJ (ed) Management of aquifer recharge for sustainability, Proc. 4th Internat. Symp. on Artificial Recharge, Adelaide, Australia, 22–26 Sept 2002, Balkema, 77–82

Stuyfzand PJ (2002b) Modelling the accumulation rate and chemical composition of sludges in recharge basins. In Dillon PJ (ed) Management of aquifer recharge for sustainability, Proc. 4th Internat. Symp. on Artificial Recharge, Adelaide, Australia, 22–26 Sept 2002, Balkema, 221–224

Stuyfzand PJ, De Ruiter CJ (1999) Quality changes of drinking water upon deep well injection near St. Jansklooster; final report about the monitoring from Apr 1996 till Jul 1998. Kiwa-report KOA 99.001, 93 p (in Dutch)

Stuyfzand PJ, Lüers F (2000) Balance of environmental pollutants in nature terrains with and without artificial recharge. Kiwa-Meded, 126, 241 p (in Dutch)

Stuyfzand PJ, Juhasz-Holterman MHA, de Lange WJ (2006a) Riverbank filtration in the Netherlands: well fields, clogging and geochemical reactions. In: Hubbs SA (ed) Riverbank filtration hydrology: impacts on system capacity and water quality, Proc. NATO Advanced Research Workshop on Riverbank Filtration Hydrology, Bratislava, 7–10 Sept 2004, Springer NATO Science Series IV, Earth and Environmental Sciences 60:119–153

Stuyfzand PJ, Wakker JC, Putters B (2006b) Water quality changes during Aquifer Storage and Recovery (ASR): results from pilot Herten (Netherlands), and their implications for modeling. Proc. 5th Intern. Symp. on Management of Aquifer Recharge, ISMAR-5, Berlin 11–16 June 2005, UNESCO IHP-VI, Series on Groundwater No. 13, 164–173

Stuyfzand PJ, Kortleve M, Olsthoorn TN, Rolf H (2007a) Accumulation of pollutants and aquifer leaching during artificial recharge compared to a natural recharge system, in the Netherlands. In: Fox P (ed) Management of aquifer recharge for sustainability, Proc. ISMAR-6, 28 Oct–2 Nov 2007, Phoenix AR USA, Acacia Publ. Inc., Phoenix, 174–184

Stuyfzand PJ, Segers W, Van Rooijen N (2007b) Behavior of pharmaceuticals and other emerging pollutants in various artificial recharge systems in the Netherlands. In: Fox P (ed) Management of aquifer recharge for sustainability, Proc. ISMAR-6, 28 Oct–2 Nov 2007, Phoenix AR USA, Acacia Publ. INc., Phoenix, 231–245

Van Beek CGEM, Breedveld RJM, Juhász-Holterman M, et al (2009) Cause and prevention of well bore clogging by particles. Hydrogeol J 17:1877–1886

Zheng C, Wang PP (1999) MT3DMS: a modular three-dimensional multispecies model for simulation of advection, dispersion and chemical reactions of contaminants in groundwater systems, Documentation and User's Guide, Contract Report SERDP-99-1, U.S. Army Engineer Research and Development Center, Vicksburg, MS

Chapter 8 Potential of Riverbank Filtration to Remove Explosive Chemicals

Chittaranjan Ray[1,2,*], Weixi Zheng[1,3], Matteo D'Alessio[1,2], Joseph Lichwa[2], and Rico Bartak[2,4]

[1] Civil & Environmental Engineering Department, University of Hawaii at Manoa, Honolulu, Hawaii 96822, USA. E-mail: matteo@hawaii.edu

[2] Water Resources Research Center, University of Hawaii at Manoa, Honolulu, Hawaii 96822, USA. E-mail: lichwa@hawaii.edu

[3] Presently at Environet, Inc., 650 Iwilei Road, Suite 204, Honolulu, Hawaii 96817, USA. E-mail: wzheng@environetinc.com,

[4] Dresden University of Technology, Dresden, Germany. E-mail: ricobartak@gmail.com

Abstract Riverbank filtration (RBF) is a low-cost and efficient water treatment technology for the removal of many surface water pollutants. It is widely used by water utilities in developed as well as developing countries to produce drinking water from surface water which is often polluted. In this research, the presence of explosive chemicals in riverbed sediments or in flowing water is considered as a potential threat to the quality of filtrate produced from RBF systems. For this, degradation experiments were conducted to examine the persistence of these compounds in river sediments. In addition, a model RBF system was setup to examine the breakthrough of the major explosive chemicals and their metabolites. Results show that HMX was the most mobile and compound followed by RDX. TNT and DNT degraded quickly. Thus, the presence of RDX and HMX could produce breakthroughs in high capacity collector wells located along riverbanks.

Keywords: Riverbank filtration, explosive chemicals, RDX, TNT

1. Introduction

When wells are placed in an alluvial aquifer adjacent to a river or a surface water body and pumped, a portion of the surface water is induced to flow to the wells. As the water passes through the riverbed sediments and underlying aquifer material, most contaminants in surface water are removed. Physicochemical and biological processes play important roles in removing suspended and dissolved chemicals.

---

* Chittaranjan Ray, Civil & Environmental Engineering Department, University of Hawaii at Manoa, Honolulu, Hawaii 96822, USA, Water Resources Research Center, University of Hawaii at Manoa, Honolulu, Hawaii 96822, USA, e-mail: cray@hawaii.edu

C. Ray and M. Shamrukh (eds.), *Riverbank Filtration for Water Security in Desert Countries*, DOI 10.1007/978-94-007-0026-0_8, © Springer Science+Business Media B.V. 2011

As most of the removal action occurs in the naturally occurring soil and aquifer materials, the process is also referred to as "natural filtration." Other terms used for this process are "riverbank filtration (RBF)" or "bank filtration." RBF systems have been in operation since the 1870s in Europe and for more than half-a-century in the United States for supplying potable water to communities (Eckert and Irmscher 2006).

Through RBF, a wide range of suspended solids (measured in terms of turbidity), pathogenic organisms and indicator bacteria, dissolved chemicals, and natural organic matter are removed (Gollnitz et al. 1997, 2003; Ray 2004). Although the river temperature fluctuates significantly in temperate climates, the riverbank filtered water is equilibrated during soil and aquifer passage and the seasonal variations in temperature are lower than that observed for the river. If RBF is used as a pretreatment before a full-scale treatment, a relatively cleaner (lower turbidity) water with an equilibrated temperature requires less chemicals to be added for coagulation and flocculation as well as for disinfection. In some locations, RBF is used as the sole treatment before disinfection (Ray 2008). A major benefit of RBF is the reduction in the formation of disinfection by-products such as trihalomethanes and haloacetic acids which are formed during and following chlorination of the filtered water (Singh et al. 2010). As the natural organic matter is removed during RBF, the formation potential for disinfection by-products is also reduced. Recent studies in Europe and United States show the efficacy of RBF to remove many pharmaceutical and endocrine disrupting chemicals (Ray 2004; Drewes et al. 2006; Wu et al. 2007).

Explosive chemicals can be a source of contamination of surface water in regions of military conflict—areas that have been long used for live ammunition training or from past disposal of military munitions into surface or groundwater. Corrosion of casings initiates dissolution of explosive chemical fillers which eventually reach the surface water. During bombing operations of bridges or structures along waterways, certain munitions do not explode and remain buried in sediments. Cracks and leaks contribute to the release of munition chemicals. Also, in certain locations, intentional disposal of explosive chemicals to sabotage water sources cannot be ruled out.

Conventional military explosives contain TNT (2,4,6-trinitrotoluene), RDX (1,3,5-trinitro-1,3,5-triazacyclohexane), and HMX (1,3,5,7-tetranitro-1,3,5,7-tetrazocane). Recently used munitions contain Composition B (called COMP B), which is a mixture of 59.5% RDX, 39.5% TNT, and 1% paraffin. These chemicals are toxic to humans and have adverse impacts to the ecosystem (Robidoux et al. 2003; Yoon et al. 2005). The objective of this article is to elucidate the degradation behavior of explosive chemicals in riverbed sediments and the transport of these chemicals from surface water to wells.

TNT, RDX, and HMX can adsorb to soil and aquifer material and degrade microbially with TNT degrading faster than the other two compounds (Pennington and Brannon 2002). Studies have shown that TNT also adsorbs to soil strongly compared to RDX and HMX (Dontsova et al. 2006). TNT can degrade under both

RIVERBANK FILTRATION TO REMOVE EXPLOSIVE CHEMICALS   131

aerobic and anaerobic conditions, while RDX and HMX appear to degrade mostly under anaerobic environments. The first objective of this study was to determine how TNT, RDX, and HMX would degrade in stream sediments in polluted stream water reflecting conditions existing in many parts of the stream. The second objective was to show if any of these chemicals would produce a breakthrough in modelled RBF conditions in a laboratory setting. Selected metabolites of these chemicals were monitored in the degradation and breakthrough studies.

2. Experimental Methods

Sediments from a relatively pristine stream flowing near the campus of the University of Hawaii in Honolulu, Hawaii (USA) were collected and sieved though US#10 sieve (1.981 mm opening) to collect the fine fractions. The sediments were wet sieved and stored in stream water at 4°C until use. Clean Ottawa sand (approximately 0.25-mm in diameter) was used to represent the aquifer material. The stream water was filtered with a 1.6 μm pore size glass fiber filter. Wastewater influent from a membrane biofiltration unit was filter sterilized by passing through a 0.45-μm membrane filter and was added to the stream water to provide organic carbon. The wastewater constituted 7% of the total volume. The 5-day biochemical oxygen demand (BOD_5) of the wastewater influent ranged between 200 and 400 mg/L.

Four different experiments were conducted. In the first, aerobic test filter experiments were conducted to examine the degradation of the three chemicals in aerobic conditions. The second experiment was setup to examine the degradation of the explosive chemicals in sediments saturated with wastewater-amended stream water naturally as oxygen was depleted in the sealed containers. In the third experiment, adsorption of the explosive chemicals to stream sediments was examined. Finally, the breakthrough of these three chemicals in a model RBF system was evaluated. Details of the four experiments are presented in Zheng et al. (2009).

In the aerobic test filter experiment, the sieved sediments from the stream were packed into a small column (25 mm diameter and 100 mm long). Wastewater spiked stream water was pumped from a reservoir and passed through the column for two weeks to establish biological equilibrium in the column. The effluent was put back into the reservoir. The flow velocity through the filter was on the order of 3 m per day (typical value for slow sand filtration). TNT, RDX, and HMX were added to the reservoir to have an initial concentration of 1 mg/L each. The concentrations of these chemicals in the reservoir were monitored as a function of time.

The degradation of TNT, RDX, and HMX was carried out in dark at room temperature. Forty-milliliter vials were used to store the sediment and wastewater amended stream water (1:3 in sediment:solution volume ratio). The three chemicals were presented in the amended stream water at a concentration of 1 mg/L, each.

Duplicate vials were opened on set intervals and the residual concentration of the chemicals was measured. Residual oxygen content of the water was also measured, as was the degradation products of the chemicals.

Batch adsorption of the chemicals in the amended stream water to sediments was measured to examine the relative retention of these chemicals. Sediment:solution ratio of 1:10 was maintained in the sorption test. Equilibrium time from a kinetic study was achieved within 18 h. The concentrations of residual chemicals in the solution were measured. The amount adsorbed to the solids was estimated from the differences in initial and final concentrations.

The column experiments were conducted using two flow rates: one representing flow to vertical wells located along river banks (slow infiltration rate, about 0.05 m/day) and the other representing flow to horizontal collector wells where the laterals can be under the riverbed (high infiltration rate, about 0.5 m/day). Four columns, three stainless steel and one glass, were used for the experiment. The first three columns were 25 mm in diameter and 100 mm long and the fourth column was of same diameter but 400 mm long. Column 1 contained stream sediments passing through US#10 sieve (1.981 mm opening), columns 2 and 3 contained stream sediments passing through US#16 sieve (1.18 mm opening) and column 4 contained washed Ottawa sand (0.25 mm diameter). From a reservoir, TNT, RDX, and HMX spiked stream water (which was also amended with filter sterilized wastewater) was pumped through these columns in an upward flow manner. Samples were collected from the exit points of columns 1, 3, and 4 and analyzed for the primary chemicals and their metabolites. A sketch of the experimental apparatus is shown in Zheng et al. (2009).

Chemical analysis of the samples was carried out employing EPA Method 8330B (USEPA 2006) using a Thermo Finnigan HPLC system. The three primary chemicals (TNT, RDX, and HMX) were obtained from commercial sources. However, selected metabolites of these chemicals were obtained from SRI International at Menlo park, CA, USA.

3. Results

Degradation tests for TNT, RDX, and HMX showed that oxygen in the system depleted within two weeks of the incubation. The concentration of TNT dropped below the detection limit in two weeks. Once the anaerobic conditions kicked in, RDX and HMX started to degrade. The pseudo first-order rate constants of degradation for the three compounds were 0.33, 0.055, and 0.033 day^{-1}, respectively. The major degradation products identified were 2-Am-4,6-DNT; 4-Am-2,6-DNT; 2,4-Diam-6-DNT; MNX; DNX (dinitroso-RDX), and mononitroso- HMX. The evolution of these metabolites as a function of time is shown in Figure 8.1.

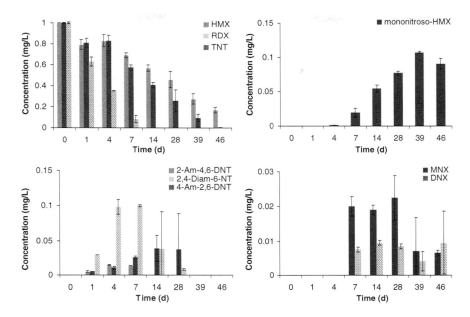

Figure 8.1 Degradation of TNT, RDX and HMX and the evolution of their metabolites as a function of time (modified from Zheng et al. 2009).

Batch sorption experiments conducted with a local streambed sediment showed the sorption coefficient (K_d) for TNT was about 50 mL g^{-1} ($R^2 = 0.89$), for RDX was about 2.2 mL g^{-1} ($R^2 = 0.97$), and for HMX was about 2.9 mL g^{-1} ($R^2 = 0.98$). Sorption equilibrium can be reached in 12 h for RDX and HMX. For TNT, however, due to its rapid degradation, the sorption equilibrium could not be fully reached. It is speculated that both biotic and abiotic processes contributed to the degradation of TNT.

In a month-long aerobic test filter experiment, TNT degraded with a pseudo first-order rate constant of 0.33 day^{-1}. However, RDX and HMX did not degrade within the month. Thus, it is clear that RDX and HMX are more persistent under aerobic conditions. It has also been showed that TNT can inhibit the degradation of RDX and HMX when present together (Moshe et al. 2009).

Three sampling points (called herein as collectors 1, 2, and 3) were located at the downstream ends of columns 1, 3, and 4. With low flow velocity, TNT was not detected in any of the collectors. The peak concentration of RDX was about 20% of the input concentration at the first collector while the concentrations were below detection at the second and third collectors. HMX breakthrough was significant in all three collectors. In collector 1, the peak concentration was about 95% of input concentration. When the flow rate was increased by a factor of 10, TNT breakthrough still did not occur. However, RDX and HMX appeared in the breakthrough quickly. HMX and its metabolites persisted during low and high flow experiments (Figure 8.2).

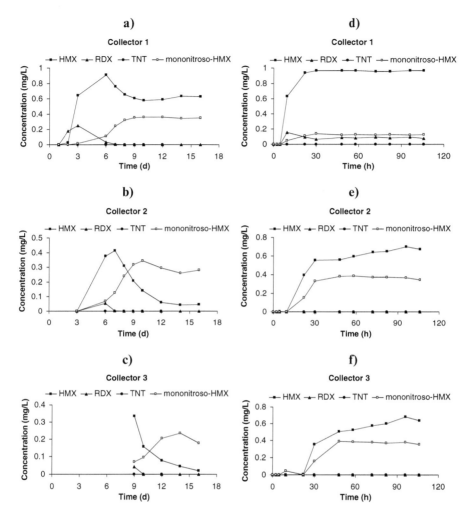

Figure 8.2 Breakthrough curves of HMX, RDX, TNT and mononitroso – HMX at low flow rate (**a–c**) and high flow rate (**d–f**).

4. Conclusions

HMX remains to be a primary concern in RBF systems. In organic-rich waters, RDX and TNT may degrade quickly. In locations where large-capacity wells with horizontal screens are used (case of high flow rate), it is still possible that some RDX might move from the river to the well. For planning and design purposes, if the river water is expected to contain HMX and RDX, appropriate caution should be taken in terms of well design/placement and additional treatment of the riverbank filtrate.

Acknowledgments   We thank Ms. Bunnie Yoneyama of the Department of Civil & Environmental Engineering at the University of Hawaii for helping in various stages of this work.

References

Dontsova KM, Yost SL, Simunek J et al. (2006) Dissolution and transport of TNT, RDX, and composition B in saturated soil columns. J Environ Qual 35(6):2043–2054

Drewes JE, Hoppe C, Jennings T (2006) Fate and transport of N-nitrosamines under conditions simulating full-scale groundwater recharge operations. Water Environ Res 78(13):2466–2473

Eckert P, Irmscher R (2006) Over 130 years of experience with riverbank filtration in Düsseldorf, Germany. J Water Supply Res 55(4):283–291

Gollnitz WD, Clancy JL, Garner SC (1997) Reduction of microscopic particulates by aquifers. J Am Water Works Assoc 89(11):84–93

Gollnitz WD, Clancy JL, Whitteberry BL, Vogt JA (2003) RBF as a microbial treatment process. J Am Water Works Assoc 95(12):56–66

Moshe SSB, Ronen Z, Dahan O et al (2009) Sequential biodegradation of TNT, RDX and HMX in a mixture. Environ Pollut 157(8–9):2231–2238

Pennington JC, Brannon JM (2002) Environmental fate of explosives. Thermochim Acta 384(10):163–172

Ray C (2004) Modeling RBF efficacy for mitigating chemical shock loads. J Am Water Works Assoc 96(5):114–128

Ray C (2008) Worldwide potential of riverbank filtration. Clean Technol Environ Policy 10(3):223–225

Robidoux PY, Bardai G, Paquet L et al. (2003) Phytotoxicity of 2,4,6-trinitrotoluene (TNT) and octahydro-1,3,5,7-tetranitro-1,3,5,7-tetrazocine (HMX) in spiked artificial and natural forest soils. Arch Environ Contam Toxicol 44(2):198–209

Singh P, Kumar P, Mehrotra I, Grischek T (2010) Impact of riverbank filtration on treatment of polluted river water. J Environ Manage 91(5):1055–1062

USEPA (2006) SW-846, Method 8330B. Nitroaromatics, nitramine, and nitrate esters by high performance liquid chromatography (HPLC). http://www.epa.gov/sw-846/pdfs/8330.pdf

Wu Y, Hui L, Wang H et al. (2007) Effectiveness of riverbank filtration for removal of nitrogen from heavily polluted rivers: a case study of Kuihe River, Xuzhou, Jiangsu, China. Environ Geol 52(1):19–25

Yoon MJ, Oliver JD, Shanks JV (2005) Plant transformation pathways of energetic materials (RDX, TNT, DNTs). In: Eaglesham A, Bessin R, Trigiano R, Hardy RWT (eds) Agricultural biotechnology: Beyond food and energy to health and the environment. National Agricultural Biotechnology Council Report 17. National Agricultural Biotechnology Council, New York

Zheng W, Lichwa J, D'Alessio M, Ray C (2009) Fate and transport of TNT, RDX, and HMX in streambed sediments: implications for riverbank filtration. Chemosphere 76(9):1167–1177

Chapter 9 Framework for Assessment of Organic Micropollutant Removals During Managed Aquifer Recharge and Recovery

Sung Kyu Maeng[1,2,3]*, Saroj K. Sharma[1], and Gary Amy[1,2,4]

[1] UNESCO-IHE Institute for Water Education, Westvest 7, 2611 DA Delft, The Netherlands.
E-mail: smaeng@kist.re.kr, s.sharma@unesco-ihe.org, gary.amy@kaust.edu.sa

[2] Technical University of Delft, Stevinweg 1, 2628 CN, Delft, The Netherlands

[3] Korea Institute of Science and Technology, P.O. BOX. 131, Cheongryang, Seoul, 130-650, South Korea

[4] Water Desalination and Reuse Center, King Abdullah University of Science and Technology (KAUST), Thuwal, Kingdom of Saudi Arabia

Abstract Managed aquifer recharge and recovery (MAR) is a reliable and proven process, in which water quality can be improved by different physical, biological, and chemical reactions during soil passage. MAR can potentially be included in a multi-barrier treatment system for organic micropollutant (OMP) removal in drinking water treatment and wastewater reuse schemes. However, there is a need to develop assessment tools to help implement MAR as an effective barrier in attenuating different OMPs including pharmaceuticals and endocrine disruptors. In this study, guidelines were developed for different classes of organic micropollutants, in which removal efficiencies of these compounds are determined as a function of travel times and distances. Moreover, a quantitative structure activity relationship (QSAR) based model was proposed to predict the removals of organic micropollutants by MAR. The QSAR approach is especially useful for compounds with little information about their fate during soil passage. Such an assessment framework for organic micropollutant removal is useful for adapting MAR as a multi-objective (-contaminant) barrier and understanding different classes of compounds during soil passage and the determination of post treatment requirements for MAR.

Keywords: Managed aquifer recharge and recovery, natural treatment systems, organic micropollutants, pharmaceuticals

---

* Sung Kyu Maeng, UNESCO-IHE Institute for Water Education, Westvest 7, 2611 DA Delft, The Netherlands, Technical University of Delft, Stevinweg 1, 2628 CN, Delft, The Netherlands, Korea Institute of Science and Technology, P.O. BOX. 131, Cheongryang, Seoul, 130-650, South Korea

C. Ray and M. Shamrukh (eds.), *Riverbank Filtration for Water Security in Desert Countries*, 137
DOI 10.1007/978-94-007-0026-0_9, © Springer Science+Business Media B.V. 2011

1. Introduction

Managed aquifer recharge and recovery (MAR) systems focused on water treatment (i.e., soil/aquifer-based natural treatment processes) can be classified into two groups: (i) bank or lake filtration (BF) in which infiltration of the source water (e.g., river or lake) is induced by a production well, and (ii) artificial recharge and recovery (ARR) or soil aquifer treatment (SAT) in which river water, lake water, or wastewater effluent is diverted to an infiltration basin or recharge well and later abstracted from a recovery well. For BF systems, more anoxic conditions often dominate the transport through the saturated zone; residence times and travel distances are generally longer than ARR and can be controlled by pumping rates. For ARR systems, more oxic conditions dominate the transport through both the unsaturated (vadose) zone and saturated zone; residence times and travel distances are often shorter compared to BF systems.

BF and ARR processes can be included in a multi-barrier treatment system for organic micropollutant (OMP) (e.g., pharmaceuticals and endocrine disruptors) removal in drinking water treatment or wastewater reuse schemes. However, a number of design and operational conditions must be met for the barrier to be effective in attenuating a range of pharmaceuticals, including non-steroidal anti-inflammatory drugs (NSAIDs), antibiotics, antiepileptic drugs, blood lipid regulators, beta-blockers, and X-ray contrast agents. Important design/operational factors include travel distances, residence times (i.e., travel times), and pumping rates. Depending on the physicochemical characteristics of the pharmaceuticals present and the water quality matrix, their principal removal mechanisms during soil/aquifer passage can be sorption, biodegradation or a combination of both (Schmidt et al. 2003). Process sustainability requires biodegradation to be the dominant mechanism. However, there must be a sufficient carbon source (i.e., electron donor), nutrients, and an adequate electron acceptor such as dissolved oxygen or nitrate (NO_3^-).

BF systems have been employed for water treatment in European countries for more than a century. Despite the long experience in the operation and maintenance of MAR systems in Europe, there are no common guidelines to aid the operation and implementation of MAR systems, especially with regard to the removal of organic micropollutants including pharmaceutical residues. In addition, there is a lack of comparative data interpretation and systematic description. Some of the challenges to the formulation of guidelines or a decision framework can be partly attributed to the complex nature of the processes and the large variations in hydro-geochemical and hydrogeological conditions of MAR sites.

Quantitative structure-activity relationship (QSAR) models have long been developed and used in environmental sciences to estimate the toxicity of pollutants as a function of their physicochemical properties. Thus, adverse effects such as toxicity from the pollutants to the environment can be predicted, and many different types of QSAR models have been developed and used by government agencies

(Bennett et al. 2009). Moreover, QSAR is an attractive alternative for a compound with very limited experimental data (Carson and Walker 2003). Therefore, it is a useful approach to estimate physicochemical properties (e.g., K_{oc}, soil sorption coefficient) of a pollutant for estimating the environmental behavior of a compound (e.g., retardation factor) in case of an unintentional release into the environment. However, there have been limited studies that use QSAR approaches to predict the fate of pharmaceutically active compounds (PhACs) and endocrine disrupting compounds (EDCs) during water treatment processes. Moreover, to the authors' knowledge no studies have been conducted on applying QSAR approaches to MAR systems.

A framework for assessment of organic micropollutant removal by MAR systems consists of two different parts: guidelines and QSAR models. First, the guidelines focus on different classes of organic micropollutants and suggest removal efficiencies of the pollutants with respect to travel times and distances. In this paper, the removal of OMPs from BF and ARR systems were collected from various field studies and incorporated into guidelines. Second, experimental results from soil column studies conducted to investigate the fate of 13 pharmaceuticals during soil passage were used to develop a QSAR model. Field studies from various BF and ARR sites were also used to develop a QSAR model for specific groups of PhACs (i.e., different therapeutic uses). The model was validated by a leave-one-out (LOO) cross validation method and an external validation using field studies data obtained from various studies.

2. Methods

Figure 9.1 shows a diagram for the development of a framework for organic micropollutant removal by MAR.

2.1. Guidelines for Estimating Removal Efficiencies

Guidelines for an assessment of organic micropollutant removal in MAR systems are based on a literature survey and include removal efficiencies of OMPs from BF and ARR field sites. Table 9.1 shows the summary of field sites used for guidelines development. The guidelines enable users to estimate the removal efficiencies based on either a known residence time or travel distance from a water source (e.g., lake or river).

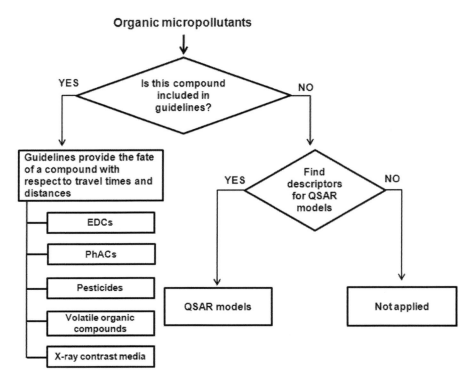

Figure. 9.1. Diagram of processing steps in the development and application of framework for assessment of organic micropollutant removal during MAR.

TABLE 9.1 A list of managed aquifer recharge and recovery sites used for guideline development.

Name	Type	Distance (m)	Time (days)	References
Lake Tegel	LBF	2–90	0–135	Grünheid et al. (2005)
Lake Tegel	LBF	90	135	Scheytt et al. (2004)
Lake Tegel	LBF	2–90	90–135	Mechlinski and Heberer (2005)
Lake Tegel	LBF	2–90	0–135	Heberer et al. (2004)
Lake Tegel	LBF	2–90	0–135	Jekel and Grünheid (2005)
Lake Tegel	ARR	2–50	50	Heberer and Adam (2004)
Lake Tegel	ARR	2–50	0–50	Grünheid et al. (2005)
Lake Wannsee	LBF	1.5–75	60–120	Heberer et al. (2008)
Lake Tegel	LBF	25–90	90–13	Verstraeten et al. (2002)
River Rhine A	RBF	160	7–20	Schmidt et al. (2007)
River Rhine B	RBF	70	12–60	Schmidt et al. (2007)
River Elbe	RBF	270	45–300	Schmidt et al. (2007)
River Ruhr	RBF	125	5–15	Schmidt et al. (2007)

2.2. QSAR

2.2.1. Dataset

All datasets of organic micropollutants collected from BF and ARR sites are heterogeneous because of different hydrogeological and hydrogeochemical conditions. In this study, the experimental data of 13 selected pharmaceuticals were collected from laboratory-scale studies, simulating a riverbank filtration system and used as a training set to develop the QSAR model. The data comprised 13 PhACs studied in soil column simulations (65 cases), with removal efficiency designated as the dependent variable. One of the limiting factors in the development of a QSAR model for BF and ARR systems is the quality of experimental data (Walker et al. 2003). Therefore, input data used for QSAR development in this study came from soil column studies that were conducted using identical experimental set-ups with a single protocol for PhACs measurements. The soil columns had an internal diameter of 100 mm, and there were two sets of column. The bottom of each column was packed with filter media support of 15 cm thick graded gravel and then filled with clean silica sands sized between 0.8 and 1.25 mm. Each set consisted of two columns, each 2.5 m in height, connected in series to simulate a 5 m depth of a one dimensional aquifer. Two sets of such soil columns were used to characterize bulk organic matter removal and the fate of PhACs. A detailed description of materials and methods used for the experimental set-ups is elaborated elsewhere (Maeng et al. 2009).

2.2.2. Molecular Descriptors

The first consideration in developing a QSAR model was to use interpretable descriptors that allowed the user to understand mechanisms (i.e., the mechanistic approach). Finding interpretable descriptors that were experimentally measured properties was expensive and time consuming. Different types of molecular descriptors were calculated and used in this study: constitutional descriptors, connectivity indices, and functional groups. The 247 descriptors calculated from software packages—including DRAGON (Talete srl 2007), EPI suite (US EPA 2009), and Chem3D (Cambridge)—were used to develop a QSAR model. For example, connectivity indices, functional groups and constitutional descriptors were calculated from DRAGON (Talete srl 2007). K_{ow} was computed using EPI suite, and model outputs from BIOWIN 1–7 were also used (US EPA 2009). Quantum-chemical descriptors such as highest occupied molecular orbital (HOMO), lowest unoccupied molecular orbital (LUMO), and heat of formation were calculated by a semi-empiric method MOPAC-PM3 (optimization geometry of a molecule) (Yangali-Quintanilla et al. 2010). Table 9.2 summarizes the types of descriptors used for the model development.

TABLE 9.2 Types of descriptors used for the development of QSAR model.

Descriptor type	No. of descriptors	Reference
Constitutional descriptors	48	DRAGON[a]
Connectivity indices	33	DRAGON[a]
Functional group counts	154	DRAGON[a]
log K_{ow}, BIOWIN 1–7	8	EPI Suite[b]
Quantum-chemical descriptors	4	Chem3D[c]

[a] DRAGON 2007 for MS windows, Version 5.5 (Talete srl 2007).
[b] US EPA 2009 Estimation programs interface suite[TM] for MS windows (US EPA 2009).
[c] Chem3D Ultra 7.0, Cambridge soft 2001 (Cambridge).

2.2.3. Model Techniques and Validation

Multiple linear regression (MLR) analysis, a common approach to develop a QSAR model, was performed. MobyDigs uses a genetic algorithm (GA) to extract the best set of descriptors and calculates a regression model by ordinary least square regression (OLS) (Yangali-Quintanilla et al. 2010). Other approaches such as principal component analysis (PCA), principal component regression (PCR), and partial least square (PLS) methods have also been used in the development of QSAR models in the environmental sciences (Eriksson et al. 2003).

A QSAR model should not be used for reproducing known data from training sets, but should provide more understanding of the behavior of new compounds. Thus, validation is a critical aspect of the QSAR model development for its reliability. LOO cross validation, the most commonly used method for internal validation, was used to estimate the robustness and predictivity of the model (Ghasemi et al. 2009). The goodness of prediction parameter, Q^2 (1-PRESS/TSS, PRESS: the predictive error sum of squares and TSS is the total sum of squares), indicates the predictive power of a model (Yangali-Quintanilla et al. 2010). During the LOO cross validation method, a single case from the training set is excluded at a time, and the remaining cases were used as the training data set to predict the single case. This process was repeated until each case in the training set had been used for prediction by developed models. Moreover, an external validation was also carried out to verify the reliability of the model. Validating QSAR models using external data greatly enhanced the prediction power of a QSAR model and its applicability. In this study, 26 cases of PhACs (removal efficiency) obtained from literature surveys (constituting a heterogeneous data set) were used as external validation data.

3. Results and Discussion

3.1. Guidelines

An assessment of organic micropollutant removal during MAR comprised not only PhACs but also other groups of organic micropollutants such as endocrine disrupting compounds (EDCs) and pesticides (all guidelines are not shown). The range of removal efficiencies for different groups of organic micropollutants, as a function of travel times and distances, were defined using scatter plots with delineation of bins, which were then summarized in tables as the proposed guidelines (Figure 9.2, Tables 9.3 and 9.4). For example, the scatter plots of PhACs data against travel times and distances at BF and ARR sites were compiled and shown in Figure 9.2. The scatter plots were required to determine the removal ranges of target compounds in the guidelines. According to Figure 9.2, PhAC concentrations were reduced gradually as travel distances and travel times increased. PhAC removal exceeds 64% when BF or ARR sites exhibit greater than 135 days of travel times or 125 m of travel distances. PhACs consist of many different compounds, which show different physicochemical properties, with respect to their usages. In contrast, EDCs are generally known to have high log K_{ow} values (indicating hydrophobicity) and are neutral compounds; thus, adsorption is an important mechanism for the removal of EDCs. Tables 9.3 and 9.4 show the analysis of scatter plots of PhAC removal with travel distances and travel times, respectively. In this study, travel times appeared to be a better parameter in estimating PhAC removal compared to travel distances in the proposed guidelines. Various removal mechanisms were responsible for the removal of PhACs (adsorption, biodegradation, etc.). The developed guidelines can be enhanced by classifying their usages; thus, there is one guideline for each type of PhAC (e.g., antibiotics and lipid regulators). Having guidelines for BF and ARR systems provides a preliminary assessment of removal or fate of compounds or different classes of compounds to users.

3.2. QSAR

Different types of molecular descriptors were used for the development of QSAR models to understand the fate of PhACs during soil passage. For example, if log K_{ow} is selected in an MLR model and has a greater influence compared to other coefficients by comparing standardized coefficients, then hydrophobic interaction plays a major role in the removal of PhACs.

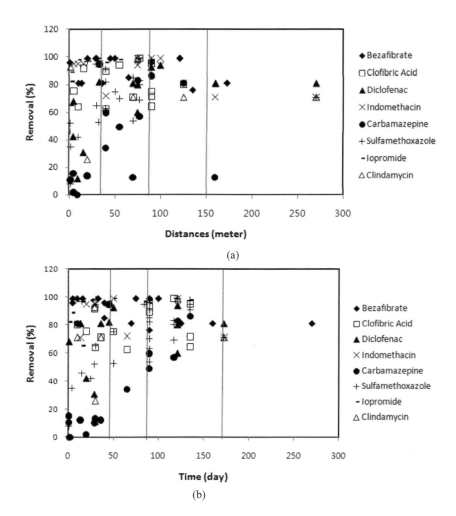

Figure 9.2. Scatter plots of removal efficiency of pharmaceutically active compounds with (a) travel distances, and (b) travel times in an aquifer from river bed.

TABLE 9.3 Analysis of scatter plot of PhAC removal with travel distances.

Distance (m)	No. of cases	Removal range (%)	Average removal (%)
0–20	39	0–60	58
20–50	45	33–99	79
50–90	45	13–99	79
>90	11	125–270	76

Sources: Grünheid et al. (2005), Scheytt et al. (2004), Mechlinski and Heberer (2005), Heberer et al. (2004), Jekel and Grünheid (2005), Heberer and Adam (2004), Heberer et al. (2008), Verstraeten et al. (2002), and Schmidt et al. (2007).

FRAMEWORK FOR ASSESSMENT OF ORGANIC MICROPOLLUTANT   145

TABLE 9.4 Analysis of scatter plot of PhAC removal with travel times.

Times (day)	No. of cases	Removal range (%)	Average removal (%)
0–25	34	0–99	63
25–50	36	11–99	69
50–100	26	33–99	78
>100	41	25–99	83

Sources: Grünheid et al. (2005), Scheytt et al. (2004), Mechlinski and Heberer (2005), Heberer et al. (2004), Jekel and Grünheid (2005), Heberer and Adam (2004), Heberer et al. (2008), Verstraeten et al. (2002), and Schmidt et al. (2007).

3.2.1. Model Development

Reliable data of PhACs are needed to develop a model that can be explained in a mechanistic manner. Having results under different spatial and hydrogeochemical conditions (i.e., field studies) will easily lead to a model with limited applicability. In this study, experimental data from column studies using four different classes of PhACs were used for developing a QSAR model for 13 selected PhACs including lipid regulators, psychostimulants, analgesics, and anticonvulsants (Table 9.5). Adding more cases from the different therapeutic usages of PhACs (e.g., antibiotics) will certainly increase the applicability of the model. GA was used to select the best descriptors followed by an OLS method to come up with the QSAR model. The model was selected based on the correlation coefficient (R^2) and the external predictivity. The following QSAR model based on four variables was selected from 247 molecular descriptors.

% removal of PhACs = 174.8($\pm$11.6)nImidazoles + 158.4($\pm$19.8)AR − 98.1($\pm$5.8)nCONN − 1830.3($\pm$193.2)ME + 1851.1($\pm$188.3)

$$R^2 = 0.84, Q^2 = 0.81 \text{ and } Q^2_{ext} = 0.64$$

Number of cases in the training set = 65, number of cases in the testing set = 26 cases.

As shown in Table 9.6, there are four variables selected in the model (i.e., NImidazloles, AR, nCONN, and ME) to estimate PhAC removal during soil passage. Selected descriptors with positive coefficients indicate that those descriptors contribute positively to the removal of PhACs, whereas descriptors with negative coefficients lead to a low degree of PhAC removal. During soil column experiments, ionic PhACs were not effectively removed under abiotic conditions using sodium azide (abiotic conditions where only sorption is influential) but were significantly removed under biotic conditions. Moreover, physicochemical properties associated with sorption and electrostatic forces (e.g., log D, dipole moment, K_{ow}, etc.) had little or no impact on the model. Therefore, it is believed

146 S.K. MAENG, S.K. SHARMA, AND G. AMY

TABLE 9.5 List of PhACs used in the training set.

Name	CAS#	Charge at pH8	Log K_{ow}[a]	Log D[b] at pH8	Therapeutic use
Gemfibrozil	25812-30-0	Anionic	4.77	2.22	Lipid regulators
Diclofenac	15307-86-5	Anionic	4.51	1.59	Analgesic
Bezafibrate	41859-67-0	Anionic	4.25	0.69	Lipid regulators
Ibuprofen	15687-27-1	Anionic	3.97	1.44	Analgesic
Fenoprofen	53-16-7	Anionic	3.9	1.11	Analgesic
Naproxen	22204-53-1	Anionic	3.18	0.05	Analgesic
Ketoprofen	22071-15-4	Anionic	3.12	0.41	Analgesic
Clofibric acid	882-09-7	Anionic	2.88	−1.08	Lipid regulators
Carbamazepine	298-46-4	Neutral	2.45	2.58	Anticonvulsant
Phenacetine	62-44-2	Neutral	1.67	1.68	Analgesic
Paracetamol	103-90-2	Neutral	0.27	0.23	Analgesic
Pentoxifylline	6493-05-6	Neutral	0.29	0.48	Analgesic
Caffeine	58-08-2	Neutral	−0.07	−0.45	Psychostimulants

[a] KOWWIN v.1.67 (US EPA 2009).
[b] ADME/Tox WEB software (http://www.pharma-algorithms.com/webboxes/).

that the descriptors may correlate to biodegradation. The four descriptors included in the model are reported in Table 9.7 in increasing order of significance (i.e., standardized regression coefficient values). The significances of the selected descriptors ME (mean atomic Sanderson electronegativity) (−0.8) and AR (aromatic ratio) (0.8) in the model are relatively low compared to nCONN (−1.6) and nImidazoles (2.5). ME was selected in the model because it indicates the negative influence of electronegativity of PhACs. On the other hand, an increase in the number of Imidazole group (i.e., functional groups) leads to an increase in biodegradation.

TABLE 9.6 Summary of selected descriptors.

Descriptor		Type	
nImidazoles	number of Imidazoles	Functional groups	
AR	aromatic ratio (number of aromatic bond over the total number of non-H bonds)	Constitutional descriptors	
nCONN	number of urea (-thio) derivatives	Functional group	
ME	mean atomic Sanderson electronegativity (scaled on carbon atom)	Constitutional descriptors	

According to the Organization for Economic Co-operation and Development (OECD) criteria, imidazole is a readily biodegradable compound that degrades between 90% and 100% within 18 days (OECD 2003). nCONN was probably selected because carbamazepine, the most persistent compound during the soil passage, contains the urea derivatives.

TABLE 9.7 Regression coefficients for selected descriptors.

Variable	Regression coefficient	Error regression coefficient	Confidence interval (0.95)	Standardized regression coefficient
Intercept	1851.1	188.3	376.5	
ME	−1830.3	193.2	386.4	−0.8
ARR	158.4	19.8	39.5	0.8
nCONN	−98.1	5.8	11.6	−1.6
nImidazoles	174.8	11.6	23.3	2.5

3.2.2. Model Validation

R^2 is often used as a measure of goodness-of-fit of a QSAR model. However, validation has to be carried out to determine the robustness and predictivity of the model. LOO cross validation, the most commonly applied for an internal validation, was used to predict the reliability of the model (Gramatica 2007). Thus, if the cross validation coefficient Q^2 is greater than 0.5, then the model can be attributed a high predictive power (Ghasemi et al. 2009). The developed QSAR model presented a Q^2 of 0.81. Therefore, the model was acceptable by analyzing LOO cross validation. Figure 9.3 shows the experimental versus predicted removal efficiencies.

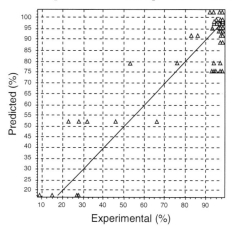

Figure 9.3. Predicted removal efficiencies versus observed removal efficiencies for selected PhACs in Table 9.5.

External data collected from various field studies were used for external validation of the QSAR model. According to Gramatica (2007), only externally validated models are applicable to both external prediction and regulatory purposes. A Q^2_{ext} value of 0.64 was obtained suggesting the prediction power of the model by external validation was lower than that of internal validation.

4. Conclusions

The framework for assessment of organic micropollutant removal for MAR systems (guidelines and QSAR models) developed from this study could be a useful tool to provide more understanding of the behavior of organic micropollutants during soil passage. The proposed guidelines deal with different classes of organic micropollutants and suggest removal efficiencies of the pollutants as function of travel times and distances. In this study, travel times and distances appear to be good parameters in estimating PhACs removal. Travel times seems slightly better compared to travel distances to estimate the removal of organic micropollutants. Different types of molecular descriptors (constitutional descriptors, connectivity indices, and functional groups) were used for the development of QSAR models to understand the fate of PhACs during soil passage. Sorption properties associated with chemical, physical, and electrostatic forces (e.g., log D, dipole moment, log K_{ow}, etc.) had little or no impact on the model. Therefore, it is likely that the selected descriptors are correlated to biodegradation. More data on removal of different types of PhACs (e.g., antibiotics) will enhance the applicability of the model. Moreover, different QSAR models for ionic and non-ionic PhACs will further improve the prediction power of the models. The framework for assessment of organic micropollutant removal would be useful to water utilities for adapting MAR as a multi-objective (-contaminant) barrier and understanding different classes of compounds, and it would also help to determine post treatment requirements for MAR.

Acknowledgments We would like to acknowledge the help of Mr. Emmanuel Ameda for the support on data collection for developing guidelines. This study was financially supported by K-WATER (Korea Resources Water Resources Corporation).

References

Bennett ER, Clausen J, Linkov E, Linkov I (2009) Predicting physical properties of emerging compounds with limited physical and chemical data: QSAR model uncertainty and applicability to military munitions. Chemosphere 77:1412–1418

Cambridge Soft Corporation, CS Chem3D Ultra 7.0, USA. http://www.cambridgesoft.com/

Carson L, Walker JD (2003) QSAR for prioritizing PBT substances to promote pollution prevention. QSAR Comb Sci 22:49–57

Eriksson L, Jaworska J, Worth AP et al. (2003) Methods for reliability and uncertainty assessment and for applicability evaluations of classification- and regression-based QSARs. Environ Health Persp 111(10):1361–1375

Ghasemi JB, Abdolmaleki A, Mandoumi N (2009) A quantitative structure property relationship for prediction of solubilization of hazardous compounds using GA-based MLR in CTAB micellar media. J Hazard Mater 161:74–80

Gramatica P (2007) Principles of QSAR models validation: internal and external. QSAR Comb Sci 26(5):694–701

Grünheid S, Amy G, Jekel M (2005) Removal of bulk dissolved organic carbon (DOC) and trace organic compounds by bank filtration and artificial recharge. Water Res 39:3219–3228

Heberer T, Adam M (2004) Transport and attenuation of pharmaceutical residues during artificial groundwater replenishment. Environ Chem 1:22–25

Heberer T, Massmann G, Fanck B et al. (2008) Behaviour and redox sensitivity of antimicrobial residues during bank filtration. Chemosphere 73:451–460

Heberer T, Mechlinski A, Fanck B et al (2004) Field studies on the fate and transport of pharmaceutical residues in bank filtration. Ground Water Monit R 24:70–77

Jekel M, Grünheid S (2005) Bank filtration and groundwater recharge for treatment of polluted surface waters. Water Sci Technol 5:57–66

Maeng SK, Abel CDT, Sharma SK et al. (2009) Impact of biodegradability of natural organic matter on removal of pharmaceutically active compounds during riverbank filtration. In: van der Helm AWC, Heijman GSJ (eds) High quality drinking water conferences (9–10 June, 2009), Delft, The Netherlands

Mechlinski A, Heberer T (2005) Fate and transport of pharmaceutical residues during bank filtration. In: Proceedings of 5th international symposium on management of aquifer recharge, (10–16 June, 2005), Berlin

OECD (2003) SIDS Initial Assessment Report for SIAM 17. UNEP PUBLICATIONS. http://www.inchem.org/documents/sids/sids/288324.pdf. Accessed 07 January 2010

Scheytt T, Mersmann P, Leidig M et al. (2004) Transport of pharmaceutically active compounds in saturated laboratory columns. Ground Water 42:767–773

Schmidt CK, Lange FT, Brauch HJ (2007) Characteristics and evaluation of natural attenuation processes for organic micropollutant removal during riverbank filtration. In: Proceedings of the Regional IWA conference on groundwater management in the Danube River Basin and other large River Basins, (7–9 June, 2007), Belgrade

Schmidt CK, Lange FT, Brauch HJ, Kühn W (2003) Experiences with riverbank filtration and infiltration in Germany. International symposium on artificial recharge of groundwater. K-WATER, Daejon, Korea, pp. 117–131

Talete srl, DRAGON, for Windows (Software for Molecular Descriptors Calculations). Version 5.5 – 2007. http://www.talete.mi.it/

Talete srl, MobyDigs – software for multilinear regression analysis and variable subset selection by genetic algorithm, in: Version 1.1 – 2009. http://www.talete.mi.it/

US EPA (2009) Estimation Programs Interface SuiteTM for Microsoft Windows, v 4.00, United States Environmental Protection Agency, Washington, DC, USA

Verstraeten IM, Heberer T, Scheytt T (2002) Occurrence, characteristics, and transport and fate of pesticides, pharmaceutical active compounds, and industrial and personal care products at bank-filtration sites. In: Ray C, Melin G, Linsky RB (eds) Riverbank filtration: Improving source-water quality. Kluwer, Dordrecht, 175–227

Walker JD, Jaworska J, Comber MH et al. (2003) Guidelines for developing and using quantitative structure-activity relationships. Environ Toxicol Chem 22(8):1653–1665

Yangali-Quintanilla V, Sadmani A, McConville M et al. (2010) A QSAR model for predicting rejection of emerging contaminants (pharmaceuticals, endocrine disruptors) by nanofiltration membranes. Water Res 44(2):373–384

Chapter 10 Dissolved Organic Carbon as an Indicator Parameter for Groundwater Flow and Transport

Dagmar Schoenheinz*

University of Applied Sciences Dresden, Faculty of Civil Engineering and Architecture, Friedrich-List-Platz 1, 01069, Germany.

Abstract The bulk parameter dissolved organic carbon (DOC) is often successfully exploited in the context of investigations into the collective behavior of dissolved organics contained in water. Though being an unspecific parameter, increased DOC concentrations can indicate anthropogenic influences (such as bank filtration or artificial recharge) on groundwater quality. To this end, DOC as an organic surrogate parameter was investigated for its potential to indicate and characterize such influences. The advantages of using DOC as an indicator parameter are the low time and cost investments required for analysis of DOC concentrations but also the fact that DOC reflects the collective behavior of the dissolved organics contained in water. Existing models for DOC characterization are limited to the identification of one or two compound groups with different degradability. Based on first order degradation kinetics, a multi-component approach is mathematically derived. The approach is demonstrated for the example of four compound groups that are (i) easily, (ii) moderately, (iii) poorly and (iv) under the given conditions not degradable as a function of time scale and flow path length. By means of the conceptual model, a decision tool is given for finding the necessary contact time as a function of hydraulic conditions and flow path length, such as in the case of bank filtration or groundwater recharge site design, as well as in finding the required distance between infiltration areas and points of interest. The discussion of contact time and time scales is a step towards a comparable interpretation of laboratory and field conditions in terms of biodegradation rate constants for DOC as well as for single organic compounds.

Keywords: Dissolved organic carbon, groundwater recharge, degradation concept, modeling

---

* Dagmar Schoenheinz, University of Applied Sciences Dresden, Faculty of Civil Engineering and Architecture, Friedrich-List-Platz 1, 01069 Dresden, Germany, e-mail: schoenheinz@htw-dresden.de

C. Ray and M. Shamrukh (eds.), *Riverbank Filtration for Water Security in Desert Countries*, DOI 10.1007/978-94-007-0026-0_10, © Springer Science+Business Media B.V. 2011

1. Introduction

Bank filtration from lakes and rivers, as well as artificial groundwater recharge using surface waters or treated effluents, or infiltration processes such as from unsealed dumping sites, often result in increased concentrations of organic compounds in the affected groundwater (Helmisaari et al. 2006; Schoenheinz et al. 2002; Grischek 2003; Kolehmainen et al. 2007).

The global change evident in such trends as population growth and the expansion of the western life style will increase the stress on water resources both quantitatively and qualitatively. Additionally, the anticipated higher frequency of extreme hydrological conditions as floods and droughts due to potential climate change will be accompanied by higher concentrations of organics in surface waters (Schoenheinz and Grischek 2010), which will affect the groundwater as well. However, the predominance of single trace organic compounds in substantial concentrations that might help to track groundwater flow paths and processes is often not given (Schoenheinz et al. 2002). Moreover, the identification of organic compounds such as persistent organic pollutants (POPs) requires knowledge of these compounds and their stable metabolites, which is never fully available due to continuous industrial creation of new POPs. In particular, the behavior and complex interactions of these compounds are often unknown. Besides the high analytical and cost requirements, the discovery of single organic compounds implies their occurrence in appropriately high concentrations compared to background concentrations but also compared to analytical detection limits. While the occurrence of specific organic pollutants is often site dependent, a general concept for groundwater flow indication based on a specific single organic compound is a contradiction in terms. Organic surrogate parameters as the bulk parameter dissolved organic carbon (DOC) are often easier to determine and the concentration is mostly elevated at anthropogenically influenced sites compared to background values. Therefore and in the particular context of riverbank filtration, soil aquifer treatment, artificial groundwater recharge, the research into monitoring of DOC behavior, characterization and transport modeling has a long history (Sontheimer and Völker 1987; Gimbel and Mälzer 1987; Arnold et al. 1996; Nestler et al. 1998; Drewes and Jekel 1998; Drewes and Summers 2002).

The DOC behavior can be considered reflective of the collective behavior of the contained dissolved organics in water. The hypothesis of the present article states that DOC is a useful indicator parameter for the identification and evaluation of groundwater flow and transport processes during aquifer passage, in particular under conditions of increased non-specific pollution of surface waters. Extending existing DOC modeling concepts, a more general concept to DOC degradation modeling will be introduced. The model considers not only the significant DOC removal within the first centimeters of infiltration (Quanrud et al. 1996; Grischek 2003; Drewes and Fox 1999a; Maeng et al. 2008), but also an ongoing degradation of more poorly degradable organics. The requirements and restrictions of this concept shall be discussed in the following.

2. Requirements of Indicator Parameters

To define a compound or compound group as an indicator parameter for groundwater flow at sites impacted by anthropogenic activities requires that the parameter satisfies the following criteria:

1. Relevance to anthropogenic impairment
2. Adequate methods exist to detect the compound or compound group with passable analytic efforts
3. Sensitivity to anthropogenic impairment beyond natural background of groundwater, i.e. measurably ascertainable concentration differences
4. Ability to characterize the parameter's transport behavior
5. Applicability in transport simulations

Looking at the bulk parameter DOC as a potential indicator parameter, the requirements (1) and (2) are normally fulfilled: (1) Naturally, the DOC is a rather generic parameter that does not give information about quality, quantity and harmfulness of individual organic compounds. Nevertheless, DOC plays an important role in raw water quality evaluation not only in terms of its potential to form disinfection byproducts during drinking water treatment but also as a measure for bacterial growth in the distribution system. (2) The analytical determination of dissolved organic compounds as DOC is nowadays a standard method which is inexpensive and quickly applied (Kölle 2003).

(3) While for unaffected groundwater low concentrations of DOC are typical, anthropogenic impairment is often accompanied by increased DOC concentrations (Kolehmainen et al. 2007). However, the sensitivity of the parameter DOC is only given if the differences between original and impaired groundwater are in the range of milligrams per liter. While usual DOC concentrations in groundwater range from 0.5 and 1.5 mg/L, impacts by surface waters, wastewaters or polluted water of other origin are mostly seen in higher concentrations. To characterize (4) and simulate (5) the transport behavior of DOC, many efforts have been undertaken which show the general applicability of DOC to groundwater flow indication and transport modeling. However, the focus and model approaches of the performed investigations were quite different (Hobby and Gimbel 1988; Mälzer et al. 1992; Drewes and Jekel 1998). As a foundation for a more generalized conceptual model of DOC as an indicator parameter, these investigations shall be discussed and systematized.

3. Review of DOC Characterization and Transport Research

The endeavor to characterize the behavior of DOC during soil passage in Germany was motivated by the direct or indirect potable use of river water influenced by wastewater discharge resulting in increasing organic loads during the 1970s. Here, the development of the testfilter method by Sontheimer and Völker (1987) was a significant contribution to quantification of the removal efficiency during bank filtration and subsequent activated carbon filtration in terms of organics removal. Therefore, the bulk organics were characterized by distinguishing between compounds that are not removed during bank filtration (relevant for waterworks) and compounds that are not removed during neither bank filtration nor during activated carbon filtration (relevant for drinking water). While the compound group 'relevant for waterworks' is considered to be hardly or not at all biodegradable, the compound group 'relevant for drinking water' is additionally not adsorbable to activated carbon.

To determine the degradable portion of the bulk organics, Sontheimer and Völker (1987) designed column experiments, the so-called testfilters. From a container, the water under investigation was pumped continuously in a circulating flow through a column filled with biologically active carbon. The column outflow discharged back into the inflow container (Figure 10.1).

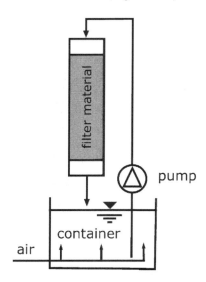

Figure 10.1. Scheme of an experimental testfilter set-up patterned after Sontheimer and Völker (1987).

Composite samples with time increments from hours to days were taken from the container and the concentration decrease of organics measured as DOC versus time was observed. Hobby and Gimbel (1988) applied a 1st order kinetic to describe the observed DOC degradation behavior in wastewater within experimental periods of approximately two weeks,

$$c(t) = c_1 \cdot e^{-\lambda t} + c_r, \qquad (10.1)$$

c_1 concentration of the degradable portion (mg/L)
c_r concentration of the residual (nondegradable, thus waterworks relevant) portion (mg/L)
λ degradation rate constant of the degradable portion (days^{-1}).

The rate constant λ is determined by non-linear regression analysis. This approach is in agreement with findings where the majority of organic carbon is removed within the most biologically active uppermost infiltration zone.

Mälzer et al. (1992) extended the interpretation of testfilter results by distinguishing between two DOC fractions of different degradability, an easily and a poorly degradable portion,

$$c_{ges}(t) = c_1 \cdot e^{-\lambda_1 t} + c_2 \cdot e^{-\lambda_2 t} + c_r, \qquad (10.2)$$

c_1, λ_1 concentration (mg/L) and degradation rate constant (days^{-1}) of the easily degradable portion
c_2, λ_2 concentration (mg/L) and degradation rate constant (days^{-1}) of the poorly degradable portion

In order to determine the easily and poorly degradable DOC portions experimentally, the column set-up was enlarged by combining a column system of four 2-m columns operated in flow-through and a downstream column operated in circulating flow similar to the testfilter (Figure 10.2). A defined outflow from the flow-through columns served as inflow volume of the subsequent testfilter operated in the circulating flow regime.

While the concentrations c_1 and c_2 in Eq. (10.2) were determined by non-linear regression based on the observation data in the flow-through column, the residual concentration c_r was defined as the concentration of the composite sample from the circulating system after 7-days of operation. The two different degradation rate constants λ_1 and λ_2 were reasoned by a decreasing concentration of biomass along the flow path as a consequence of reduced organic supply along the flow path, which was supposed to be a ratio of 10:1 for the comparison of biomass concentration at the beginning and the end of the flow-through column system (Mälzer 1993).

Drewes and Jekel (1998) determined the concentrations of the two differently degradable DOC fractions as a function of retention time in flow-through columns. They defined the easily degradable DOC fraction as compounds degraded within 3 days and the less degradable fraction as compounds eliminated within a retention time of 16–20 days. Different experimental set-ups and boundary conditions of laboratory column experiments performed in different research projects are summarized in Table 10.1. Experiments in circulating flow regime that consider the sum of contact time in the column and residence time in the container as experimental time are indicated by *. The term retention time refers to the contact time in the column only.

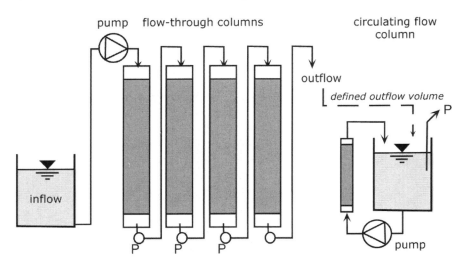

Figure 10.2. Experimental column set-up with sampling points P patterned after Mälzer et al. (1992).

Equations (10.1) and (10.2) rely on assumptions with limited validity. For example, the residual concentration c_r is defined as equivalent to the concentration observed at the end of the realized investigation time. This time is considered to be sufficient for the elimination of the degradable compounds even though it varied between 7 days (Sontheimer and Völker 1987) and up to three weeks (Gimbel and Mälzer 1987; Sontheimer 1991; Drewes and Jekel 1998).

Also, Eqs. (10.1) and (10.2) imply that all degradable compounds can be eliminated within a few days or weeks, at least within a time scale that can be realized at the laboratory scale. The remaining concentration c_r is considered to be non-degradable. This is inconsistent with the observations of different authors who documented ongoing degradation processes in the field (Grischek and Nestler 1998; Sacher et al. 2000; Lindroos et al. 2002; Kolehmainen et al. 2007). The consideration of c_r as a nondegradable portion was also put into perspective by Drewes and Fox (1999b) who proved an ongoing DOC degradation of less degradable compounds in the field where longer retention times up to months were realized. Obviously, c_r seems to be only apparently non-degradable under certain boundary conditions and relatively short contact times.

Another approximation in Eqs. (10.1) and (10.2) is the choice of the rate constants λ_1 and λ_2 that are considered to be representative for the whole group of degradable dissolved organic compounds. Compounds that are very slowly mineralized are essentially neglected. Comparing the results documented in Table 10.1, this assumption is invalidated by the wide range of the data.

TABLE 10.1. Overview of experimental set-ups for the DOC characterization.

	Sontheimer and Völker (1987)	Hobby and Gimbel (1988)	Mälzer et al. (1992)	Drewes and Jekel (1998)	Drewes and Fox (1999a)	Drewes and Fox (1999b)
water type	treated effluent	treated effluent	River Rhine	treated effluent	treated effluent	treated effluent
material	carbon	carbon	pumice	aquifer	aquifer	aquifer
flow regime	circulating	circulating	flow-through + circulating	flow-through	flow-through	flow-through
redox conditions	aerobic	aerobic	aerobic	aerobic and anoxic	anoxic	anoxic
column length	n.i.	0.5 m	8 m + 0.5 m	2 m	4 m	4 m
experimental*/ retention time	1–7 days*	14–21 days*	3–10 days + 7 days*	16 days	21 days	21 days
v_a	n.i.	n.i.	0.06–0.24 m/h	0.025–0.05 m/h	n.i.	n.i.
DOC fractions	2	2	3	3	2	3
determination c_1	$c_0 - c_{circ}$	$c_0 - c_{circ}$	$c_0 - c_{thr}$	$c_0 - c_{thr}$; $t_A = 3$ days	nonlinear regression	$c_0 - c_{thr}$ $t_A = 21$ days
determination c_2	n.a.	n.a.	$c_1 - c_{circ}$	$c_0 - c_1 - c_{thr}$; $t_A = 16$ days	n.a.	extrapolation laboratory – field
determination c_r	c_{circ}	c_{circ}	c_{circ}	c_{thr}; $t_A = 16$ days	extrapolated to 30 days	field
rate constant λ_1	2–13 days^{-1}	0.5 days^{-1}	3–11 days^{-1}	0.9–1.3 days^{-1}	0.08–0.13 days^{-1}	0.09 days^{-1}
rate constant λ_2	n.a.	n.a.	0.3–1.9 days^{-1}	0.1–0.15 days^{-1}	n.a.	0.005 days^{-1}

n.i. – no information provided; n.a. – not applicable; t_A – retention time; c_{circ} – end concentration in circulating flow; c_{thr} – end concentration in flow-through

Close to the infiltration source containing increased concentrations of organics, fast growing organisms are dominant. With far distances, slower growing bacteria dominate that are able to metabolize more poorly degradable compounds of DOC (Kaplan and Newbold 2000). Also, the biomass concentration decreases along the flow path (Lehtola et al. 1996; Preuß and Nehrkorn 1996; Rauch and Drewes 2005).

Degradation Rate Constants depend on both compound-specific features and the biological activity of the system. The rate constants of single substances determined in laboratory and field studies often vary up to one order of magnitude.

158   D. SCHOENHEINZ

These variations are explained by differences in experimental boundary conditions as initial concentrations (Wesnigk et al. 2001), redox conditions (e.g. Alexander 1991; Brauner and Widdowson 2001), reactor temperature (e.g., Palmisano et al. 1991; Jurado-Exposito and Walker 1989; Veeh et al. 1996), and solubility and interactions among organic compounds. For aerobic conditions, rate constants for selected organic compounds under aerobic and anoxic/anaerobic conditions are compared in Table 10.2.

TABLE 10.2. Examples of determined DOC degradation rates.

| Compound group | Reactor | Rate constant λ (day^{-1}) | | Scale | References |
		Aerobic	Anoxic/anaerobic		
BTEX	n.i.	on average $8 \cdot 10^{-2}$	on average $9 \cdot 10^{-3}$	Field/ lab	Suarez and Rifai (1999)
BTEX	Aquifer	n.i.	$1.3 \cdot 10^{-1} - 1.4 \cdot 10^{-2}$	Field	Bockelmann et al. (2001)
Benzene, paraxylene, naphthalene	Aquifer	$1.6 \cdot 10^{-2} - 7 \cdot 10^{-3}$	n.i.	Field	Boggs et al. (1993)
Linear alkylbenzene-sulfonates	Treatment plant, river, aquifer	$7 \cdot 10^{-1} - 3.5 \cdot 10^{-2}$	n.i.	Lab	Krüger et al. (1998)
	Aquifer	$7 \cdot 10^{-2} - 3.5 \cdot 10^{-3}$	n.i.	Field	
PAH	Aquifer	n.i.	$3.1 \cdot 10^{-2} - 3.7 \cdot 10^{-3}$	Field	Bockelmann et al. (2001)
Halogenated hydrocarbons		$3.6 \cdot 10^{-3} - 1.6 \cdot 10^{-4}$	n.i.	n.i.	Matthess (1994)
CHC		$1.6 \cdot 10^{-1}$	$3.3 \cdot 10^{-2}$	field/ lab	Suarez and Rifai (1999)

n.i. – no information provided, CHC – chlorinated hydrocarbons

The assignment of rate constants determined in the lab to site-specific field conditions frequently fails. Experimentally determined degradation is very often significantly faster than those occurring in the field. An example is given for linear alkylbenzenesulfonates in Table 10.2. Possible reasons are changes in the microbial community in the field and in the lab, but also often lower temperatures under field site conditions. In column experiments performed in circulating flow, a source of error may be an incorrect time reference for the data analysis as discussed below.

Contact Time and Degradation Rates   Both Hobby and Gimbel (1988) and Mälzer et al. (1992) determined the degradation rate constants with respect to the total experimental time (sum of contact time in the column and retention time in the container, indicated as experimental time in Table 10.1). However, in this paper it is assumed that only the retention time in the reactor but not the retention time in the container contributes to the removal of organic substances and thus to the effective contact time. This is justified by the fact that the majority of biological degradation occurs by sessile microorganisms attached in the biofilms (e. g. Hazen et al. 1991; Preuß and Nehrkorn 1996; Momba et al. 2000). The degradation during the water retention in the container is negligible if the container volume is large compared to the container surface in contact with the water body (Schoenheinz 2004).

In this context, column experiments operated in flow-through or circulating flow differ with respect to the effective contact time of dissolved organics and the solid phase. While in flow-through regime the effective contact time, i.e., the retention time, is equivalent to the experimental time, in circular flow regime the effective contact time is a function of the ratio of water volume filling the pore space of the sediment in the column and the total volume of the fluid in both the column and the inflow container,

$$t_K = t \cdot \frac{V_P}{V_T}, \quad V_T = V_P + V_C \qquad (10.3)$$

t_K effective contact time (days),
t total experimental time (days),
V_P water volume filling the pore space of the column (m^3),
V_C water volume in the container (m^3),
V_T total water volume in the system (m^3).

This consideration is important for the interpretation of degradation kinetics. Neglecting the effective contact time of column experiments performed in recirculating flow systems results in the underestimation of the degradability of organics. This is also evident in Table 10.1, showing that only compounds with relatively high degradation rates are determined by recirculating flow systems compared to the results from flow-through experiments. The observation is explained by the circulating flow regime that did not provide sufficient contact time to determine the portion of more slowly degradable compounds.

4. Modeling DOC Degradation

4.1. Mathematical Model

The limitation of Eqs. (10.1) and (10.2) to explain the ongoing organics degradation in the field compared to observations in the lab shall be approached by a generalized mathematical model and a subsequent conceptualization.

As the bulk parameter DOC describes the sum of all dissolved organic compounds contributing to the organic carbon concentration, the total concentration of DOC equals to

$$c_T = \sum_{i=1}^{n} c_i \tag{10.4}$$

with $i = 1, 2, \ldots$ n.

Suggesting a first order degradation kinetics results for each single compound in

$$c_i(t) = c_{0i} \cdot e^{-\lambda_i \cdot t} \tag{10.5}$$

where c_{0i} is the carbon concentration of the compound i at time $t = 0$,

$$c_{0i} = c_i(0), \tag{10.6}$$

with $i = 1, 2, \ldots$, n–1,

all non-degradable compounds are combined in the fraction

$$c_n = c_r = constant. \tag{10.7}$$

The total concentration c at the time t is defined by

$$c_T(t) = \sum_{i=1}^{n-1} c_i(t) + c_r. \tag{10.8}$$

Abandoning the common suggestion that the non-degradable DOC portion is equivalent to the lowest observed value, the initial condition of

$$c_{0T} = \sum_{i=1}^{n-1} c_{0i} + c_r, \text{ i.e., } c_r = c_{0T} - \sum_{i=1}^{n-1} c_{0i}, \tag{10.9}$$

results under consideration of Eqs. (10.5) and (10.8) in

$$c_T(t) = \sum_{i=1}^{n-1} c_{0i} \cdot (e^{-\lambda_i \cdot t} - 1) + c_{0T}. \tag{10.10}$$

This formulation contains both the models according to Eqs. (10.1) and (10.2) but also enables the consideration of even more organic compound groups of different degradability. Each fraction i combines organic compounds with similar degradation behavior characterized by concentrations c_{0i} and rate constants λ_i. The parameters c_{0i} and λ_i are assessed from observation data by non-linear curve fitting.

4.2. Model Assumptions

The spatial biomass distribution decreases along the flow path (Rauch and Drewes 2005). Nevertheless, by applying a 1st order kinetic to biodegradation processes, the local biomass concentration is thought to be balanced (i.e., growth and death rate are in equilibrium and the local biomass density is constant in time). By this assumption, changes in the degradation rate due to fluctuating concentrations of organics are neglected.

Furthermore, biological degradation within soil sediment systems is assumed to be quantitatively more significant compared to abiotic adsorption. With respect to the premise of equilibrated conditions between the organics concentration in the infiltrate and organics concentration, the solid phase sorption is mostly neglected.

Consequently, the main model assumption is to postulate equilibrium conditions between the water phase and the biofilm and solid phase, respectively. Effects of changing organics concentration in the water phase on the degradation rate constant and on the sorption conditions are neglected.

4.3. Conceptual Model

The model in Eq. (10.10) is the basis of the DOC concept introduced in this work. Theoretically, Eq. (10.10) can be broken down in an arbitrary number of DOC groups with different biodegradability. However, an optimum needs to be found between the simplest approximation in Eq. (10.1) and the complete expression in Eq. (10.10). An appropriate differentiation based on measurably ascertainable differences would be reasonable. The literature review showed that two degradable fractions with λ_1 and λ_2 can be reliably distinguished in the laboratory so far (see Table 10.1). Furthermore, field observations suggested an ongoing relatively slow degradation. Combining these observations results in the definition of four hypothetical organic compound groups. Introducing these groups to Eq. (10.10) with n = 4 gives the formulation

$$c_T = c_{01} \cdot \left(e^{-\lambda_1 \cdot t} - 1\right) + c_{02} \cdot \left(e^{-\lambda_2 \cdot t} - 1\right) + c_{03} \cdot \left(e^{-\lambda_3 \cdot t} - 1\right) + c_{0T}, \quad (10.11)$$

where c_{01}, c_{02}, c_{03} are the concentrations of hypothetical compounds with different degradation rate constants.

The ratio of λ_1 for easily to λ_2 for moderately degradable organic compounds was found to be about 10:1 for water of the River Rhine (Sontheimer 1991). The same ratio was chosen by Mälzer (1993) and also Table 10.1 documents differences of about an order of magnitude for the degradation rates λ_1 and λ_2. This means to measure a degradation of a defined percentage of the moderately degradable portion takes ten times as long as the same percentage degradation of the easily degradable portion.

The necessary contact time to degrade a compound by a certain percentage can be calculated accordingly to the calculation of the half life time as

$$t_\alpha = -\ln(1-\alpha)/\lambda, \qquad (10.12)$$

where α is the aspired percentage of degradation and λ is the degradation rate.

Considering a contact time of 21 days as realized in different laboratory studies and a reduction of organics by 95% enables the elimination of organics with rate constants ≥ 0.14 days^{-1},

$$\lambda \geq -\ln(1-0.95)/21. \qquad (10.13)$$

Consequently, to measure concentration differences for the elimination of slowly degradable organics, very long retention times are required. These retention times can be realized practically only by observing the degradation along longer flow paths in the field. At river bank filtration sites in Europe, retention times of 30–600 days are typical (Grischek et al. 2002). Analogous to the ratio for λ_1 and λ_2 as observed in laboratory experiments, for the poorly degradable fraction c_{03} the rate constant λ_3 shall be an order of magnitude smaller compared to λ_2. An order of magnitude is already given by Drewes and Fox (1999b) where the rate constant for degradation in the field was determined to be 0.005 days^{-1}.

Taking this observation into account for the conceptual model, the easily and moderately degradable DOC fraction shall be defined by rate constants λ_1 and λ_2 in the range of $\geq 10^{-1}$ and 10^{-2} days^{-1}, respectively, and the slowly degradable fraction shall be characterized by values for λ_3 in the range of 10^{-3} days^{-1}. Necessary retention times to eliminate compounds with rate constants of 10^{-2} days^{-1} and 10^{-3} days^{-1} by 95% are 300 and 3,000 days. Compounds with smaller degradation rate constants than 10^{-3} days^{-1} are considered to be non-degradable within the typical retention times between the infiltration zone and abstraction well at bank filtration sites. Figure 10.3 visualizes the explanations and presents a decision tool for the planning laboratory experiments and the necessary time frames.

The diamonds indicate degradation rate constants as function of different contact times determined in the laboratory (Schoenheinz 2004).

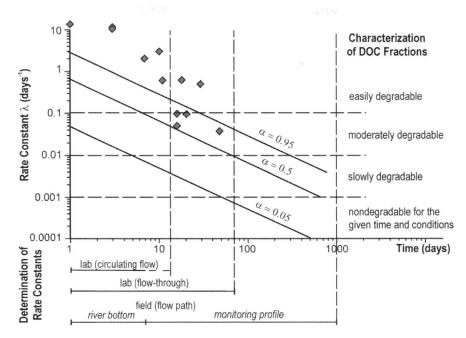

Figure 10.3. Methods for the parameter determination following the Conceptual Model.

5. Application of the Conceptual Model to the Evaluation of Anthropogenically Impacted Aquifers

The evaluation of field sites in terms of groundwater flow and transport requires first of all sufficient hydrogeological information regarding groundwater dynamics and travel times. To apply DOC as an indicator and evaluation parameter, data in sufficiently fine spatial and temporal resolution are necessary. With respect to the conceptual model, a logarithmic scale of space and time is suggested. Methods to obtain these data are shown in Figure 10.4.

In the case of the existence of well-designed monitoring cross sections, the laboratory studies can be abandoned completely. Depending on number, location and quality of existing monitoring wells, the methods for the parameter determination can be combined.

Monitoring Cross Section Well-designed cross sections allow the description of the flow path between the infiltration or pollution area and the abstraction well or other protection areas along a stream line. This enables the calculation of travel times and the extraction of extensive hydro-geochemical information. With respect to the concept for DOC modeling, observations immediately beneath the river bottom (interstitial) facilitate the detection of relatively easily degradable DOC compounds, observations in a monitoring well located close to the river bank allow the determination of moderately degradable compounds, and in one or more monitoring wells at a further distance from the river the elimination of poorly degradable compounds can be detected.

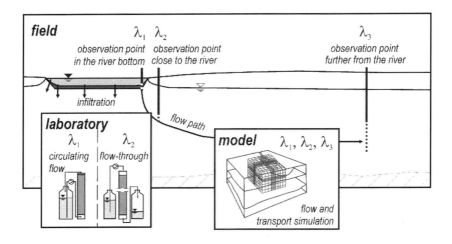

Figure 10.4. Methods for the parameter determination following the Conceptual Model.

Laboratory Experiments Though well-designed cross sections represent the ideal case, this situation is rarely given. Nevertheless, usually there is at least one monitoring well between the infiltration zone and the abstraction well/protection zone, even though it is often further from the bank or infiltration area. Therefore, the evaluation of easily and moderately degradable DOC compounds can be reasonably done by carrying out laboratory experiments as described above. Rate constants of the easily degradable portion are defined to be >0.1 days^{-1} with half-lives shorter than 7 days. Mineralization of compounds belonging to this group occurs in the water itself or in the first centimeters of the river bottom. The methodical determination is reasonable in the laboratory by exploring short time column experiments with contact times of approximately 7 days. The moderately degradable fraction consists of organics with degradation rates between 0.1 days^{-1} and 0.01 days^{-1} which is similar to half-life times between 7 and 70 days. Those rate constants can be derived from column studies with accordingly long retention times. The laboratory results can then be linked to the field observations of the monitoring well and allow conclusions about the poorly degradable DOC fraction's concentration and degradation rate.

Flow and Transport Modeling Site investigation under tough financial and technical restrictions normally results in a lack of representative hydrochemical information. Despite this, the implication of simulation models may support the evaluation. By site-relevant and task-adequate flow simulation, water balances, flow paths and travel times can be estimated. In the accompanying transport simulation, DOC fractions in accordance to Eq. (10.11) and Figure 10.3 will be included.

Conventional tools for the transport simulation require a separate modeling of one or more single substances whose specific behavior is well known. Theoretically, each group of organic compounds with similar degradation characteristics is considered as a single compound, which allows their separate investigation in the model. The results will be combined by superposition at the end of the calculation.

In reality, modeling is wise only for compounds whose concentrations are changing during the considered spatial and temporal scale. Dealing with short retention times of days or weeks, a finer discretization is needed and the behavior of all DOC fractions might warrant consideration. However, focusing on long flow paths the elimination process of the easily and moderately degradable fractions might be out of interest. Therefore, only the poorly degradable with λ_3 on the order of 10^{-3} days^{-1} will be simulated and the non-degradable fraction under given conditions will be considered a tracer in the model. The parameter identification for λ_3 and the concentrations of the poorly and the non-degradable portion is performed indirectly by model calibration based on field observations.

The criteria for the relevance of a hypothetical DOC compound for the model is the theoretical time $t_\alpha = f(\lambda)$ of the fractions in comparison to the simulated retention time t_A. A handy way to realize the conceptual model of DOC as an indicator parameter for field sites is the application of groundwater flow modeling software, such as MODFLOW.

6. Conclusion

The conceptual model of this work is based on the simplified consideration of 1st order degradation kinetics for all organics. The model implies equilibrium conditions with respect to the biomass concentration. Degradation is considered to be dominated by sessile microorganisms. A limitation of the application of DOC as an indicator parameter is given by its non-specific character and the low sensitivity of DOC measurements to concentration changes below 0.2 mg/L.

Nevertheless, the introduced conceptual model provides a helpful tool for site characterizations and flow and transport modeling, particularly at sites or in regions with limited means for hydrochemical analysis or with a lack of specific single trace organics. The parameter determination following the conceptual model helps to design laboratory experimental set-ups as well as field sites aimed at abstraction of bank filtrate or artificially recharged groundwater.

The four-component approach to the conceptual model of DOC degradation and transport was successfully applied by Schoenheinz (2004) to three different field sites, (i) a bank filtration site at River Elbe in Torgau, Germany, (ii) a treated effluent recharge site in Arizona, U.S.A., and (iii) an unintended infiltration site from pulp and paper mill sludge storage basins in Sjastroj, Russia.

The conceptual model is applicable for different climate conditions if the degradation rate constants are determined for site-specific temperatures. If the model is calibrated, the effect of extreme floods and droughts on the DOC removal can be predicted if the retention time in the equations is adapted according to the resulting groundwater flow velocities.

Acknowledgments The author thanks Wolfgang Nestler, Jörg E. Drewes, Thomas Grischek, Hilmar Börnick, and Eckhard Worch for their continuous discussions, significant comments, and ongoing support in developing the demonstrated concept.

References

Alexander M (1991) Introduction to soil microbiology. Krieger, Malabar

Arnold RG, Quanrud DM, Wilson LG et al. (1996) The fate of residual wastewater organics during soil aquifer treatment. Proceedings American Water Works Association Water Reuse '96, San Diego, CA:505–530

Bockelmann A, Ptak TH, Teutsch G (2001) An analytical quantification of mass fluxes and natural attenuation rate constants at a former gasworks site. J Contam Hydrol 53:429–453

Boggs JM, Beard LM, Waldrop WR et al. (1993) Transport of tritium and four organic compounds during a natural-gradient experiment (MADE-2). EPRI TR-101998. Research Project 2485-05. TVA Eng. Lab., Norris. Tennessee

Brauner JS, Widdowson MA (2001) Numerical simulation of a natural attenuation experiment with a petroleum hydrocarbon NAPL Source. Ground Water 39(6):939–952

Drewes JE, Fox P (1999a) Behavior and characterization of residual organic compounds in wastewater used for indirect potable reuse. Water Sci Technol 40(4–5):391–398

Drewes JE, Fox P (1999b) Fate of natural organic matter (NOM) during groundwater recharge using reclaimed water. Water Sci Technol 40(9):241–248

Drewes JE, Jekel M (1998) Behavior of DOC and AOX using advanced treated wastewater for groundwater recharge. Water Res 32(10):3125–3133

Drewes JE, Summers RS (2002) Natural organic matter removal during riverbank filtration: Current knowledge and research needs. In: Ray C, Melin G, Linsky RB (eds) Riverbank filtration: Improving source-water quality. Kluwer, Dordrecht, The Netherlands:303–309

Gimbel R, Mälzer H-J (1987) Testfilter experiments to evaluate drinking water relevance of organic compounds in running waters (in German). Vom Wasser 69:139–153

Grischek T, Schoenheinz D, Worch E, Hiscock K (2002) Bank filtration in Europe – an overview of aquifer conditions and hydraulic controls. In: Dillon PJ (ed) Management of aquifer recharge for sustainability. A.A. Balkema Publ., Sweets & Zeitlinger, Lisse:485–488

Grischek T (2003) Management of bank filtration sites along the Elbe River. PhD thesis, Faculty of Forestry, Geo and Hydro Sciences, Dresden University of Technology

Grischek T, Nestler W (1998) Behaviour of DOC during bank filtration (in German). University of Applied Sciences, Research report 6, 02WT9454/0

Hazen TC, Jiménez L, de Victoria GL, Fliermans CB (1991) Comparison of bacteria from deep subsurface sediment and adjacent groundwater. Microb Ecol 22:293–304

DISSOLVED ORGANIC CARBON AS AN INDICATOR PARAMETER   167

Helmisaari H-S, Derome J, Hatva T et al. (2006) Artificial recharge in Finland through basin and sprinkling infiltration: Soil processes, retention time and water quality. UNESCO IHP-VI, Series on Groundwater 13:280–285

Hobby R, Gimbel R (1988) Results from testfilter experiments with sewage water of municipalities and pulp and paper mills (in German). DVGW-Schriftenreihe Wasser 60:117–151

Jurado-Exposito M, Walker A (1989) Degradation of isoproturon, propyzamide and alachlor in soil with constant and variable incubation conditions. Weed Res 38:309–318

Kaplan LA, Newbold JD (2000) Surface and subsurface dissolved organic carbon. In: Jones JB, Mulholland PJ (eds) Streams and ground waters. Academic Press,San Diego:237–248

Kolehmainen RE, Langwaldt JH, Puhakka JA (2007) Natural organic matter (NOM) removal and structural changes in the bacterial community during artificial groundwater recharge with humic lake water. Water Res 41:2715–2725

Kölle W (2003) Water analyses – correct evaluation (in German). Wiley, VCH Weinheim

Krüger CJ, Radakovich KM, Sawyer TE et al. (1998) Biodegradation of the surfactant linear Alkylbenzenesulfonate in sewage-contaminated groundwater: A comparison of column experiments and field tracer tests. Environ Sci Technol 32(24):3954–3961

Lehtola M, Miettinen I, Vartiainen T et al. (1996) Changes in microbiology and water chemistry during slow sand filtration. Proceedings of International Symposium on "Artificial Recharge of Groundwater," NHP Report 38:197–202

Lindroos A-J, Kitunen V, Derome J, Helmisaari H-S (2002) Changes in dissolved organic carbon during artificial recharge of groundwater in a forested esker in Southern Finland. Water Res 36:4951–4958

Mälzer H-J (1993) Investigation into transport and degradation processes during bank filtration considering effects of shock loads (in German). PhD thesis, Faculty of Engineering, University Duisburg

Mälzer H-J, Gerlach M, Gimbel R (1992) Development of testfilters to simulate shock loads for bank filtration (in German). Vom Wasser 78:343–353

Maeng SK, Sharma SK, Magic-Knezev A, Amy G (2008) Fate of effluent organic matter (EfOM) and natural organic matter (NOM) through riverbank filtration. Water Sci Technol 75(12):1999–2007

Matthess G (1994) The quality of groundwater (in German). Gebr. Borntraeger, Berlin, Stuttgart

Momba MNB, Kfir R, Venter SN, Cloete TE (2000) An overview on biofilm formation in distribution systems and its impact on the deterioration of water quality. Water SA 26(1):59–66

Nestler W, Walther W, Jacobs F, Trettin R, Freyer K (1998) Water production in alluvial aquifers in the Elbe river catchment (in German). UFZ Research report 7, Leipzig

Palmisano AC, Schwab BS, Maruscik DA, Ventullo RM (1991) Seasonal changes in mineralization of xenobiotics by stream microbial communities. Can J Microbiol 37(12):939–948

Preuß G, Nehrkorn A (1996) Succession of microbial communities during bank filtration and artificial groundwater recharge. Proceedings of International Symposium on "Artificial Recharge of Groundwater," NHP Report 38:215–221

Quanrud DM, Arnold RG, Wilson LG et al (1996) Fate of organics during column studies of soil aquifer treatment. J Environ Eng 122(4):314–321

Rauch T, Drewes JE (2005) Quantifying biological organic carbon removal in groundwater recharge systems. J Environ Eng 131(6):909–923

Sacher F, Brauch HJ, Kühn W (2000) Fate studies of hydrophilic organic micro-pollutants in riverbank filtration. In: Jülich W, Schubert J (eds) Proceedings of the International Riverbank Filtration Conference. IAWR:139–148

Schoenheinz D, Grischek T, Worch E et al .(2002) Groundwater pollution near the pulp and paper mill Sjasstroj at Lake Ladoga, Russia. In: Hiscock KM, Davison RM, Rivett MO (eds) Sustainable groundwater development. Geological Society Special Publ 193, London:277–291

Schoenheinz D (2004) DOC as control parameter for the evaluation and management of aquifers with anthropogenic influenced infiltration. PhD thesis, Faculty of Forestry, Geo and Hydro Sciences, Dresden University of Technology

Schoenheinz D, Grischek T (2010) Behavior of dissolved organic carbon (DOC) during bank filtration under extreme climate conditions. In: Ray C, Shamrukh M (eds) Riverbank filtration for water security in desert countries. Springer, Dordrecht, The Netherlands

Sontheimer H (1991) Drinking water from river Rhine? (in German) Academia Verlag Sankt Augustin

Sontheimer H, Völker E (1987) Characterisation of waste water discharges considering drinking water supply (in German). University of Technology Karlsruhe 31

Suarez MP, Rifai, HS (1999) Biodegradation rates for fuel hydrocarbons and chlorinated solvents in groundwater. Biorem J 3(4):337–362

Veeh RH, Inskeep WP, Camper AK (1996) Soil depth and temperature effects on microbial degradation of 2,4-D. J Environ Qual 25(1):5–12

Wesnigk JB, Keskin M, Jonas W, Figge K, Rheinheimer G (2001) Predictability of biodegradation on the environment: Limits of prediction from experimental data. In: Beek B (ed) Biodegradation and persistence. Springer Verlag, Berlin

Chapter 11 Planning, Design and Operations of Collector 6, Sonoma County Water Agency

Jay Jasperse*

> Deputy Chief Engineer, Sonoma County Water Agency, Engineering and Resource Planning Division, 404 Aviation Boulevard, Santa Rosa, CA, USA

Abstract   The Sonoma County Water Agency (SCWA) has operated water supply facilities along the Russian River in California to deliver potable water since the late 1950s. These facilities, consisting of radial collector and vertical wells, utilize natural filtration processes within the alluvial aquifer surrounding the Russian River to remove impurities from the water and provide a source of high quality drinking water.

Keywords: Aquifer, caisson, collector well, groundwater, laterals, MODFLOW

In May 2006, the SCWA completed construction of a sixth collector well (Collector 6). This case study describes the process employed by the SCWA to plan, design, construct, and operate Collector 6. Due primarily to potential impacts on riparian habitat for sensitive terrestrial and fish species, it was necessary to locate this collector well about 107 m from the edge of the Russian River. Consequently, Collector 6 required a different design than SCWA's existing collector wells. The collector well was constructed with conventionally installed laterals in addition to longer and larger diameter laterals installed using horizontal drilling methods. This design not only enhanced pumping capacity but also distributed pumping stress over a larger area of the aquifer, thereby increasing the efficiency of extraction.

1. Introduction

SCWA's facilities provide water on a wholesale basis to retail water agencies serving over 600,000 people in Marin and Sonoma Counties of California (Figure 11.1). Water pumped from SCWA's water supply facilities are delivered to its retail municipal customers via SCWA's transmission system, which consists of approximately 90 miles (145 km) of pipelines, 129 million gallons (488 million liters) of storage, and several pumping stations.

---

* Jay Jasperse, Deputy Chief Engineer, Sonoma County Water Agency, Engineering and Resource Planning Division, 404 Aviation Boulevard, Santa Rosa, CA, USA, e-mail: jay.jasperse@scwa.ca.gov

C. Ray and M. Shamrukh (eds.), *Riverbank Filtration for Water Security in Desert Countries*,   169
DOI 10.1007/978-94-007-0026-0_11, © Springer Science+Business Media B.V. 2011

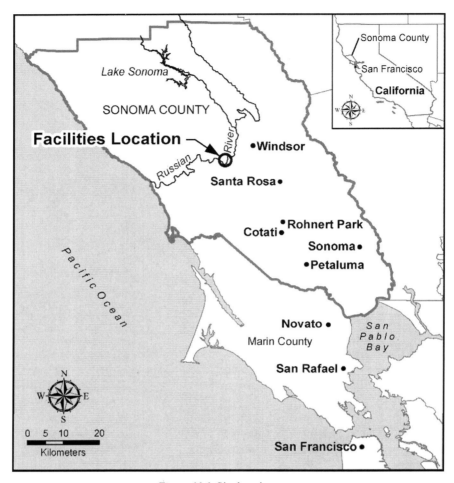

Figure 11.1. Site location map.

2. Purpose of Collector 6 Project

The objective of the Collector 6 project is to provide up to 106 mld (million liters per day) of additional water supply capacity for the SCWA's transmission system. The additional production capacity resulting from Collector 6 helps to meet water supply requirements during periods of peak demand and provide for backup collector well capacity for increased system reliability. The development of backup capacity is important because the peak production capacity is primarily dependent on the use of infiltration ponds to enhance the capacity of SCWA's collector wells. In August 1996 and June 1997, SCWA's total storage declined to about a 12-h supply due to reduced infiltration rates in the ponds. These two events raised concern about the adequacy and reliability of SCWA's Russian River facilities to meet water supply demands. To address this issue, SCWA implemented operational changes related to the infiltration ponds and proceeded with the Collector 6 project.

3. Background

This section provides background information regarding SCWA's water supply facilities by describing the hydrologic and hydrogeologic setting of the Russian River basin.

3.1. Hydrology of the Russian River

The Russian River is the primary surface water resource in Sonoma County (Figure 11.1). The river begins in Mendocino County, which borders Sonoma County on the north. Dams on two main tributaries, Warm Springs Dam on Dry Creek and Coyote Valley Dam on the East Fork of the Russian River, create the reservoirs of Lake Sonoma (46,947 ha-m) and Lake Mendocino (15,098 ha-m), respectively, for water-supply storage in the basin. Russian River flows are augmented by diversions operated by Pacific Gas & Electric from the Eel River to the East Fork of the Russian River above Lake Mendocino. The United States Geologic Survey (USGS) has been operating stream gauges on the Russian River since 1939 and specifically working with SCWA since 1984 to provide continuous stream gauging and peak and sediment discharge information at numerous locations throughout the watershed. Thus the surface water flow patterns of the Russian River system are well documented.

The Russian River drains a watershed of nearly 3,885 km^2 centered approximately 97 km northwest of San Francisco, and empties into the Pacific Ocean near the town of Jenner. The watershed is bordered on the west by the Coast Range and on the east by the Mayacamas Mountains. The watershed is divided into a fog-influenced coastal region and an interior region with hot, dry summers. The basin-wide mean annual precipitation is 104 cm, with a range of 56–203 cm. Approximately 93% of the annual runoff occurs from November to April during Pacific frontal storms. This weather pattern, combined with summer releases by SCWA from Warm Springs and Coyote Valley Dams, result in the winter season experiencing variable short-term high river flow events; while low flow conditions are consistently observed in the summer and fall seasons.

Median winter flows range from 23.5 to 39.75 m^3/s at Hopland, 33.98 to 62.3 m^3/s at Healdsburg and 50.97 to 110.45 m^3/s at Guerneville. The Russian River watershed responds rapidly to variations in rainfall. For example, on February 17, 1986, peak flows were 740 m^3/s at Hopland, 2,013 m^3/s at Healdsburg, and 2,888

m^3/s at Guerneville. As previously mentioned, summer flows in the Russian River are less variable than those during the winter. Summer flows are consistent due to the absence of major storm events and controlled releases from Warm Springs and Coyote Valley Dams. Median summer flows at Hopland, Healdsburg and Guerneville are 4.73 to 6.54 m^3/s. Figure 11.2 presents a hydrograph that illustrates this flow pattern.

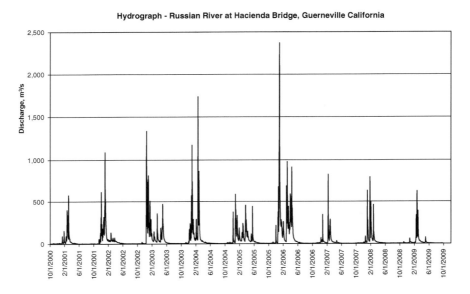

Figure 11.2. Hydrograph of Russian River near Town of Guerneville illustrating typical flow conditions.

A number of management practices have altered historical channel characteristics in the Russian River. These practices include: (1) stream terrace and in-channel gravel mining; (2) channelization; (3) flood control projects; (4) removal of riparian vegetation; (5) operation of dams; and (6) inter-basin water transfers.

3.2. Hydrogeologic Setting

The Russian River, in the general vicinity of the project site, is situated within the northwest slope of the Santa Rosa Plain and is bounded by hills to the east, west and south. The Russian River is bounded by a highly transmissive alluvial aquifer, which in the immediate vicinity of the project site, is relatively narrow in extent (typically less than 610 m wide) as the river has incised into the surrounding low permeability metamorphic rocks of the Franciscan Formation. This formation is comprised of assemblages that include complexly folded and faulted greywacke, shale, and sandstone (PES 1999). The Franciscan Formation exhibits materials of very low hydraulic conductivity and is characterized by fracture flow. The contact between the Russian River alluvial aquifer and the Franciscan Formation is distinct and usually characterized by a weatherized zone of clays and siltstone.

Approximately 0.85 km upstream of the project site the alluvial aquifer is bounded by the Wilson Grove and Glen Ellen Formations. The weakly to moderately consolidated Pleistocene aged Glen Ellen Formation contains stratified but poorly sorted alluvial fan and floodplain deposits comprised of sands, clays, and gravels (Cardwell 1965). The Pliocene-Pleistocene aged Wilson Grove Formation is composed of fine- to medium-grained fossiliferous marine sand and sandstone (PES 1999). The region is structurally complex and exhibits numerous faults. Recent investigations conducted by SCWA near the project site have identified a fault splay approximately 0.85 km upstream (east) of the project site.

Along the Russian River, most pumped groundwater is from the recent alluvium (comprised primarily of poorly sorted sands and gravels). These highly permeable river deposits exhibit transmissivity values ranging from 3,902 m^2/day to 11,148 m^2/day. SCWA diverts water from these alluvial deposits surrounding the river under its appropriative water rights authorized by the State of California.

3.3. SCWA Water Supply Facilities and Operations

With the construction of Collector 6, the SCWA now operates six collector wells and seven vertical production wells in the Wohler and Mirabel areas adjacent to the Russian River (Figure 11.3). Combined, these facilities have a peak production capacity of at least 424 mld. Water treatment is accomplished by utilizing the naturally occurring alluvial aquifer to remove impurities from the water. Once water is diverted from the subsurface, it is chlorinated and softened (to prevent pipeline corrosion) and then conveyed via a transmission system to SCWA's retail water supply customers. SCWA also operates three vertical production wells in the Santa Rosa Plain groundwater basin not associated with the Russian River alluvial aquifer that provide a supplemental supply of water.

SCWA's water supply facilities are permitted through the California Department of Health Services (DHS). SCWA conducted detailed studies in 1993 and 1998, in consultation with DHS, that examined the relationship of river flow and water quality with collector well water quality throughout full seasonal cycles. The results of these studies showed that water produced by Collectors 1 through 4 was of continuously high quality throughout the sampling program, including periods of high river flow. Based on these results, DHS determined that Collectors 1 through 4 were not classified as "groundwater under the direct influence of surface water," (http://www.epa.gov/safewater/disinfection/lt2/regulations.html) and instead are considered having a groundwater source. The studies also showed that Collector

5 had excellent water quality throughout the sampling program with the exception of short periods of time during high river flow events. During times of high river flow events, the bacterial quality of Collector 5 was found to change prior to chlorination. As a result, DHS determined that Collector 5 also operates as a groundwater source unless river flows are greater than 141.6 cm s^{-1} and until flows drop to below 56.64 cm s^{-1}. During these high flow events, water produced from Collector 5 is considered to be groundwater under the direct influence of surface water. SCWA does not operate Collector 5 under such conditions.

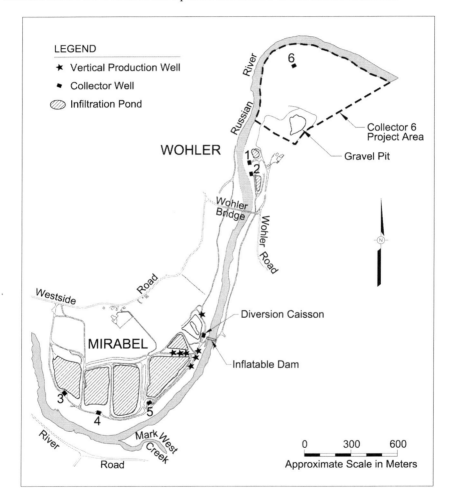

Figure 11.3. Map of the Wohler and Mirabel water supply facilities along the Russian River.

Collectors 1 and 2 were constructed in the late 1950s in the Wohler area. Between 1975 and 1983, Collectors 3, 4, and 5 were constructed in the Mirabel area (Figure 11.3). Each collector well consists of a 4–4.9 m diameter concrete caisson extending approximately 18–30.5 m into the alluvial aquifer. Horizontal perforated intake laterals (20–25 cm in diameter) extend in a radial pattern from the bottom of each caisson to a maximum of 53.3 m into the aquifer. Each collector well houses two vertical turbine pumps that are driven by electrical motors. The SCWA also operates the Russian River Well Field (RRWF) consisting of seven conventional vertical production wells located in the Mirabel area.

To increase production capacity during periods of peak demand from late spring through fall, the SCWA raises an inflatable dam at Mirabel (Figure 11.3). Water pools behind the Mirabel inflatable dam and is diverted into the infiltration ponds to recharge the aquifer near Collectors 3, 4, and 5 (Figure 11.4). In addition, the inflatable dam increases the stage of the river and the area of infiltration behind the dam, thus increasing groundwater levels and production capacity of SCWA's collector wells in the Wohler area. Studies conducted by SCWA prior to the operation of Collector 6 indicate that by operating the inflatable dam, the production capacity of the Russian River facilities was approximately doubled.

Operation of the inflatable dam creates a low velocity; relatively warm water pool extending approximately 3 km upstream. These conditions cause a decrease in the conductance of the streambed in this reach of the river from spring to fall, which in turn affects the flow mechanics within the aquifer. This issue is the subject of ongoing studies and research conducted by SCWA and its collaborating partners.

4. Collector 6 Site Selection

The 56.7-ha Collector 6 site, originally identified based on site surveys and aerial photograph review, was selected for several technical and logistical reasons (Figure 11.3). Of primary importance was the fact that the site is located along a portion of the river that is within the backwater of the SCWA's inflatable dam. As previously mentioned, water levels in the aquifer are increased when the dam is inflated between spring and fall, enhancing production capacity of water supply facilities in this area. In addition, the site exhibits thick deposits of sands and gravels and is situated on the inside bend of the river which is advantageous for locating riverbank filtration (RBF) facilities (Grischek et al. 2007). Such a configuration provides optimal recharge of the aquifer from the river in the vicinity of the radial collector well. The site is also located adjacent to other property in the Wohler area owned by SCWA, which contains power facilities, disinfection facilities, and the main transmission pipelines (thus minimizing the length of a connecting pipeline to the transmission system).

Relative negative aspects of the site were the existence of a former terrace gravel pit and the known occurrence of flooding at the site. These factors were accounted for by ensuring that the collector well was located a sufficient distance away from the gravel pit to avoid negative water quality impacts and designing pumphouse facilities above the 100-year flood elevation.

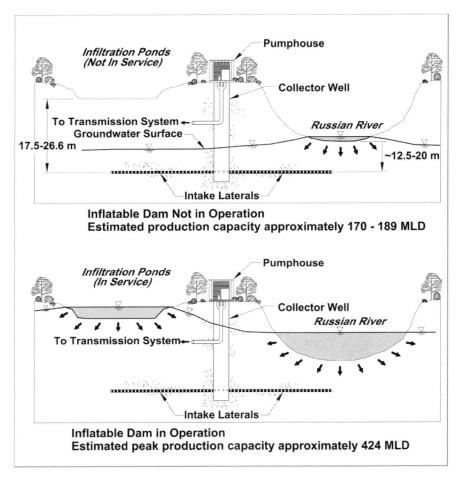

Figure 11.4. Schematic illustrating how the inflatable dam increases production capacity of SCWA's water supply facilities. This schematic does not show the occurrence of a variably saturated zone beneath the river or infiltration ponds that can develop under high pumping conditions.

5. Planning and Site Characterization Activities

The Collector 6 project involved extensive hydrogeological and engineering planning activities. Once sufficient information had been obtained through this planning process to make a determination to move forward with the project, the environmental impacts of the project were evaluated in an environmental impact report that complied with the requirements of the California Environmental Quality Act. As described below, and summarized in Table 11.1, the technical planning activities were conducted in a phased manner.

TABLE 11.1. Summary of site characterization activities.

	Phase 1	Phase 2
Objective	Evaluate site conditions to assess feasibility of a water supply facility	Refine database to develop hydrogeologic framework model
	Evaluate geologic and hydrogeologic properties to provide initial estimate of potential water production	Develop numerical model to analyze alternative designs and improve estimate of yield
		More precisely establish location and depth of a facility
Review of literature and existing site data	River flow (stream gage data)	
	Riverbed survey data	
	Hydrogeologic and water quality data (nearby sites)	
	Historical aerial photographs	
Geophysical surveys	Seismic refraction	Seismic refraction
	Electrical resistivity	
	Self-potential	
Subsurface investigations	Lithology, soil borings	Long-term water-level and temperature monitoring
	Monitoring wells	Soil sample analysis
	Aquifer testing	Water quality sampling
Analysis	Analytical solutions of aquifer test data	Development of numerical model

5.1. Phase 1—Initial Site Characterization

The primary purpose of the initial site characterization activities conducted in the 1980s was to evaluate the feasibility of the site for a water supply facility. If the site was found suitable for this purpose, the secondary objectives of these studies were to identify a general location for such a facility within the site, estimate aquifer properties (hydraulic conductivity, transmissivity, and specific yield), and provide a preliminary estimate of the potential yield of a water supply facility.

The initial site characterization activities included a review of (1) historical aerial photographs, (2) riverbed survey data, (3) river flow data, and (4) hydrogeologic studies of nearby areas for SCWA. In 1987, SCWA conducted a geophysical investigation consisting of seismic refraction, electrical resistivity, and self-potential

(SP) surveys (Norcal 1987). Geophysical methods were chosen as an initial screening level investigation because such methods are non-invasive and provide cost-effective screening level data over a large area. The seismic refraction survey was conducted to evaluate the thickness of alluvial materials and to estimate the depth to bedrock. The results of the electrical resistivity investigation were used in conjunction with the seismic refraction data to evaluate the electrical properties of subsurface materials and assess the variability of these materials. The SP survey was conducted to qualitatively assess groundwater flow for pumping and non-pumping conditions by nearby Collectors 1 and 2.

Based on the results of the geophysical investigation, a hydrogeologic investing-ation was conducted in 1988 (Harding Lawson Associates 1988). This investigation consisted of: (1) drilling fifteen exploratory borings to bedrock and installing 5-cm diameter monitoring wells in these borings; (2) drilling and installing a 30-cm diameter test well; (3) conducting a 72-hour aquifer test by pumping the test well at 13.6 mld; and (4) evaluating the test results using analytical methods to estimate aquifer properties and potential yield of a water supply well. The results of the aquifer test indicated that the site is comprised of highly transmissive alluvial materials (10,219–23,226 meters squared per day). Based on analytical solutions of the aquifer test data, the study also concluded that in order to achieve a production capacity of about 106 mld, 16 ha of infiltration ponds would need to be constructed at the site to enhance production capacity. Based on these activities, the property was found to be a suitable site for a water supply facility and was purchased by SCWA in 1995.

5.2. Phase 2—Detailed Site Assessment

A second phase of site investigation activities was conducted by SCWA to more precisely characterize subsurface conditions for the purpose of locating and determining the depth of a water supply facility. In addition, these activities involved obtaining supplemental data to develop a hydrogeologic framework model of the site and to better estimate the potential yield of a water supply facility. A numerical model was then developed based on the framework model in order to simulate the potential yield of a water supply facility. The additional site characterization activities included: (1) conducting additional seismic refraction surveys (Norcal 1999 and 2001); (2) drilling soil borings and analyzing soil samples in 2001; (3) obtaining water level and temperature measurements from monitoring wells; and (4) conducting numerical flow modeling.

Soil samples were analyzed for grain size distribution, expansion potential, compressive strength, and moisture density relationships. The results of the Phase 1 and 2 geophysical surveys were calibrated with the lithologic data obtained from the drilling program. Figure 11.5 presents an example of how the lithologic and geophysical data were correlated. The agreement between the geophysical data and the lithologic logs was generally good. Based on this analysis, a hydrogeologic framework model was developed which indicated that alluvial materials at the site were comprised of primarily poorly sorted sands and gravels, with zones of relatively finer-grained sands and silts. Figure 11.6 presents a cross-section illustrating the understanding of geologic conditions at the site. In addition, the contact between the alluvial material and underlying bedrock of the Franciscan Formation was mapped as shown on Figure 11.7. This analysis indicated the presence of a continuous feature of deeper alluvial deposits, likely representing a former channel of the river or paleochannel, which provides a favorable feature for groundwater extraction.

Water-level and temperature measurements were collected from spring through fall from several monitoring wells at the site and surrounding properties using temperature instruments and pressure transducers equipped with data loggers. This monitoring program provided an extensive database to characterize groundwater flow conditions. In addition, the temperature data was used to evaluate the temporal variability of riverbed conductance (Su et al. 2004). These analyses were then used to develop and calibrate a numerical flow model (MODFLOW).

6. Development of Collector 6 Conceptual Design

Based on the Phase 1 site characterization data, SCWA decided to move ahead with the design and construction of a collector well at the site. Subsequent Phase 2 activities were conducted in conjunction with the development of the conceptual design. Initially the design for Collector 6 was intended to be similar to SCWA's other collector wells. However, circumstances developed which necessitated modification of this design. The site was found to provide habitat to salmonid species listed as threatened under the State and Federal Endangered Species Act (ESA). SCWA decided to locate Collector 6 farther away from the river than its other collector wells, in part, to reduce impacts to riparian habitat along the river. Moving the collector well off the river also would reduce noise and visual impacts. Initial input from regulatory agencies also indicated that construction of recharge ponds on the property was not viable given the environmental impacts of such facilities. Combined, the increased distance to the river and the lack of recharge ponds would reduce the yield of the collector well.

180 J. JASPERSE

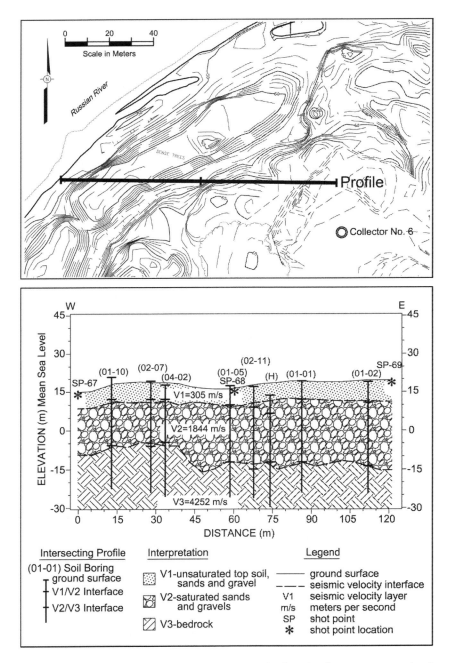

Figure 11.5. Profile view of data resulting from a seismic refraction transect correlated to lithologic data obtained from soil borings. The contact with underlying bedrock and the saturated alluvial is identified. This approach has proven to be a cost-effective means of characterizing subsurface conditions along the Russian River.

PLANNING, DESIGN AND OPERATIONS OF COLLECTOR 6

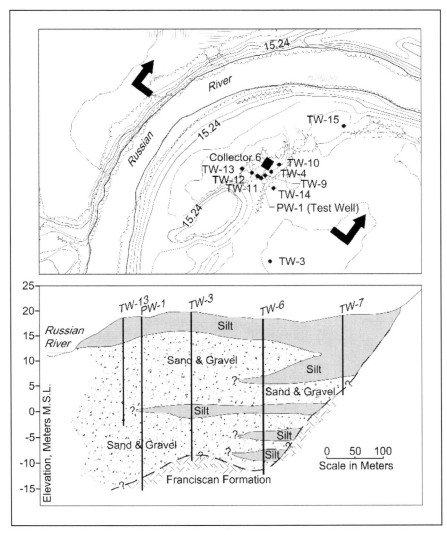

Figure 11.6. A cross section showing interpretation of site geology based on geophysical surveys and soil boring data.

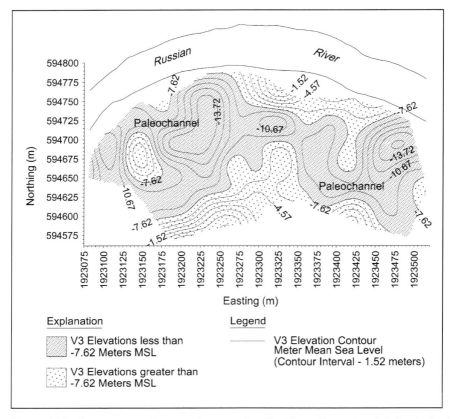

Figure 11.7. Contour map of contact between the alluvial aquifer and underlying bedrock (Franciscan Formation) indicating the presence of a paleochannel, exhibiting thicker deposits of sands and gravels.

At this time, SCWA was evaluating potential alternative designs using new technologies to the traditional collector well as part of a long-range planning effort. This evaluation focused on sites other than the Collector 6 project site that could possibly be developed in the future. The feasibility evaluation considered nine conceptual designs that utilized shaft, tunnel, microtunnel, or horizontal boring techniques.

Given the aforementioned considerations that dictated that a buffer zone between Collector 6 and the river was necessary, it was decided to apply one of the conceptual designs to the Collector 6 project with the goal of overcoming the anticipated reduction in production capacity. The conceptual design involved modifying the traditional collector well by installing two tiers of intake laterals. The lower tier of laterals consisted of conventionally installed 30-cm diameter laterals using a jacking method. An upper second tier of long and large diameter laterals (LLDLs) was then added using horizontal drilling methods. The conceptual design envisioned three LLDLs extending from 152 to 305 m from the caisson and oriented directly towards the river, in the upstream direction, and in the downstream direction. Numerical modeling of this configuration using SCWA's MODFLOW model indicated that the yield of Collector 6 would be significantly increased with the LLDLs.

7. Detailed Design and Construction of Collector 6

Although RBF facilities have many advantages compared to conventional surface water treatment technologies, one of the relative disadvantages is that the yield and water quality of the RBF facility remain uncertain until the facility is in-place and operational. This uncertainty can be reduced by a comprehensive site characterization program, however it cannot be eliminated. Compounding this uncertainty for the Collector 6 project was the additional risk of adding an experimental component to the design and construction of the facility. Consequently, the construction of Collector 6 was implemented in discrete phases and the design of the subsequent phase was reevaluated and modified as appropriate based on the most recent information. It was important that throughout the process, the status and uncertainty/risk was communicated to SCWA's Board of Directors. The following summarizes the significant steps involved with the design and construction of Collector 6.

7.1. Permitting

An important factor to consider in the planning and design phase is the regulatory constraints that may be imposed on the facility. For the Collector 6 project, all environmental permits were acquired by SCWA prior to construction (Table 11.2). Other construction related permits that were required were acquired by the contractors. In particular, the California Occupational Safety and Health Administration (Cal-OSHA) required permits for trench excavation, mining and tunneling operations, confined space, and compressed air operations.

7.2. Caisson

The caisson was the first component of Collector 6 to be constructed. The depth and location of the caisson and conventional laterals were based on the information obtained from the above-described site characterization activities. The caisson was constructed of reinforced concrete in 6.1 m segments formed at the site with an outside diameter of 7.16 m and an inside diameter of 5.5 m. As previously mentioned, the bottom segment of the caisson was constructed with two tiers of portholes (Figure 11.8). The bottom tier consisted of ten 31 cm and one 46-cm diameter portholes. The upper tier consisted of seven 46-cm diameter porthole assemblies.

TABLE 11.2. Environmental permitting for the Collector 6 project.

Permit	Jurisdiction
Clean Water Act Section 404 Nation Wide Permit #12 (Utility Line Discharges)	U.S. Army Corps of Engineers
Clean Water Act Section 404 Nation Wide Permit #18 (Minor Discharge)	U.S. Army Corps of Engineers
Clean Water Act Section 401 Water Quality Certification	California Regional Water Quality Control Board (RWQCB)
National Pollutant Discharge Elimination System (NPDES) General Permit for Discharges of Construction Dewatering Flows to Surface Water	RWQCB
NPDES General Permit for Discharges of Storm Water Runoff Associated with Construction Activity	California State Water Resources Control Board
1601 Lake and Streambed Alteration Permit	California Department of Fish and Game

Figure 11.8. Form of the bottom section of the caisson prior to installation showing configuration of lateral portholes.

A cutting shoe was constructed at the base of the first caisson segment to aid in the advancement of the caisson. A concrete stabilizing ring was also poured around the perimeter of the caisson to ensure that the caisson maintained vertical alignment during installation. A clam shell excavator was used to remove alluvial material from inside the caisson so that the caisson advanced into the subsurface under its own weight (Figure 11.9). The finished floor of the caisson was constructed to a total depth of 27.7 m below ground surface. The caisson extended another 6.4 m above ground surface to ensure that the pumphouse was located above the 100-year flood elevation.

Figure 11.9. Installation of caisson accomplished by removal of material using clamshell excavator.

7.3. *Installation of Conventional Laterals*

Upon completion of the caisson, a total of ten 30-cm diameter laterals were installed through the bottom tier of porthole assemblies using conventional installation methodology. The orientation of the laterals was selected to take advantage of the deeper deposits of alluvial materials identified by the site characterization program (Figure 11.10). These laterals were installed about 85 cm above the floor of the caisson at an approximate depth of 26.8 m below ground surface. This method employed jacking a solid casing into the aquifer until maximum pressures were achieved. The length of the conventional laterals ranged from 21.3 to 51.8 m, which is similar to the length of laterals in SCWA's other collector wells. The jacking equipment was sealed at the porthole assembly to prevent excessive water from entering the caisson. Stainless steel wire screen wrapped well with an end cap was then advanced into the blank casing and the outer blank casing was then removed. The screen size of the intake lateral ranged from 0.254 to 0.445-cm based on grain size analyses of soil samples collected during the initial pipe jacking activities. No drilling fluids other than water were used to install the laterals.

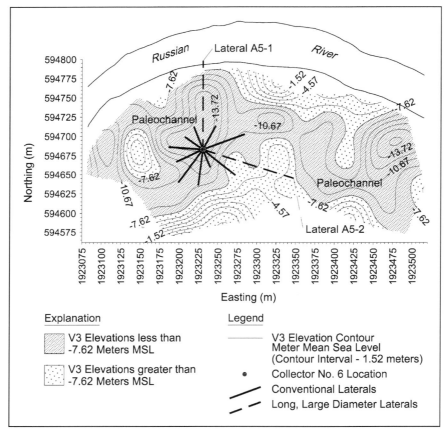

Figure 11.10. Plan view of location and orientation of intake laterals. The contour map of the contact between the alluvial aquifer and underlying bedrock as established by the seismic refraction surveys and the soil boring program is shown for reference.

7.4. Development of Conventional Laterals

The conventional laterals were developed by air lifting using compressed air and then surging or flushing the laterals. Each lateral was developed by isolating the well screen in 1.8-m sections using a double packer isolation tool. Flow rate, pH, electrical conductivity, turbidity, and sand content were measured during development. This process was repeated for each section of the lateral until water quality parameters reached stable levels and could not be improved by further development.

7.5. Performance Testing

After installation of the caisson and conventional intake laterals, a 72-h performance test of the conventional laterals was conducted to obtain information regarding production capacity, aquifer response, and water quality (Ranney 2002). Prior to the constant rate test, a 10-hour step-drawdown test was performed consisting of 2-h pumping steps at 0.27, 0.44, 0.66, 0.88, and 1.16 m^3/s. Based on the results of the step-drawdown test, it was decided to conduct the test at a constant rate of 0.78 m^3/s to ensure that a constant pumping rate could be maintained through the duration of the test. Water level and temperature monitoring devices equipped with data loggers were deployed in the river, the caisson, and in seven nearby monitoring wells to continuously collect monitoring data prior to, during, and after the test.

Hydrographs from the performance test show stable water levels were achieved in the caisson and monitoring wells during the test (Figure 11.11). The stabilized drawdown in the caisson during the constant rate test was 3.8 m.

As Table 11.3 illustrates, there is good agreement between the aquifer properties estimated in: (1) the performance test; and (2) previous aquifer tests conducted during characterization activities at the site and nearby sites. Water quality analyses conducted during the performance testing indicated no measurable sand content or turbidity.

7.6. Design and Installation of Long, Large Diameter Laterals

Because intake laterals had not previously been installed at the length and diameter envisioned for Collector 6, there was a high degree of uncertainty regarding successful installation of the LLDLs. The initial attempts to install an LLDL were unsuccessful, and as a result, the goals of the project were revised based on the experience gained during these initial efforts. The installation methodology was also significantly modified based on the lessons learned from the unsuccessful attempts. Detailed documentation of the LLDL installation is provided by a report prepared by Underground Construction Managers (2007). The following describes how the LLDL program evolved from the initial design to what was eventually constructed.

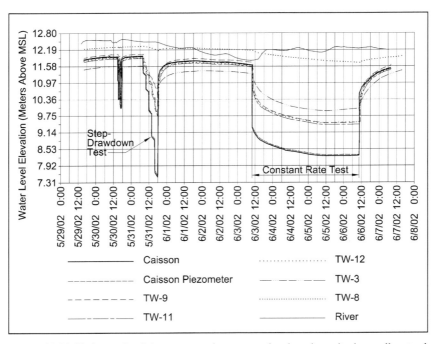

Figure 11.11. Hydrograph of river stage, caisson water level, and monitoring well water levels during the performance test.

TABLE 11.3. Comparison of aquifer and performance test results.

	Site characterization aquifer test	Collector 6 performance test
Hydraulic conductivity (mpd)	487.73 to 792.57	470.18 to 854.47
Transmissivity (m^2pd)	10,220 to 13,935	9,647 to 17,375
Specific yield	0.14 to 0.32	0.15 to 0.20

Note: mpd = meters per day, m^2pd = meters squared per day.

7.6.1. Initial Design and Drilling System Configuration

As mentioned in Section 7.2, an array of eight 45-cm diameter porthole assemblies located at specific orientations and depths was included in the caisson design. The orientation of porthole assemblies were selected based on subsurface lithogy and depth of alluvial sands and gravels (Figure 11.10). Multiple LLDL portholes were installed to provide for redundancy in case initial attempts were unsuccessful. One of the LLDL portholes was installed in the lower tier at the same depth as the conventional portholes approximately 26.8 m below ground surface (85 cm above the caisson floor) while the remaining LLDL portholes were installed at an approximate depth of 24.4 m below ground surface in the upper tier.

A drill rig was designed and fabricated for the specific requirements of the project (Figure 11.12). The drill rig was made to fit within the limited space in the caisson on a rail and sled and is capable of producing approximately 385,554 kg of thrust to the lateral casing. Ancillary equipment included a hydraulic motor, gearbox, hydraulic power unit, augers, top hammer, and a cutting shoe. Support equipment included lifting cages, ladders, a rig mounting platform, and discharge pumps, tanks, and piping.

Figure 11.12. A drill rig was fabricated specifically for the installation of the LLDLs.

The lateral casing consisted of 1.5-m long, 45-cm diameter by 1.3-cm thick steel pipe. Prior to installation, 0.95-cm by 10-cm keystoned axial slots were cut using a computer guided plasma torch. Slot size was based on the slot sizes of intake laterals from other SCWA collector wells. The perforation design had to balance having sufficient open area to provide an adequate hydraulic capacity versus (1) maintaining casing strength given the anticipated high pressures from hammers and jacking and (2) prohibiting excessive sanding issues.

7.6.2. Initial Unsuccessful Installation Attempts

The initial attempts to install an LLDL involved driving the perforated casing into the aquifer. This was done through the use of a top hammer in conjunction with the jacking cylinders of the drill rig that kept drive pressure on the casing while the hammer pulses drove the perforated casing into the aquifer. A cutting shoe was welded on the front of the casing. This installation method did not use drilling mud or fluid. The water management system consisted of dewatering pumps, plugs within the casing, and valves. The first two attempts to install an LLDL via this method proved unsuccessful mainly because the top hammer drive drilling method was ineffective in overcoming bearing pressures encountered in the alluvial materials and the water management system was unable to effectively control water flow into the caisson. These laterals, referred to as PSL#1 (lower tier) and PSL#2 (upper tier) extend only 21.3 and 18.3 m into the aquifer, respectively.

7.6.3. Modified Installation Methodology

Substantial modifications to the drilling system and methodology were made based on the experience of the initial attempts to install the LLDLs. Modifications to the drilling system included adding a new hydraulic power unit, incorporating the use of drilling mud, and the use of the top hammer and drive cylinders in conjunction with the augers and cutting head which were fitted inside the 46-cm diameter perforated casing. Spoils were collected below the drill rig platform in a spoils bin and periodically removed from the caisson.

A significant change in methodology was the use of compressed air to the drilling operation. Compressed air operations required several modifications to the system including:

- A sealed bulkhead was constructed 9.1 m from the bottom of the caisson to isolate the work area from atmospheric conditions. The seal of the bulkhead was tested at 2.1 kg/cm^2 (kilogram-force per square centimeter) .
- A decompression chamber was installed on top of the bulkhead with a second decompression chamber placed at ground surface next to the caisson as a backup.
- A 2.4 m diameter steel dome was installed in the center of the bulkhead to allow for equipment and materials access.
- An extensive communication system was installed including telephones and closed circuit TV.
- An air delivery system with redundancy was added to the system.

Figure 11.13 shows a schematic of the modified drilling system using compressed air. A supervising physician with extensive experience in compressed air operations was intimately involved in the planning activities and was at the site or on call during actual drilling operations. All drilling personnel were trained for compressed air operations prior to commencing with operations. Cal-OSHA was involved in the design and planning activities related to the compressed air operations and were also frequently at the site to ensure compliance with safety regulations. One of the primary reasons that the compressed air drilling operation led to successful installation of the LLDLs is that oxygen rather than air was used for decompression of the workers. This dramatically reduced the time for decompression and lengthened the shift time for drilling. At no time did any worker experience decompression sickness.

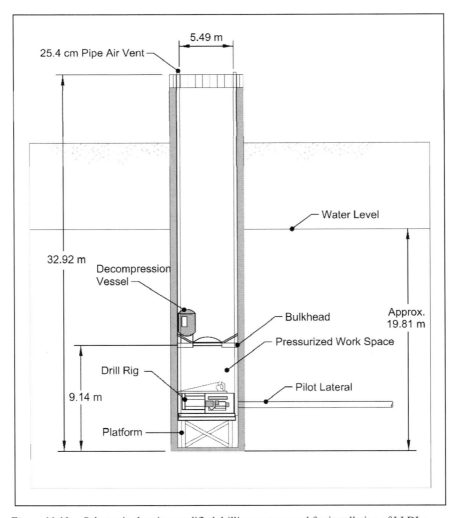

Figure 11.13. – Schematic showing modified drilling system used for installation of LLDLs.

The system modifications described above resulted in successful installation of two LLDLs, referred to as A5-1 and A5-2. Figure 11.14 shows drilling operations under the modified system. During installation of laterals A5-1 and A5-2, operating pressures within the bulkhead were maintained between 1.55 and 1.69 kgf/cm^2 depending on the groundwater level, as measured in a nearby monitoring well. The modified drilling methodology successfully reduced inflow of water through the casing and porthole to a trickle, which was easily handled with a relatively small dewatering pump.

Figure 11.14. Drilling operations during LLDL installation.

7.7. Successful Installation of Lateral A5-1

The first successfully installed LLDL (lateral A5-1) was constructed using compressed air and drilling mud to its target length of 106 m in a porthole oriented directly towards the river (Figure 11.10). It was decided to limit the lateral to this length because of the proximity to the river. A plug of grout was poured at the end of the lateral and a specially fabricated locking end cap was then installed to complete the lateral.

7.8. Analytic Element Modeling and Installation of Lateral A5-2

Upon completion of lateral A5-1, SCWA developed an analytic element model (AEM) to examine whether a second LLDL would be beneficial and to better understand at what length would head losses reduce the operating effectiveness of the lateral. An AEM was selected because it is capable of simulating both pipe and aquifer hydraulics. The AEM was also developed to be integrated into the regional MODFLOW model. The results of the AEM analysis indicated that another 46-cm diameter lateral would be beneficial in distributing pumping stresses and more efficiently extracting water by reducing drawdown within the caisson for a given flow rate in addition to increasing the overall production capacity of Collector 6. The modeling also helped to select which of the remaining portholes to attempt a second LLDL.

Results of the AEM analysis using estimated pipe friction factors for rough pipe also indicated diminishing returns in terms of increased capacity relative to increased length of lateral. For example, a 183-m lateral would have only a 30% increase in yield compared to a 91.5-m lateral (Bakker et al. 2005). Consideration was given to a second lateral as long as 305 m, however based on this analysis, the design of the second LLDL (lateral A5-2) was envisioned to be approximately 183 m long, consisting of 60.9 m of blank casing nearest the caisson to reduce interference with the conventional laterals. Additional research regarding head loss in laterals would improve the analysis of collector well hydraulics and enhance the future design of lateral systems.

Lateral A5-2 was successfully installed to a length of 114 m in the general upstream direction parallel to the river (Figure 11.10). At this point dramatically increased drilling pressures, possibly from a large boulder or tree stump, dictated that drilling should stop to avoid breaking the drill string or a casing weld. Lateral A5-2 was then completed by fitting a cap to the end of the casing and installing a shut off valve at the penetration with the caisson.

The risk of losing the entire lateral by continuing drilling operations outweighed the potential benefit of a longer lateral. This was an illustration of how the uncertainty associated with constructing the below ground surface components of RBF systems can lead to differences between design and actual as-built facilities. In this case, although lateral A5-2 was shorter than designed, it was still long enough to significantly enhance the production and pumping efficiency of Collector 6.

7.9. Development of LLDLs

Laterals A5-1 and A5-2 were developed by surging isolated sections of the lateral. Statistical criteria for turbidity and sand content for the laterals were developed prior to commencing development activities. These criteria were met for each section of the lateral before development was deemed complete. Initial attempts to develop lateral A5-1 involved initially spinning the augers followed by air lifting in isolated 1.5-m segments using a double packer system, surging of the entire lateral, and finishing with the 1.5-m packer system for segments that required additional development. A camera was periodically sent into the lateral to video the conditions between stages of the development process. Based on the experience of developing lateral A5-1, the development methodology was modified for lateral A5-2 to emphasize the techniques that were most effective. The development of lateral A5-2 was accomplished by isolating and surging 15.2-m sections of the lateral using a hydraulically activated knife valve. This approach resulted in a more time-efficient development of lateral A5-2.

7.10. Embankment

Due to the 100-year flood stage elevation of the river, the finished floor elevation of the Collector 6 pumphouse was set approximately 6.4 m higher than the existing ground surface. Therefore, similar to other SCWA collector wells, an embankment was constructed around the pumphouse (Figure 11.15). Though there are several benefits, the most obvious benefit provided by the embankment is access to the pumphouse to facilitate operation and maintenance activities. Nonetheless, the day to day benefits of having the embankment at Collector 6 were partially offset by some increased risk of soil liquefaction occurring in deeper soil strata during a significant seismic event.

Construction of the embankment included 7,864 m^3 of subgrade excavation (1.5 m deep) and placement of 20,665 m^3 of engineered fill. An additional 2,743 m^3 of temporary overfill material was also placed and removed to accelerate initial settlement. Soil material for the embankment was excavated from nearby on-site areas or existing stockpiles. The soil material was acquired from within the same reach and overbank area of the river, which helped to mitigate adverse flood flow impacts that may have resulted from placement of the embankment.

7.11. Pumphouse Facility and Associated Equipment

The design and construction of the pumphouse represented the final construction stage of the project. The major components of the pumphouse and associated equipment are described below.

PLANNING, DESIGN AND OPERATIONS OF COLLECTOR 6

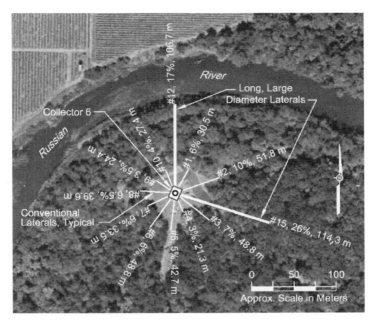

Figure 11.15. Aerial view of Collector 6 showing embankment and projections of laterals. Relative contribution of each intake lateral for a pumping rate of 1.62 m³/s. Note that lateral A5-1 is lateral # 12 and lateral A5-2 is lateral #15 (ERM 2010).

7.11.1. Pumphouse Building and Foundation

The Collector 6 pumphouse structure is a pre-engineered metal building, 11.6 by 14.6 m in plan dimension, with a height of approximately 10.7 m. Personnel access is provided through two man-doors and a roll-up door for vehicular access. An additional roll-up door provides access to the electrical switchgear and motor control centers. The structure also supports a 15.24 t overhead traveling crane. To help mitigate noise related environmental and recreational impacts along the river, the building walls and roof are fitted with noise-attenuating acoustical panels.

7.11.2. Pumping System

The Collector 6 pumping facility consists of two vertical turbine pumps, 919.4 kW and 1,471 kW in size. The pump design points are 29,905 lpm(liters per minute) and 49,560 lpm at 159 m of head, respectively, making the combined nominal design flow approximately 110 mld. The pumping head was determined by hydraulic modeling of the SCWA's transmission system.

Variable speed units were considered for this pumping application, but were later rejected due to increased cost, increased footprint requirements, increased heat generation, and overall decreased pumping efficiency based on anticipated use. Provision for future regulation of flow was provided by constructing a control valve vault for possible installation of a future throttling control valve in the discharge of the larger pump.

7.11.3. Surge Control System

A 28.32 m^3 surge control tank was provided to mitigate the adverse affects of transient pressures in the system (typically encountered upon power failures). The tank is partially filled with water and partially filled with compressed air, which is fed by an oil-less compressor. The surge tank is fitted with level controls and solenoid valves to maintain the water to near the mid-height of the of the 2.4 m diameter, 6.7 m long tank.

7.11.4. Chlorination System

Initial chlorination occurs inside the wet well of the caisson with final chlorination occurring at SCWA's Wohler disinfection facilities. However, it was determined early in the design that chlorine gas would not be stored at the Collector 6 facility. This necessitated a system of mixing and conveying a highly concentrated chlorine solution between the Collector 6 facility and the SCWA's chlorine gas storage facility, located approximately 1,067 m away. The system is a compound loop system that feeds chlorine at a rate based upon the flow of water and the measured chlorine residual at a downstream location.

7.11.5. Water Quality Sampling System

Two separate water quality sampling systems were provided for the Collector 6 facility. The caisson sampling system samples the composite water in the caisson (i.e. combined inflow of all open well laterals), and the lateral sampling system samples water from each individual well lateral. Both systems use a commercially available 1/4 kW submersible pump to pump water (at a rate of 38 lpm) up to the water quality testing equipment located in the pumphouse. The lateral sampling system is remotely controlled from the pumphouse. It utilizes compressed air from the surge control system to operate pneumatic ball valves to open and close individual sampling valves from each well lateral.

8. Startup Testing

SCWA is currently working with DHS to identify the permit status of Collector 6. SCWA developed a monitoring plan for the startup of Collector 6 consisting of two distinct phases, pre-production and post-production. DHS reviewed and approved the monitoring plan in November 2005. As described in Section 7.11.5, the design of the laterals allowed water quality sampling from each individual lateral. In addition to water quality testing, water level, temperature and pumping data are collected from the collector well and surrounding monitoring wells. This data is being used to refine SCWA's MODFLOW model and AEM.

After initial startup testing activities were completed in the spring of 2006 to ensure proper functioning of the pumping and control systems, the start-up monitoring plan was implemented as described below.

8.1. Pre-Production Monitoring

Pre-production monitoring was intended to demonstrate that the water quality from Collector 6 met all applicable regulatory requirements. The monitoring was designed to be completed over a short time-frame in order to place Collector 6 on-line as quickly as possible. Pre-production monitoring included the following:

- Discrete sampling of each lateral for 12 h or until the turbidity of water being produced from the lateral dropped to 0.2 NTU. Turbidity was measured continuously for each lateral.
- Coliform bacteria was analyzed for each lateral at the end of the sampling period.
- Source chemical monitoring of the composite water from Collector 6 in accordance with Title 22 of the California Code of Regulations (Title 22). This program includes general physical parameters, general minerals, metals, volatile and synthetic organic chemicals (including pesticides), radiological constituents, and microbiological constituents.

Pre-production monitoring was completed in two days. While there was some initial concern regarding the rate in which the turbidity in the laterals would drop below 0.2 NTU, turbidity in each of the laterals fell below the required limit within two hours of their specific discrete sampling period. There was no differentiation between the turbidity of the LLDLs and the conventional laterals.

Source chemical monitoring from the composite water of Collector 6 was also completed per Title 22 requirements. All chemical monitoring results were below their respective maximum contaminant levels.

The pre-production monitoring results were submitted to DHS for review and approval of the new water source. In May 2006, DHS approved the use of Collector 6 as a source of supply for SCWA and production immediately commenced.

8.2. Post-Production Monitoring

At this time, the post-production monitoring program is ongoing. All post-production monitoring was specified in the plan to be conducted on the composite Collector 6 water. The DHS approved monitoring plan includes the following:

- Continuous measurements of turbidity reported on an hourly basis;
- Weekly measurements of temperature, conductivity, and pH;
- Monthly raw water bacteriological sampling analyzed using a method that enumerates bacteria density up to 2,400 organisms per 100 ml of sample; and
- Weekly particle counts.

In addition to the above-described monitoring, supplemental monitoring of Russian River water is being conducted to compare to data obtained from Collector 6. Specifically, temperature, conductivity, pH and particle counts are collected weekly from the Russian River as well as from Collector 6. In addition, continuous turbidity measurements from each of the individual laterals are collected.

9. Sustainable Operations

SCWA conducts several activities to ensure that its facilities (including Collector 6) continue to reliably provide high quality water to its retail customers. These activities include routine monitoring and data collection activities in addition to studies and applied research focused on specific aspects of the river and alluvial systems. Table 11.4 summarizes some of the activities SCWA is involved with to ensure sustainable operations of its water supply facilities.

TABLE 11.4. Summary of activities to support sustainable operations.

Routine monitoring
- Continuous river flow monitoring.
- Annual riverbed surveys of dedicated river cross sections to ensure stability of the river bed.
- Monitoring of temperature and water levels to assess flow paths and streambed conductance.
- Monitoring of surface water and groundwater quality (including trace organics, pharmaceuticals, and endocrine disrupting compounds) through a network of monitoring stations and wells.

Studies and applied research
- Focused studies related to infiltration rates along the riverbed and infiltration ponds (e.g., seepage meters and geophysical investigations).
- Characterization of alluvial materials including grain size distribution and grain chemistry (e.g., metal oxides).
- Characterization of dissolved organic matter.
- Assessment of pathogen transport potential (column studies and in-situ mircrosphere experiments).
- Investigation of variably saturated zones beneath the river.

Analysis and Modeling
- MODFLOW model – regional planning level model.
- Analytic Element Model – detailed analysis of lateral and aquifer hydraulics for Collector 6.
- TOUGH 2 – Detailed evaluation of saturated and variably saturated conditions including the development of unsaturated areas beneath the river under certain conditions.

10. Capacity Analysis of Collector 6

A capacity analysis was conducted to evaluate overall collector production capacity trends and to investigate intake lateral hydraulics (Environmental Resources Management [ERM] 2010). Table 11.5 provides a summary of the results of specific capacity estimates for Collector 6 for tests conducted in 2002, 2006 and 2008. The 2002 event evaluated only the conventional laterals as described in Section 7.5 using the data shown in Figure 11.11 (Ranney 2002). Tests conducted in 2006 and 2008 reflect the operation of both the conventional laterals and LLDLs (ERM 2010). A review of the specific capacity estimates indicates that the overall specific capacity has not decreased with time. The specific capacity of the collector appears to be approximately the same over 72 h of testing in the 2002 test (only the conventional wells operating) and the 2008 test (conventional and LLDLs operating). However, the overall production of the collector at that specific capacity is significantly higher for the 2008 test than the 2002 test. As shown in the step-drawdown test data for the 2002 test event (Figure 11.11), a pumping rate of 1.16 m^3/s resulted in an unstable (significantly increasing) drawdown of over 7.5 m. Therefore, the specific capacity at this pumping rate using only the conventional laterals would be less than 0.15 m^3/s/m, whereas the results of the 2006 test at a higher pumping rate (1.53 m^3/s) resulted in a drawdown of only 5.8 m (specific capacity of 0.27 m^3/s/m). Therefore, it is apparent that while the specific capacity of the collector is about the same with and without the LLDLs, the operation of the LLDLs allows a significantly greater production rate to be achieved while maintaining the specific capacity.

TABLE 11.5. Summary of specific capacity estimates for Collector 6.

	2008 Test		2006 Test	2002 Test
Duration of test (h)	72	24	24	72
Pumping rate (m^3/s)	0.82	0.82	1.53	0.78
Specific capacity conventional and LLDL's (m^3/s/m)	0.22	0.27	0.27	N/A
Specific capacity conventional laterals (m^3/s/m)	N/A	N/A	N/A	0.21

The study conducted by ERM also evaluated the flow characteristics of the individual laterals (including the LLDLs). While operating the collector, flow measurements were made at the outlet of each lateral. Figure 11.15 compares the percent contribution of each lateral for a given pumping rate (1.62 m^3/s). As expected, the LLDLs provide a significant portion of the total production (~43%). The orientation of the laterals relative to the river does not appear to influence the flow contribution.

In addition to measuring the overall flow characteristics of the intake laterals, a flow meter was inserted and advanced into 4 of the conventional laterals (# 2, 3, 6, and 8) and both of the LLDLs to measure the flow profile along the length of the lateral. Figure 11.16 illustrates the total cumulative flow for a given distance from the caisson. These data indicate that the contribution of flow from the outer extent of the laterals represents a significant portion of the total flow of the lateral. ERM estimates that about 16.8% of the flow is derived from the inner third of the lateral length and 62.4% of the flow comes from the outer third of the lateral length.

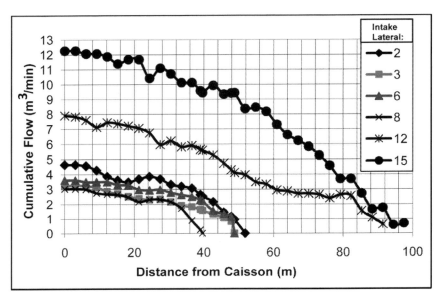

Figure 11.16. Plot of flow versus distance from caisson. Note that LLDL A5-1 is #12 and A5-2 is # 15. (ERM 2010).

11. Conclusions and Recommendations

RBF systems can offer several advantages relative to conventional surface water treatment technologies in terms of cost and water quality. However, the uncertainty of RBF facilities is greater than conventional systems because they depend on and operate more closely within the natural environment, especially if these systems are utilized as the primary means of treatment. Consequently, there will always be inherent uncertainty regarding the final production capacity and water quality of RBF facilities until they are constructed and begin operating. The following strategies helped to make the Collector 6 project a success even with these uncertainties and the risk of attempting a new design.

- Conduct a comprehensive and phased site characterization that builds upon each successive investigation.
- Understand the inherent uncertainty in the natural system throughout the planning, design, and construction phases.
- Communicate the risk and uncertainty to decision makers throughout the process.
- Incorporate redundancy into the design (e.g., intake laterals) to allow for unforeseen conditions to alter the design.
- Be flexible throughout the process, learn and adapt while progressing through the project.

To promote efficient operations, SCWA intends to conduct condition assessments of Collector 6 on a periodic basis. These assessments will include: (1) estimation of specific capacities under similar hydrologic conditions of past evaluations; (2) measurement of the distribution of flow from each of the intake laterals and comparison to past measurements; and (3) measurement of the flow contribution within specific laterals and comparison with prior measurements to assess trends.

Additional applied research is needed to better understand the natural processes that control the performance of facilities such as Collector 6. In particular, the following areas of investigation should be a priority:

Surface Water – Groundwater Interactions The development of cost-effective and reliable field methods that allow utilities to evaluate the spatial and temporal variations in surface water and groundwater interactions is an important topic of research. Ideally, these new methods could be incorporated into a monitoring program to promote more effective operations in terms of production capacity and water quality.

Mechanisms of Natural Filtration The investigation of the physical, chemical, and biological processes that promote natural filtration is a topic of extensive research. It is important to continue to support this area of investigation to better understand the dominant mechanisms that control natural filtration and the environmental conditions that enhance or inhibit effective filtration.

Interaction of Aquifer and Intake Lateral Hydraulics It is important to better understand the inter-relationship of aquifer and pipeline hydraulics as new RBF facility designs employing horizontal drilling methods are developed to address various constraints (e.g., land use, environmental). Areas of investigation include the measurement of flow distribution versus length of laterals, estimation of head loss coefficients for intake laterals, and improved modeling tools.

Acknowledgments The author acknowledges the contributions of Kent Gylfe, Jeanette Wilson, Lori Armburst, and Nathan Basket of SCWA to the preparation of this document.

References

Bakker M, VA Kelson, KH Luther (2005) Multilayer analytic element modeling of radial collector wells. Ground Water 43(6):926–934

Cardwell GT (1965) Geology and ground water in Russian River valley areas and in Round, Laytonville and Little Lake Valleys Sonoma and Mendocino Counties California. US Geological Survey Water-Supply Paper 1548

Environmental Resources Management (2010) Capacity analysis of Sonoma County Water Agency Wohler Radial Collector Wells 1, 2, and 6. February

Grischek T, Schubert J, Jasperse JL, Stowe SM, Collins MR (2007) What is the appropriate site for RBF? International Symposium on Managed Aquifer Recharge. 466–474

Harding Lawson Associates (1988) Hydrogeologic investigation Wohler aquifer study Sonoma County California. December

Norcal Geophysical Consultants Inc (1987) Geophysical survey of the Wohler aquifer study area Russian River California. September

Norcal Geophysical Consultants Inc (1999) Seismic refraction survey pump and collector capacity project Wohler Collector No. 6 Sonoma County California. June

Norcal Geophysical Consultants Inc (2001) Supplemental seismic refraction survey Wohler Collector 6 Russian River Sonoma County California. February

PES Environmental Inc (1999) Hydrogeologic evaluations Westside Farms & Lazy "W" Ranch Healdsburg California. January

Ranney Division of Layne Christensen Company (2002) Wohler Ranney Collector Well No. 6 new lateral design, installation & capacity testing, Sonoma County Water Agency, Santa Rosa California

Su GW, Jasperse J, Seymour D, Constantz J (2004) Estimation of hydraulic conductivity in an alluvial system using temperatures. Ground Water 42(6):890–901

Underground Construction Managers (2007) Report pilot study installation of long, large diameter laterals Russian River diversion facilities Wohler Collector 6 Sonoma County California. January

Chapter 12 Evaluation of Bank Filtration for Drinking Water Supply in Patna by the Ganga River, India

Cornelius Sandhu[1]*, Thomas Grischek[1], Dagmar Schoenheinz[1], Triyugi Prasad[2], and Aseem K. Thakur[3]

[1] University of Applied Sciences Dresden, Faculty of Civil Engineering & Architecture, Friedrich-List-Platz 1, 01069 Dresden, Germany. E-mail: grischek@htw-dresden.de, schoenheinz@htw-dresden.de

[2] Integrated Hydro Development Forum, Patna 800 001, India. E-mail: drt.prasad@gmail.com

[3] Department of Science & Technology, Government of Bihar, Patna 800 001, India. E-mail: aseem_thakur@yahoo.co.in

Abstract   Hydrogeological investigations in Patna by the Ganga River, together with water quality investigations were conducted in 2005–2006 to evaluate riverbank filtration (RBF) wells in the city. A groundwater flow model was used to obtain a better characterization of the groundwater flow conditions. The investigations showed that RBF is an efficient treatment technique for the removal of coliform bacteria (>4 log). The increase in Ganga water level during monsoon helps improve the surface water – groundwater interaction by scouring a 10 m thick sediment layer deposited on the river bed during the pre-monsoon. The dissolved organic carbon concentration was found to be low in Ganga water and groundwater (both <3 mgL^{-1}), except in monsoon the river water showed an increase (4.9 mgL^{-1}). The investigated RBF wells in Patna were found to provide sustainable drinking water in terms of quality and quantity throughout the year.

Keywords: Bank filtration, Ganga, Patna, drinking water, coliform removal

1. Introduction

Surface water bodies (mainly rivers) are the main source for drinking for most riparian urban areas in India. To cope with the large increase in water demand, especially for irrigation and drinking to serve the growing Indian economy, groundwater abstractions have also increased significantly. This has resulted in a

---

* University of Applied Sciences Dresden, Faculty of Civil Engineering & Architecture, Friedrich-List-Platz 1, 01069 Dresden, Germany, e-mail: sandhu@htw-dresden.de

C. Ray and M. Shamrukh (eds.), *Riverbank Filtration for Water Security in Desert Countries*, DOI 10.1007/978-94-007-0026-0_12, © Springer Science+Business Media B.V. 2011

widespread lowering of the groundwater table by more than 0.2 myear^{-1} from 1981–2000, especially in the regions to the south of the Ganga and Sutlej rivers (Indo-Gangetic alluvial plain) up to the southern margin of the hard-rock Deccan Plateau (Nagaraj et al. 1999, Jain et al. 2007). In the case of many cities along the Ganga River in India, the direct abstraction of surface water for drinking appears problematic during periods of low-flow or when the surface water receives large quantities of untreated or partially treated domestic sewage and industrial wastewater (Ray 2005). However, the hydrogeology of many riparian cities in India makes riverbank filtration (RBF) both feasible and attractive as an alternative to direct surface water abstraction, as it is possible to readily extract large quantities of water from river banks.

Patna, having a population of more than 1.7 million (Census of India 2001), is located along the Ganga River (Figure 12.1) and, by simple visual inspection, also appears to be a suitable RBF site. There, six wells of varying depths (150–200 m below ground level) are located 9–236 m from the river and supply water directly to the public distribution network. Analyses of river water and abstraction wells conducted in November 2005 and February 2006 indicated possible bank filtrate being abstracted (Sandhu et al. 2006), though its proportion and travel-time could not be determined then. Additionally, the presence of a confining clay layer and seasonal high sediment deposition interferes, to a varying extent, with the direct hydraulic connection between the aquifer beneath the city and the Ganga River (Sandhu and Thakur 2006). In conjunction with the previous studies, further field and laboratory investigations have been conducted on the water levels, water quality and potential of using bank filtration in Patna (Prasad et al. 2009).

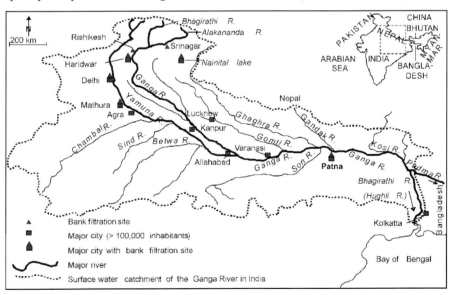

Figure 12.1. Ganga River catchment in India and major bank filtration sites.

In this study, results from field and laboratory investigations (2005–2006) and groundwater flow modeling are used to characterize some of the existing RBF wells in Patna. No investigations on these wells were conducted before 2005, and thus no knowledge about the RBF system in Patna was available. This study expands upon the previous work by Sandhu et al. (2006) and Prasad et al. (2009). An improved understanding of the hydrogeological boundary conditions by groundwater flow modeling is presented. Additionally, the effect of the high sediment deposition during the pre-monsoon and scouring by the Ganga during the monsoon season are discussed.

2. Study Area

2.1. Physiography and Hydrogeology

2.1.1. South Ganga Plain

The city of Patna lies at the northern edge of the South Ganga Plain (SGP) in the Indo-Gangetic Alluvial Plain (Figure 12.2). The SGP is a physiographic unit referring to the linear tract of land bounded by the Ganga in the north, the Rajmahal Hills in the east and the Precambrian Highlands in the south (Saha et al. 2007). The area of Patna city covers a 25 km stretch along the right (south) bank of the Ganga River, and is approximately 5 km in width. The elevation of the city from south to north varies between 48 and 54 m above mean sea level (MSL). According to Saha et al. (2007), the SGP varies from hilly and undulating topography in the south to flat monotonous terrain in the central part and in the north. The flat alluvial terrain in the centre and north displays an average regional slope of 0.63‰. More than 80% of the average annual rainfall $(1,010 \text{ mm year}^{-1})$ from the SW monsoon occurs from June to September (Government of Bihar 1994).

The drainage of the SGP is controlled by the Ganga, flowing from west to east along its northern edge and a number of ephemeral streams originating from the Precambrian Highlands and flowing towards the Ganga in a NNE direction. Just upstream of Patna, a few large perennial rivers such as the Ghaghra, Son and Gandak join the Ganga. Thus the mean annual flow of the Ganga at Patna is significant around the year, and on an average is around $364 \times 10^9 \text{ m}^3\text{a}^{-1}$ (Jain et al. 2007). Even in the dry season (October–May) the flow of the river remains substantial.

A comprehensive description of the geology and hydrogeology of the SGP is provided by Saha et al. (2007). The alluvium deposited by the Ganga and ephemeral streams originating from the Precambrian highlands (Bisaria 1984) is underlain by the northerly sloping basement of crystalline rocks belonging to the Precambrian period (Mathur and Kohli 1963). Deep seismic refraction studies by Bose et al.

(1966) indicate an increase in thickness of the alluvium from south to north, with the maximum thickness of 500–550 m recorded along the course of the Ganga.

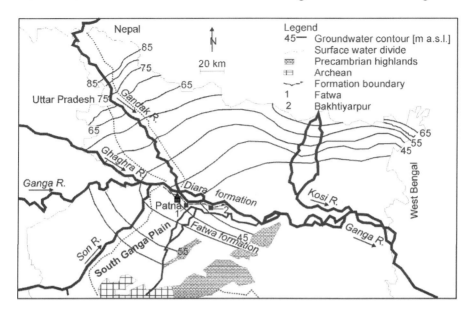

Figure 12.2. Geology and hydrogeology of the South Ganga Plain (after CGWB 1997 and Saha et al. 2007).

The alluvium in the SGP within the study area can be classified into two main formations, namely the Diara and Fatwa formations (Figure 12.2). The Diara formation of the late Holocene age covers the present course of the Ganga (adjacent to Patna) and consists of fine sand, silt and minor clay. The Fatwa formation of the middle to upper Holocene age forms a 10–25 km wide belt to the south of the Diara formation (including Patna and the area to the south). It consists of sand, silt, sandy clay and clay. In this formation, a clay layer is encountered immediately beneath the ground surface along the stretch from Patna to the east bordering the Diara formation. While in Patna the clay layer is usually 50 m thick, towards the east it reduces to 20–40 m. As presented by Saha et al. (2007), lithological information gathered by the Central Ground Water Board (CGWB 1990–1994) and Saha (1999) from three drinking water production wells at the towns of Fatwa, Bakhtiyarpur and Barh to the east of Patna along the Ganga indicate the presence of a 30–50 m thick upper clay layer followed by a 200–220 m thick sand aquifer. These wells are situated on the boundary of the Diara and Fatwa formations.

Within the Quaternary alluvium, the aquifers are semi-confined to confined and are made of unconsolidated, fine to coarse grained sands with occasional gravel beds. The groundwater level decreases from more than 60 m MSL to the south of the SGP to 45 m MSL along the south and north bank of the Ganga at Patna (Figure 12.2). The regional groundwater flow direction in the SGP is to the north with an average gradient of 0.77‰. The hydraulic parameters of the SGP as presented by Saha et al. (2007) were determined based on pumping test data of 19 deep tube wells (CGWB 1990–1994, Saha 1999). Accordingly, the transmissivity and storativity were determined by Jacob's Straight Line Method and by the Theis Curve Matching Method. In the SGP, the hydraulic conductivity (K) gradually increases towards the north, with the highest K ($>1.7 \times 10^{-3}$ ms^{-1}) observed along the Fatwa-Bakhtiarpur-Barh stretch on the northern edge of the Fatwa formation along the Ganga (Saha et al. 2007).

2.1.2. Patna

The aquifer geometry of an approximately 6 km long and 3 km wide stretch of Patna city located along the south bank of the Ganga (Figure 12.3) has been interpreted from borehole logs of the drinking water production wells situated in the study area (T3, T6 and T7), as well as from logs of boreholes drilled during the construction of the *Mahatma Gandhi Bridge* (Public Works Department 1966). These have enabled the construction of two cross-sections, ABC and DEF as shown in Figures 12.3 and 12.4, respectively.

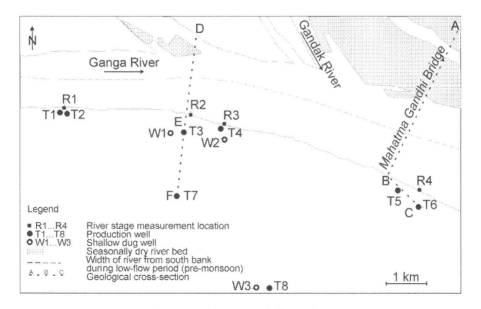

Figure 12.3. Study area in Patna city.

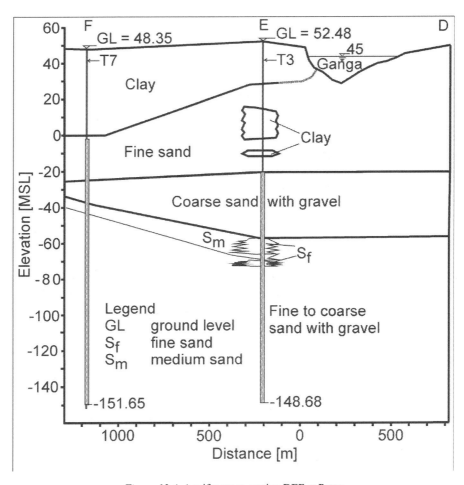

Figure 12.4. Aquifer cross-section DEF at Patna.

In general, the geometry of the aquifer corresponds to the description of Saha et al. (2007). The hydrogeology map of Bihar describes the aquifer below Patna as fairly thick and regionally extensive extending to a depth of 300 m (CGWB 1997). This aquifer depth is derived from the drilling depth of various wells in the city, and does not necessarily indicate the presence of an impermeable base. Hence, as described by Bose et al. (1966), it is probable that the aquifer is deeper (>500 m below ground level). The interpretations of the production well logs reveal a top confining clay layer below the city having a thickness of 24–59 m and extending from the ground level to a depth of 29 to −5 m MSL (Figures 12.4 and 12.5). However, the lateral extent of this confining clay beneath the Ganga is difficult to ascertain. Figure 12.5 indicates that the confining clay layer nearly extends beneath the entire width of the Ganga towards the north bank. However, it is not possible to reach the same conclusion from Figure 12.4 (cross-section DEF), since no boreholes exist in the river bed or across it on the north bank to indicate the extent of the confining clay layer beneath the Ganga. Below the clay layer, the aquifer comprises fine to coarse sand and gravel. However, a second clay layer separating an upper fine to medium sand layer from the main lower lying medium to coarse sand and gravel layer is visible at some locations. At these locations, where a double layered aquifer is visible, the well screens of the production wells are positioned only in the main lower fine to coarse sand and gravel layer at an elevation of −2 m MSL and below. In all the production wells, the groundwater levels rise above the base of the upper clay layer, thus exhibiting confining conditions.

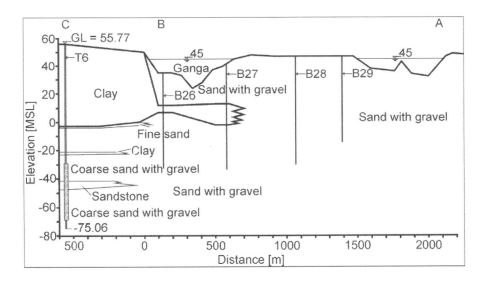

Figure 12.5. Aquifer cross-section ABC at Patna.

Aquifer transmissivities, and yields of existing drinking water production wells located more than 2 km to the south of the Ganga and parallel to it have been determined in previous studies conducted by the Central Ground Water Board (Maitra and Ghosh 1992). Transmissivities obtained from Jacob's straight line and Theis recovery methods were in the range of 5×10^{-2} to 2.2×10^{-1} m^{2}s^{-1} and 4.4×10^{-2} to 1.5×10^{-1} m^{2}s^{-1}, respectively. No indications of the saturated aquifer thickness for these wells on which the tests were conducted could be derived, and thus the same was assumed to be between 90 and 151 m (coinciding with the length of the filter-sections in the aquifer) in order to derive the hydraulic conductivity. This varied by an order of magnitude between 1.4×10^{-3} and 5.6×10^{-4} ms^{-1} depending on the aquifer thickness and presence of fine to coarse sand (and gravel).

2.2. Drinking Water Supply

The entire rural, and more than 90% of the urban water supplies of the SGP is met from the underlying Quaternary aquifers (Saha et al. 2007). In Patna, the drinking water is supplied by groundwater from 137 vertical production wells operated by the Patna Water Board (PWB) and Public Health Engineering Department (PHED) supplemented by numerous private production wells (Prasad et al. 2009). Each of these wells abstract 500–3,200 $m^3 day^{-1}$ from the fine to medium sand aquifer beneath the city. Each well operates discontinuously on an average for a total duration of around 12 hours per day. The wells are located throughout the city. The depth of these wells is 150–200 m. Ganga River water is not abstracted for domestic water supply. Ground water is preferred to surface water abstraction, as the cost of constructing, operating and maintaining a surface water treatment plant exceeds the financial resources available to the PWB (PWB 2005). Fifteen overhead reservoirs that were once used for storing drinking water are now not operational.

Prasad et al. (2009) report that the population of Patna has been increasing in recent years, necessitating increasing supply of drinking water mainly by the installation of deeper production wells. Out of 85 production wells operated by the PWB, more than 60% were installed from 2000 to 2004. The PHED (2006) reports that on an average the groundwater table declines at a rate of approximately 0.3 $myear^{-1}$, forcing them to install the impeller of the pumps at greater depths or even installing higher-capacity pumps. Many private production wells are installed by numerous housing cooperative societies and individual estate owners. These are not registered, and hence neither their production rates nor construction details are known. Within the city, many shallow dug wells exist in local aquitards within the top clay layer. Apart from irrigating small vegetable plots, these wells are also used for drinking and watering domestic animals.

3. Data and Methods

3.1. Field and Laboratory Investigations

The methodology used to investigate RBF in Patna consisted of field investigations, laboratory analyses and groundwater flow modeling (Figure 12.6). No previous site-specific data were available, and hence the investigation programme consisted of basic, but essential tasks. Geodetic surveys to determine the elevation of the ground surface and bench marking of the monitoring sites was carried out by the Integrated Hydro Development Forum Patna (IHDF) in 2005. A review of existing literature, including unpublished data and borehole logs was also done at Patna by the authors. Unpublished data on the Ganga River level and river bed-elevation was obtained by the authors from the river stage monitoring station of the Central Water Commission (CWC) in Patna.

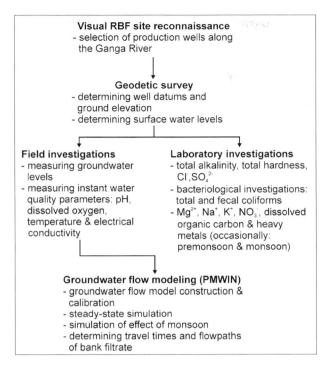

Figure 12.6. Methodology for investigating RBF at Patna.

The water levels and water quality of eight production wells (T1…T8), three dug-wells (W1…W3) and of the Ganga River at four sites (R1…R4) were investigated in this study (Figure 12.3) according to the methodology described by Prasad et al. (2009). Six wells (T1…T6), which abstract bank filtrate, are located within 10–236 m from the riverbank (Table 12.1). Two production wells are located further inland (T7 and T8), in the central part of the city. These were used to investigate the flow of groundwater from the south.

TABLE 12.1. Details of production wells used to investigate RBF in Patna.

Drinking water production well	Total production ($m^3 day^{-1}$)	Daily operation (h)	Distance from river bank (m)
T1	838	10.5	10
T2	1,094	12	70
T3	3,211	13	236
T4	2,736	12	40
T5	2,408	13	100
T6	503	11.5	92
T7	2,736	12	>1,000
T8	2,052	12	>2,000

Ground and surface water levels and water quality were investigated by IHDF at monthly intervals during the pre-monsoon (November 2005–May 2006) and at weekly intervals during the monsoon (June–October 2006). One set each of pre-monsoon and monsoon samples from all locations were analyzed at the laboratory of the water company Stadtwerke Duesseldorf AG (Germany) for dissolved organic carbon (DOC), major anions and cations and trace metals. The groundwater samples were taken from taps (for domestic use; i.e., drinking, bathing, washing) attached to the main supply pipe at each well. A plastic container (10 cm diameter) was used to collect the water from the taps to measure the instant field parameters for groundwater and collect samples for laboratory analyses. Samples to be analyzed for DOC were filtered into a sterilized glass bottle with a 45 μm one-way filter attached to a syringe. One drop of nitric acid was added with a pipette for preserving the sample. Two 100 mL bottles (one each for cation and anion analyses separately), were filled and three drops nitric acid were added as preservative to the sample for cation analyses. The groundwater for cation and anion analyses was not filtered as no suspended particles were visible and the water was clear, however surface water samples for cations, anions and DOC were always filtered. Additionally, one sample for isotope analyses was collected from each site in a glass bottle and sent to the University of East Anglia for analyses.

3.2. Groundwater Flow Model Geometry and Boundary Conditions

To get an improved understanding of the boundary conditions and the hydrogeological regime of the drinking water production wells, a two-layered groundwater flow model (Figure 12.7) using Processing Modflow Version 5.3.0 (Chiang and Kinzelbach 2001) was constructed from field observations of water levels, interpreting data and information provided in the borehole logs and hydrogeological map of Bihar (CGWB 1997 and Saha et al. 2007).

Model Geometry and Initial Conditions   The two-layer groundwater flow model with a discretisation of 62 rows and 44 columns was constructed for the study area covering 30 km^2. A uniform thickness of 53 and 300 m was assigned to the upper and lower layers, respectively. A cell size of 200 × 200 m was used initially. The cell sizes of the production wells were refined to 20 × 50 m. An initial hydraulic head of 45 m MSL was used, which corresponds to the 45 m groundwater contour passing through Patna as shown in the hydrogeology map of Bihar (Figure 12.2). Although this resembles a fictive scenario considering the current high groundwater abstraction in Patna, it however would be a reasonably realistic scenario in case of no or limited groundwater abstraction. During the study period (2005–2006), the groundwater levels measured at rest in the production wells T1…T8, were in the range of 36–45 m MSL.

Boundary Conditions   The Ganga River was simulated using the river boundary condition. The pre-monsoon surface water level (Ganga River) measured in April 2006 at the river stage measuring locations (R1, R2, R3, and R4) was interpolated

EVALUATION OF BANK FILTRATION FOR DRINKING WATER SUPPLY 213

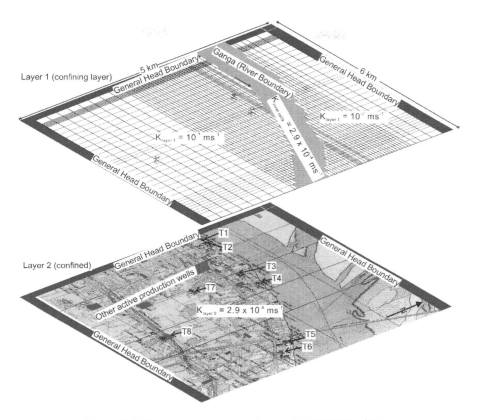

Figure 12.7. Numerical groundwater flow model (PMWIN) of Patna.

and assigned to the river cells. The hydraulic conductance of the river bed (leakage factor in m^{2}s^{-1}) was calculated for each river cell based on its dimensions and using a uniform hydraulic conductivity of 1×10^{-6} ms^{-1}. This conservative value is based on investigations by Grischek et al. (2002) that proposed a hydraulic conductivity of 10^{-6} ms^{-1} for application in groundwater flow modeling to simulate low surface flow conditions of the Elbe River. The current groundwater flow model for Patna also simulates a low discharge scenario for the Ganga. Grain size distribution analyzes the Ganga bedload sediments conducted by Singh et al. (2007) determined that fine to very fine sand having a grain size diameter between 63 and 500 μm (hydraulic conductivity ~10^{-5} ms^{-1}) are the dominant fractions (up to 80 %) of an approximately 300 km long stretch of the Ganga passing by Patna. Grischek et al. (2002) concluded that the hydraulic conductivity of the clogging layer was at least one order of magnitude lower than that of the bed sediments. Additionally, the thickness of the river bed bottom was taken to be equivalent to the difference between the pre- and post-monsoon elevations of the river bed. Thus the hydraulic conductance of the river cells ranged from 0.001 to 1 m^{2}s^{-1}. A general head boundary (GHB) condition was assigned to both layers in the north, south and west of the model domain. The hydraulic head of the GHB corresponds to the groundwater contour of 45 m (Figure 12.2) in the north, and to 50 m in the south and west of the model domain respectively. Initially only the eight production wells investigated during the study (T1...T8) were assigned to the lower model layer. However, with the current abstraction rates (Table 12.1) no significant drawdown occurred. In reality, a significant number of additional production wells (Figure 12.7, dark shaded rectangles in layer 2) exist within the 30 km^2 model domain, of which the location of at least 18 wells operated by the PWB could be identified. The total daily abstraction of these 18 wells (average operation of 12 h/day) is 34,300 m^3. In comparison, the total daily abstraction of the wells T1...T8 is 15,500 m^3 (45%). Hence the 18 wells operated by PWB were additionally assigned to the model domain. The production rates of all the wells are in the range 0.05–0.13 m^{3}s^{-1}.

4. Results and Discussion

4.1. Ground and Surface Water Levels

The water levels of all the production wells, and that of the Ganga River at the farthest downstream gauging site (R4), are shown in Figure 12.8. It is observed that the Ganga River level is higher than the production wells throughout most of the year. The pattern of groundwater levels resembles the Ganga River stage, indicative of a behavior for a confined aquifer. Thus, it can be inferred that the river is influent and in hydraulic contact with the aquifer. Due to the very high abstractions in Patna, a local groundwater trough is created, thus inducing surface water infiltration from the Ganga. However, this conclusion only applies at the local scale of Patna, and cannot be inferred from the regional hydrogeology map of Bihar in Figure 12.2.

4.2. Ganga River Morphology

The highest recorded Ganga River stage by the CWC field station in Patna (since 1965) is 50.27 m MSL in August 1994, which is nearly 2 m higher than observed in August 2006 (Figure 12.8). The corresponding channel width of the Ganga was 2520 m (measured approximately 200 m upstream of R2).

The data for the post-monsoon (December 2004) and the pre-monsoon (June 2005) river bed elevation profile have been combined with the lithological profile of production well T3 (Figure 12.9). Accordingly, the lowest post-monsoon elevation of the river bed was 29 m MSL In the pre-monsoon along the same profile, the lowest elevation measured was 34 m MSL, recorded 250 m from the bank. By the end of the pre-monsoon season in June 2005, the river bed elevation had risen up to 10 m.

Figure 12.9 indicates a dynamic change in the river bed morphology. Similar to the observation by Singh et al. (2007), an increase in the river bed elevation commences after the monsoon ends (post-monsoon) and continues up to the onset of the next monsoon the following year (pre-monsoon 2005) due to the maximum amount of deposition of bed load sediments with decreasing flow. By comparing the elevation of the sand layer in the adjacent aquifer to the south and the Ganga River bed, the extent of the hydraulic contact is increased due to scouring as a result of the monsoon. In the pre-monsoon (April–June), during the period when the Ganga water level is lowest up to when it rises as a result of increased snow melt, a 2–5 m thick sequence of bed load sediments deposits is formed that constitutes mainly fine sand (69–87%), medium and very fine sand with fractions (2%) of silt and clay (Singh et al. 2007).

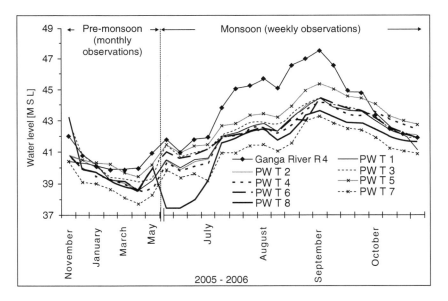

Figure 12.8. Water level observations for the Ganga River (R4) and production wells (T1...T8) at Patna.

4.3. Water Quality

The summary of the water quality data (Table 12.2) indicates that all physicochemical parameters from the production wells are within the maximum permissible limits set by the Bureau of Indian Standards (IS 10500-1991), and thus the abstracted water from the production wells is of good quality. It can be observed that the water in the shallow dug-wells exhibits greater mineralization than the deeper groundwater of the production wells. This is probably due to the much higher residence time of groundwater in the aquitards within the upper confining clay layer.

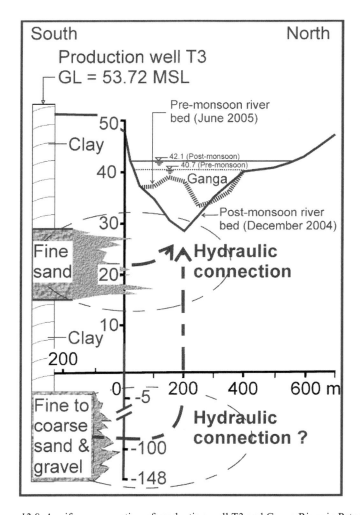

Figure 12.9. Aquifer cross-section of production well T3 and Ganga River in Patna.

The observed high total coliform count in the Ganga River (Table 12.2) is largely due to the discharge of untreated domestic sewage. The presence of total coliform in some of the production wells is most likely also due to the contamination of landward side groundwater from the south, as a result of many open unlined drains carrying domestic sewage. Nevertheless, the more than four log removal of total coliform is a considerable advantage of pre-treatment by RBF. The high total coliform count in the dug-wells can be attributed to the fact that these wells are open at the top, and only surrounded by a parapet. Additionally, the bathing of animals is also often undertaken adjacent to these wells. The wastewater is usually discharged into open unlined drains or septic tanks.

EVALUATION OF BANK FILTRATION FOR DRINKING WATER SUPPLY   217

TABLE 12.2. Water quality during non-monsoon and monsoon periods.

Parameter	Ganga River (R1…R4)		Production wells (T1…T8)		Dug wells (W1…W3)	
	Pre-monsoon	Monsoon	Pre-monsoon	Monsoon	Pre-monsoon	Monsoon
pH	6.9–8.4	7.8–8.2	7.1–7.8	7.5–7.9	7.0–8.0	7.4–7.7
Water temperature (°C)	18.0–33.6	29.4–30.7	18.0–29.0	28.3–28.7	18.1–28.3	28.0–28.5
Electrical conductivity (μScm^{-1})	303–549	168–296	505–681	219–940	734–1547	210–1532
Dissolved oxygen (mgL^{-1})	4.8–9.0	2.9–5.6	0.7–7.4	0.0–6.6	1.8–6.9	3.5–5.5
Total alkalinity (mgL^{-1})	110–230	57–284	151–238	57–277	194–332	287–294
Total hardness (mgL^{-1} as $CaCO_3$)	105–506	90–123	84–364	142–218	148–503	356–446
Ca^{2+} (mgL^{-1})	43–46[3]	22[1]	67–84[9]	70–74[4]	97–104[3]	n. d.
Mg^{2+} (mgL^{-1})	13–18[3]	5[1]	16–20[9]	15–16[4]	26[1]	n. d.
Na^+ (mgL^{-1})	17–29[3]	10[1]	25–35[9]	27–29[4]	83[1]	n. d.
K^+ (mgL^{-1})	4–5[3]	3[1]	2–7[9]	3–5[4]	55[1]	n. d.
Cl^- (mgL^{-1})	3–21	4–9	3–69	3–41	39–83	52–114
SO_4^{2-} (mgL^{-1})	12–29	9–42	2–31	3–74	18–57	29–60
NO_3^- (mgL^{-1})	<1[2]	n. d.	0.7–7.7[5]	1–15[4]	55–83[2]	n. d.
DOC (mgL^{-1})	1.9–2.1[2]	4.9[1]	0.2–2.8[4]	0.6–1.6[4]	0.6[1]	n. d.
Total coliform (MPN/100 mL)	24,000–160,000	90,000–160,000	8–170	8–300	170–2200	800–5000

MPN most probable number, n = 180, [1…9] n = 1…9, n. d. = not determined

Some analyses in the pre-monsoon and monsoon, of iron and manganese, and certain trace metals were conducted in surface and groundwater. The analyses showed that the values were within the permissible limits in most cases. For a few exceptions, such as manganese, the values were within the tolerance limits (IS 10500-1991). While the concentration of iron and barium was <0.1 mgL^{-1} in all cases, that of manganese was in the range of 0.02–0.2 mgL^{-1}. Other heavy metals such as arsenic and copper were <0.01 mgL^{-1} in all samples. Nickel, lead and chromium were <2 μgL^{-1}. Cadmium was <0.5 μgL^{-1}. The concentration of zinc was <0.05 mgL^{-1} in all samples. The values of certain parameters such as the *pH* and concentrations of dissolved oxygen, total hardness, sodium, chloride, sulphate, nitrate, potassium, iron and the total coliform count of the river water (Table 12.2) corroborate well with results of the studies on the Ganga River water by Tiwary et al. (2005) and Ram and Singh (2007).

4.4. Groundwater Flow Modeling

A steady-state calibration of the groundwater flow model was done for two scenarios, one where all the production wells are inactive (Figure 12.10a) and the other where they are all active (Figure 12.10b).

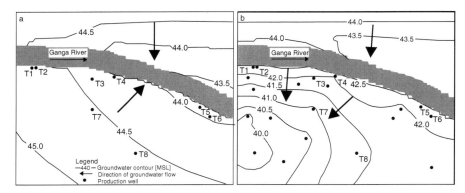

Figure 12.10. Direction of groundwater flow for (**a**) inactive production wells, and (**b**) active production wells in Patna.

For the fictive scenario where all the production wells are inactive (Figure 12.10a), the groundwater flow pattern and contours resemble the 45 m groundwater contour passing through Patna as seen in Figure 12.2. This is a logical scenario, considering that the Ganga is effluent for most of the year, except during monsoon when it becomes influent (Chaturvedi and Srivastava 1979, Saha et al. 2007). For the realistic scenario where all the production wells are active, the groundwater flow direction is reversed to the south of the Ganga (Figure 12.10b). To the immediate north of the Ganga, the groundwater flow direction remains unchanged, since there are no substantial groundwater abstractions in that area. The groundwater flow direction to the south corroborates with the results of the monthly groundwater level field measurements (Figure 12.8), where lower groundwater levels are observed throughout the year (compared to Ganga water levels), as a result of the high groundwater abstractions to meet the drinking water supply of Patna. Thus, as seen in Figure 12.2, the aquifer is effluent on a regional scale. At a local scale, it is reasonable to expect a groundwater trough in Patna resulting in an influent river as seen in Figure 12.10b.

Considering the significant aquifer thickness and storage, and the fact that most of the recharge to the aquifer occurs to the south of the SGP bordering the Precambrian highlands, the GHB to the south and west is a sensitive model parameter that affects the water balance and groundwater table of the model. With the current boundary conditions and model parameters, a good water balance with an error of −0.08% and −0.01% is achieved for both scenarios depicted in Figure 12.10a and b (Table 12.3). The discrepancy between the observed and simulated groundwater

EVALUATION OF BANK FILTRATION FOR DRINKING WATER SUPPLY 219

heads of the production wells was 0.13–0.62 m. Thus considering the scale of the model, a good calibration was achieved.

TABLE 12.3. Water balance for groundwater flow model for Patna.

Flow term	Production wells inactive		Production wells active	
	Inflow into model (m^3s^{-1})	Outflow from model (m^3s^{-1})	Inflow into model (m^3s^{-1})	Outflow from model (m^3s^{-1})
River leakage	0.003	0.32	0.53	0.002
GHB	0.31	0.0	0.62	0.0
Production wells	0.0	0.0	0.0	1.145
Sum	0.313	0.320	1.150	1.147

4.5. Isotope Analyses

One set of pre-monsoon samples of the Ganga River and groundwater was analyzed for the ^{18}O isotope, and one set of monsoon samples was analyzed for ^{18}O and ^2H isotopes (Table 12.4). The relative isotopic abundance (δ) indicates that in the pre-monsoon the $\delta^{18}O$ signature for the Ganga River (−6.35‰) is slightly greater than that of water from the production wells (−6.44 to −6.72‰). During monsoon the $\delta^{18}O$ signature of the Ganga water shows greater depletion (−8.70‰) compared to water from the production wells (−6.61 to −6.68‰). Likewise, the δ^2H signature of the Ganga River water is greater (−62.05‰) during the monsoon compared to production well water. The more depleted isotopic signature can be attributed to the very high proportion of precipitation in the Ganga River. The decrease in the $\delta^{18}O$ signature of the production well water during monsoon indicates a contribution of more depleted isotopic water (such as due to recharge) which could be due to mixing with bank filtrate.

TABLE 12.4. Relative isotopic abundance in Ganga River and groundwater in Patna (Hiscock 2006).

Relative isotopic abundance (‰)	Ganga River		Production wells	
	Pre-monsoon	Monsoon	Pre-monsoon	Monsoon
$\delta^{18}O$	−6.35[b]	−8.70[a]	−6.44 to −6.72[c]	−6.61 to −6.78[d]
δ^2H	n.d.	−62.05[a]	n.d.	−47.17 to −48.46[d]

[a...d] n = 1...4, n.d. = not determined

5. Conclusion

The RBF wells investigated in this study in Patna were found to be efficient for the removal of coliform bacteria. Even during the monsoon, no turbidity was observed in the production wells. Water quality results show that the mineral content of the Ganga at Patna is low, but decreases further during the monsoon. Except for the dug wells, the mineral content of the bank filtrate wells and Ganga River water do not differ significantly. The organic carbon concentration in the river and well water is very low. The low presence of coliform bacteria in some wells is probably due to contamination from land side groundwater due to infiltration of waste water from open and unlined drains. Yet the sufficiently thick clay layer at the surface above the aquifer serves as an effective barrier against potential widespread contamination due to anthropogenic activities. Due to the confining nature of the aquifer, a sharp increase in water levels of the production wells is observed during monsoon which corresponds to the dynamic rise in the Ganga water levels indicating a good hydraulic connection.

The RBF wells in Patna provide significant quality improvements and are sustainable in terms of water quality throughout the year. Nevertheless, a final disinfection of the distributed water is recommended. Although a lowering of the observed groundwater levels in the production wells abstracting bank filtrate over a study period of one year is not observed, this observation cannot apply to all the production wells in the city. Thus, the RBF wells also appear to be sustainable in terms of water quantity abstracted. An abstraction strategy that incorporates the conjunctive operation of bank filtrate and groundwater production wells could arrest the declining groundwater levels in the city to some extent. This would be a useful tool in future.

According to Prasad et al. (2009), evidence from the records of the archives of the Bihar State suggest that the old course of the Son River passed parallel to the south of the Ganga River and through the city in 750 A.D., before joining the Ganga River near Fatwa approximately 10 km to the east of Patna (Public Works Department 1966). Currently the confluence of the Son with the Ganga is approximately 30 km to the west of Patna (Figure 12.2). The abandoned bed of the river in Patna city is characterized by an east-west aligned belt of coarse sand 2 m below ground level. The coarse sand is typical of the current river bed of the Son. Thus, at the points of confluence of the former course of the Son with the current course of the Ganga, the coarse sand aquifers found at a shallow depth would have a good hydraulic contact with the Ganga River. These could serve as potential RBF sites (Prasad et al. 2009).

Acknowledgements Financial support for this study was partly received from the German Ministry of Education and Research (BMBF) for the project 'Riverbank Filtration Network (BMBF IND 08/156),' as part of the BMBF programme 'India and Germany – Strategic Partners for Innovation.' It was also partly supported by the Cooperation Centre for Riverbank Filtration Haridwar (India) and the EU-India River Bank Filtration Network (2005–2006) funded by the

EVALUATION OF BANK FILTRATION FOR DRINKING WATER SUPPLY 221

European Union Economic Cross Cultural Programme (ECCP). The authors are also very grateful to Mr. A.K. Varma and Mr. B.P. Singh for their valuable assistance. The cooperation of the Patna Water Board and Public Health Engineering Department of the Government of Bihar for this study is also appreciated.

References

Bisaria, BK (1984) Geology and geomorphology of a part of Son-Ganga alluvial belt, Nalanda, Nawada and Patna districts, Bihar. Unpubl. Rep. Geol. Survey of India, Kolkata. In: Saha D, Upadhyay S, Dhar YR, Singh R (2007) The aquifer system and evaluation of its hydraulic parameters in parts of South Ganga Plain, Bihar. J Geol Soc India 69(5):1031–1041

Bose PK, Saikia BC, Mukherjee BB (1966) Refraction seismic survey of ground water in Patna and Gaya districts, Bihar. Unpubl. Rep. Geol. Survey of India, Kolkata. In: Saha D, Upadhyay S, Dhar YR, Singh R (2007) The aquifer system and evaluation of its hydraulic parameters in parts of South Ganga Plain, Bihar. J Geol Soc India 69(5):1031–1041

Census of India (2001) List of million plus cities. 2001 Census Results, Census of India, Office of the Registrar General, New Delhi. Available: http://www.censusindia.net; accessed: January 2010

CGWB (1990–1994) Basic data reports of exploratory wells in Nalanda, Nawada, and Seikhpura districts, Bihar. Unpubl. Rep. Central Ground Water Board, Patna, India. In: Saha D, Upadhyay S, Dhar YR, Singh R (2007) The aquifer system and evaluation of its hydraulic parameters in parts of South Ganga Plain, Bihar. J Geol Soc India 69(5):1031–1041

CGWB (1997) Bihar Hydrogeology Map 1:2,000,000. Central Ground Water Board, Patna, India

Chaturvedi MC, Srivastava VK (1979) Induced groundwater recharge in the Ganges basin. Water Resour Res 15(5):1156–1166

Chiang W-H, Kinzelbach W (2001) 3D-Groundwater Modeling with PMWIN. Springer-Verlag, Berlin, Heidelberg, New York

Government of Bihar (1994) Report of the second Bihar State Irrigation Commission. In: Saha D, Upadhyay S, Dhar YR, Singh R (2007) The aquifer system and evaluation of its hydraulic parameters in parts of South Ganga Plain, Bihar. J Geol Soc India 69(5):1031–104

Grischek T, Macheleidt W, Nestler W (2002) River bed specifics and their effect on bank filtration effeiciency. In: Dillon PJ (Ed.) (2002) Management of Aquifer Recharge for Sustainability. Proceedings of International Symposium on Artificial Recharge of Groundwater, ISAR-4, Adelaide, S. Australia 22–26 September 2002. Balkema, Lisse/Abingdon/Exton (PA)/Tokyo, 59–64

Hiscock KM (2006) Application of stable isotope methods in RBF studies: examples from Europe and India. In: Grischek T (2007) EU-India River Bank Filtration Network (RBFN) Final Project Report. European Union Economic Cross Cultural Programme (ECCP). ASIE/2004/095-733

IS 10500 (1991) Indian standard Drinking water – Specification, Edition 2.2, Bureau of Indian Standards, New Delhi, India

Jain SK, Agarwal PK, Singh VP (2007) Hydrology and water resources of India. Water Science and Technology Lib. 57. Springer, Dordrecht, The Netherlands. ISBN 9781402051791. 1258 p

Maitra MK, Ghosh NC (1992) Groundwater Management and Application. Efficient Offset Printers, New Delhi, 301 pp

Mathur LP, Kohli G (1963) Exploration and development of the oil resources of India. 6. World Petroleum Cong., Section 1, New Delhi. In: Saha D, Upadhyay S, Dhar YR, Singh R (2007) The aquifer system and evaluation of its hydraulic parameters in parts of South Ganga Plain, Bihar. J Geol Soc India 69(5):1031–1041

Nagaraj N, Marshall Frasier W, Sampath RK (1999) A comparative study of groundwater institutions in the western United States and Peninsular India for sustainable and equitable resource use. Department of Agricultural and Resource Economics, Colorado State University, Fort Collins. In: Jain SK, Agarwal PK, Singh VP (2007) Hydrology and Water Resources of India. Water Science and Technology Library, 57. Springer, Dordrecht, The Netherlands. ISBN 9781402051791

PWB (2005) Patna Water Board. Written communication to Integrated Hydro Development Forum Patna

Prasad T, Varma AK, Singh BP (2009) Water supply to Patna through river bank filtration: Problems and prospects. Proc. Int. Conf. Water, Environment, Energy and Society WEES, New Delhi, 12–16 January 2009, 3, 1348–1357

PHED (2006) Public Health Engineering Department. Verbal communication

Public Works Department (1966) Detailed Project Report of Ganga Bridge at Patna. Government of Bihar. In: Prasad T, Varma AK, Singh BP (2006) EU-India River Bank Filtration Network Final Report of Patna. Integrated Hydro Development Forum Patna

Ram P, Singh AK (2007) Ganga water quality at Patna with reference to physico-chemical and bacteriological parameters. J Env Sci Eng 49(1):28–32

Ray C (2005) Drinking water supply in India and future potential for riverbank filtration. In: Ray C, Ojha CSP (2005) (eds) Riverbank Filtration – Theory, Practice and Potential for India. Proc. Int. Workshop Riverbank Filtration, 1–2 March 2004, Indian Institute of Technology Roorkee, India. Water Resources Research Center, University of Hawaii, Manoa, Cooperative Report CR-2005-01, 3–8

Saha D (1999) Hydrogeological framework and ground water resources of Bihar. Unpubl. Rep. Central Ground Water Board, Patna. In: Saha D, Upadhyay S, Dhar YR, Singh R (2007) The aquifer system and evaluation of its hydraulic parameters in parts of South Ganga Plain, Bihar. J Geol Soc India 69(5):1031–1041

Saha D, Upadhyay S, Dhar YR, Singh R (2007) The aquifer system and evaluation of its hydraulic parameters in parts of South Ganga Plain, Bihar. J Geol Soc India 69(5):1031–1041

Sandhu C, Thakur AK (2006) Report to the field investigations in Patna, 14–20 February 2006. EU-India River Bank Filtration Network Report, Cooperation Centre for Riverbank Filtration, Haridwar, India

Sandhu C, Grischek T, Schoenheinz D, Ojha CSP, Irmscher R, Uniyal HP, Thakur AK, Ray C (2006) Drinking water production in India – Bank filtration as an alternative. Water Digest 1(3):62–65

Singh M, Singh IB, Müller G (2007) Sediment characteristics and transportation dynamics of the Ganga river. Geomorphology 86(1–2):144–175

Tiwary RK, Rajak GP, Abhishek, Mondal MR (2005) Water quality assessment of Ganga River in Bihar Region, India. J Env Sci Eng 47(4):326–335

Chapter 13   Minimizing Security Risks Beyond the Fence-Line: Design Features of a Tunnel-Connected Riverbank Filtration System

S. Hubbs[1]*, K. Ball[2], and D. Haas[3]

[1] WaterAdvice Associates, 3715 Hughes Road, Louisville, KY, USA

[2] Louisville Water Company, 550 S Third Street, Louisville, KY, USA. E-mail: kball@lwcky.com

[3] Jordan, Jones, and Goulding, Inc. a Jacobs Engineering Group company, 6801 Governors Lake Parkway, Norcross, GA, USA. E-mail: David.Haas@jjg.com

Abstract   Large-capacity riverbank filtration (RBF) facilities often require large amounts of river frontage, exposing the facility to increased security risks. The design of the 230,000 m^3/day RBF facility at Louisville involved construction of four radial collector wells along 2,000 m of riverbank, much of which was not owned by the utility. Concerns regarding the security of the facility were addressed by constructing the riverside portion of the facility entirely underground, with a single pump station located within the fenced perimeter of the water treatment plant. This chapter describes the unique design features of the facility, including sub-grade collector wells, connection to a hard-rock tunnel, and the pump station.

Keywords: Riverbank filtration, construction, horizontal collector well, tunnel, cost, Louisville

1. Introduction

The Louisville Water Company (LWC) has provided water to the city of Louisville since 1860. Today, LWC provides approximately 500,000 m^3/day (130 mgd) to a community of 700,000 people in the region. Until recently, the water supply for Louisville was the Ohio River, a large Midwest US river impacted by large cities and industry upstream. In the 1950s, groundwater riverbank filtration (RBF)

---

* S. Hubbs, Louisville Water Company, 550 S Third Street, Louisville, KY, USA, e-mail: stevehubbs@bellsouth.net

C. Ray and M. Shamrukh (eds.), *Riverbank Filtration for Water Security in Desert Countries*, 223
DOI 10.1007/978-94-007-0026-0_13, © Springer Science+Business Media B.V. 2011

was investigated as a potential source for water quality reasons. Water quality improvements in the watershed between 1950 through the 1970s reduced the need for a change in water supply, and LWC continued to use the Ohio River as its primary source of water.

RBF was again considered in the 1980s as a source water for Louisville, driven by water quality concerns for disinfection by-products, and again in the 1990s driven by concerns for pesticides, microbial contaminants (*Cryptosporidium*), and MIB/Geosmin taste and odor events. In 1999, a 75,000 m^3/day (20 mgd) demonstration well was constructed, and design was initiated to convert the smaller of two plants (Payne WTP) to RBF. This was the second of three phases planned for advanced treatment technology at LWC, with the final phase involving the upgrade of the 680,000 m^3/day (180 mgd) Crescent Hill WTP.

The design capacity for the Payne WTP was set at 230,000 m^3/day (60 mgd), which would require a total length of approximately 1,800 m (6,000 ft) of riverbank. The land was privately owned, and security and aesthetics were factors to be considered in the design. The project objectives were thus set at: improved water quality, long-term cost efficiency, adequate capacity, facility security, and aesthetics. This paper focuses on the security elements of this facility.

2. Design Considerations

In the design stage of RBF facilities, several design options are available for extraction (vertical wells versus horizontal collector wells), conveyance (pressure mains, gravity conveyance via tunnels, and siphon systems), and pumping (pump stations at each well versus centralized pump station; submersible pumps versus vertical turbine pumps). Each of these options has an impact on the security of the system.

2.1. Extraction Systems

The most common method used in the US to extract water from a well is to install a pump and motor at each well, and lift the extracted water to a central treatment or distribution point. Such systems allow for the most flexibility in design, as water can be lifted from any depth and transported any distance using relatively inexpensive facilities. Individual pumps can be operated independent of the others, and can be removed for servicing without disruption to the operation of the remaining wells. Motors can be located either above ground, or installed within the well casing itself (submersible pumps), minimizing the visual impact of an elevated pump head and motor. For very large systems, such traditional systems result in a large number of visually apparent facilities dispersed over a large area, requiring occasional access for maintenance. Overhead electrical facilities and trenched pressure-discharge piping can also result in the disruption of neighborhoods during construction.

In Europe, many facilities have been built using vacuum-siphon extraction systems, with several wells connected to a central siphon point where water is then lifted to a central distribution or treatment facility. These facilities have the benefit of several wells being attached to a single siphon/pumping facility, thus reducing the number of pumping facilities and electrical service required for individually pumped wells. These siphon systems are limited, however, to a maximum water lift of approximately 10 m, due to the vapor pressure of water. Thus, a facility designed for a siphon extraction system is restricted to a total differential of 10 m between the lowest water surface in the well and the highest point of lift in the extraction system. In many systems, this highest point would be approximately the normal static water level of the aquifer. Because of the need to keep the siphon pump station at a relatively low elevation, the siphon pump stations often have to be protected from surface flooding.

Noting the desire for security, operational simplicity, and landscape aesthetics, a design was envisioned for Louisville that would connect several wells together by a tunnel system that would allow the pump station to be situated independent of the well locations. In this case, the desired location of the pump station was on secure property and adjacent to the treatment plant, approximately 600 m from the edge of the river and over 2,000 m from the farthest well. Water would flow by gravity from the wells into the tunnel, which would have to be located below the maximum drawdown level of the well. The tunnel would link all wells to a central pump station, allowing all wells to serve a single pump station.

2.2. Design Issues

Several design issues had to be resolved which were unique to the gravity-flow tunnel design concept: the type of well to use (vertical wells or horizontal collector wells), the type of tunnel to construct (hard-rock or soft soil in the saturated aquifer), and the method of sealing the pressure differential between the aquifer and the well/tunnel system. Of these issues, the seal between the well and the aquifer was of greatest concern, as this type of connection was not typical to water well installations. During maximum drawdown, as much as 15 m of water pressure would be exerted on the seal between the aquifer and the well and tunnel. Thus the design had to include an effective seal between aquifer and the well and tunnel, as any failure in the seal would result in the tunnel being flooded with sand and significant land subsidence at the surface.

Two types of wells were considered for the system: vertical wells and horizontal collector wells. Both types of wells have been used in the aquifer, which has a saturated thickness of approximately 25 m. The horizontal collector design called for four wells and had the advantage over vertical wells of allowing a greater amount of drawdown, while the vertical design called for 30 wells and allowed greater ease of maintenance.

Options for the connecting tunnel included a soft-soil tunnel in the aquifer and a hard-rock tunnel 15 m below the aquifer/bedrock surface. Considering the desire to have the tunnel extend away from the river where the aquifer was thinner, the uncertainty involved in connecting the wells to the tunnel in the saturated sand gravel aquifer, and a limited number of soft-soil tunnel contractors, the hard-rock tunnel option was selected.

Finally, considering the number of construction sites needed and the critical nature of the seal between the aquifer and the well/tunnel system, horizontal collector wells were selected as the least-risk prone design option. The final design thus featured four horizontal collector wells spaced over 1,800 m, capped at grade, and connected to a 2,300 m long, 3 m diameter hard-rock tunnel. The tunnel paralleled the river under the four collector wells, and then curved away from the river towards the pump station located on the treatment plant grounds.

3. Construction

The tunnel was constructed by first digging an access shaft through the saturated aquifer and into the bedrock to the level of the tunnel, a total depth of 57 m. This access shaft was constructed using a "diaphragm wall" technique, wherein a series of vertical rectangular steel-reinforced panels were poured into slurry-filled holes bored with a "hydro-mill" through the aquifer and anchored into rock. A total of eight panels were placed in a circular pattern with each locked into the other, resulting in a 13 m diameter concrete-reinforced structure (see Figures 13.1 and 13.2). The four primary panels were approximately 1 m thick, 9 m long, and 40 m deep. These were linked with four secondary panels each, each approximately 3 m long. After the diaphragm wall shell was completed into bedrock, the sand and gravel within the structure was excavated by clamshell, and remaining 20 m of the shaft to tunnel elevation was constructed with conventional rock blasting mining. The interior wall was then lined with shotcrete.

MINIMIZING SECURITY RISKS BEYOND THE FENCE-LINE 227

Figure 13.1. (**a**) Hydro-mill string on crane, (**b**) rebar being inserted into slurry-filled hole, and (**c**) final wall sections keyed into place.

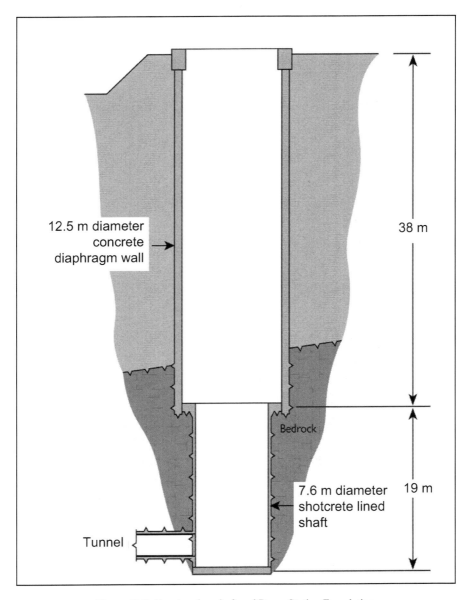

Figure 13.2. Construction shaft and Pump Station Foundation.

The tunnel was then excavated using a tunnel boring machine, with all mined material transported by small train to the access shaft and extracted. The tunnel was mined to a diameter of approximately 4 m, and the tunnel boring machine was disassembled and withdrawn through the access shaft. The tunnel was then finished to a diameter of 3.2 m with a poured-in-place liner (Figure 13.3).

While the tunnel was being excavated, the collector wells were set along the path of the tunnel to the aquifer-bedrock surface and grout-sealed to the bedrock. The collector wells were spaced at approximately 600 m intervals. Each was constructed by sinking a series of 4 m cast-in-place concrete steel reinforced rings into the aquifer by clamshell excavation of the soil inside the ring. The initial ring was fitted with a drive-shoe and pre-cast ports for laterals (Figure 13.4). When the caisson reached bedrock, a concrete floor approximately 3 m thick was poured into the caisson.

After the concrete had cured, the water was pumped out of the caisson and the laterals were set by jacking out 1.5 m sections of 0.4 m diameter steel pipe horizontally through the pre-constructed ports (typically 70–74 m into the aquifer and 4 m above bedrock). Next, 3 m sections of 0.3 m diameter screens were pushed into the steel pipe and the steel pipe was then pulled back into the caisson, exposing the screen to the aquifer. The screens were developed by surge-pumping, and capped to allow the caisson to again be dewatered. A total of eight laterals totaling approximately 550 m of wellscreen were placed in each caisson (except collector #3, which has 5 laterals totaling approximately 300 m of wellscreen).

After the tunnel was completed and the caissons and screens installed, the connection between the well and tunnel was constructed. A 1.5 m diameter hole was drilled through the bottom of the caisson and bedrock into the tunnel, and a 1.2 m diameter drop-shaft approximately 12 m long was placed in the hole and grouted in place (Figure 13.5). Each well was fitted with a flange at the top of the drop-shaft to allow each collector well to be isolated from the others for maintenance. The tunnel was then completed by constructing a cast-in-place tunnel liner.

After the wells and tunnel were completed, the system was disinfected and capped at grade. A diagram of the collector well and connection to the tunnel is shown in Figure 13.6.

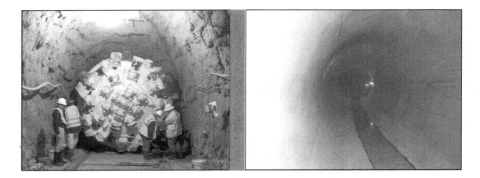

Figure 13.3. Tunnel boring machine in entrance of tunnel (*left*), and finished tunnel showing cast-in-place concrete liner (*right*).

Figure 13.4. Construction of initial caisson section showing rebar and portals (*left*) and finished section with drive-shoe (*right*) ready to be "sunk."

Figure 13.5. Drilling connection between well and tunnel (*left*), and installation of drop-shaft between the well and the tunnel (*right*).

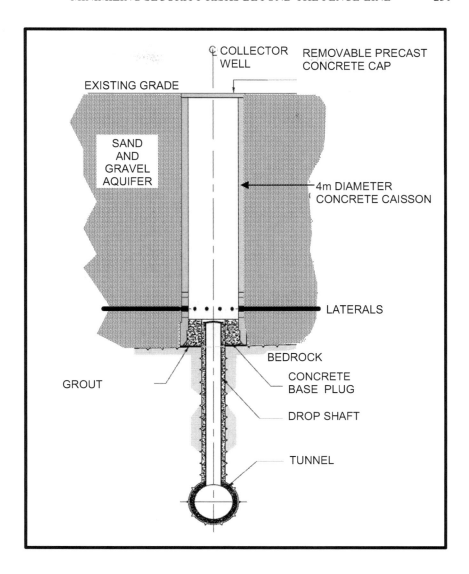

Figure 13.6. Collector well and connection to tunnel.

The pump station was constructed over the construction access shaft, using the shaft as both the foundation for the building and the wet-well for the pumps (Figures 13.7 and 13.8). Pumps were sized to allow flexibility in operation, with a continuous pumping range from 60,000 to 230,000 m^3/day. The tunnel and pump station were constructed to allow future connections to the tunnel and expansion of the pump station capacity.

Figure 13.7. Pump station foundation under construction over construction access shaft.

MINIMIZING SECURITY RISKS BEYOND THE FENCE-LINE 233

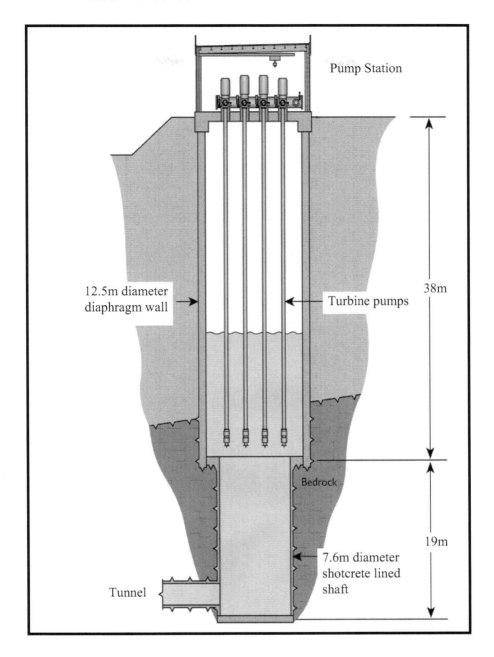

Figure 13.8. Pump station constructed over construction access shaft.

4. Summary

The total construction cost was approximately $37 million, which included $9 million for four collector wells, $22 million for the tunnel and shaft (which also served as pump station foundation and wet well), and $5 million for the pump station. The facility is under construction, with start-up anticipated in July 2010.

When completed in summer 2010, the 230,000 m^3/day RBF facility will provide access to a large expanse of riverbank constructed on privately held property, publicly owned property, and property owned by the utility. All structures located beyond the secured plant fence line are below grade and highly secured, accessible only with heavy equipment. All mechanical, electrical, and control equipment is located within the single pump station located adjacent to the treatment facility on secured utility property. The facility was constructed with minimal disruption to privately held property, and is essentially invisible beyond the fenced pump station perimeter.

Chapter 14 Removal of Iron and Manganese Within the Aquifer Using Enhanced Riverbank Filtration Technique Under Arid Conditions

Kamal Ouda Ghodeif*

Geology Department – Suez Canal University – Ismailia – Egypt.

Abstract Nowadays, riverbank filtration (RBF) technique is receiving more attention from drinking water suppliers in Egypt. An innovative design has been established to couple RBF and removal of Fe and Mn in situ. A preferential flow has been induced through constructing two sand filters crossing the upper clay layer. Continues monitoring for water quality indicators in RBF wells has revealed the potentiality of the design. The contents of Fe and Mn on average have been reduced with time. Detailed investigations of removal processes for Fe and Mn within the aquifer require long term monitoring. This innovative idea can be transferred along the Nile River where the clay layer exists and the already existing drinking water wells producing water of high iron and manganese contents.

Keywords: Riverbank filtration, Egypt, iron, manganese, water quality

1. Introduction

The population growth and high density of human activities in the Nile valley and Delta have increased the drinking water demand and lead to contamination of surface water sources. This has increased the cost of drinking water treatment with the conventional methods. At present, conventional treatment of surface water in Egypt includes pre-chlorination, coagulation, flocculation, sedimentation, filtration, and post-chlorination disinfection. The currently implemented conventional treatment processes have failed to remove several pollutants (Abdel-Shafy and Aly 2002). Moreover, pollution of source water has reduced the efficiency of sand filters due to accumulation of microorganisms and clogging. An important limitation of sand filtration is the need for high-quality source water (Logsdon et al. 2002). Cost efficient and sustainable techniques are needed to provide safe drinking water in

---

* Kamal Ouda Ghodeif, Geology Department – Suez Canal University – Ismailia – Egypt, e-mail: kghodeif@yahoo.com

C. Ray and M. Shamrukh (eds.), *Riverbank Filtration for Water Security in Desert Countries*,
DOI 10.1007/978-94-007-0026-0_14, © Springer Science+Business Media B.V. 2011

sufficient quantity and with high quality in the future. The government has used groundwater sources to supply drinking water to communities in rural areas and cover larger parts of the country with safe drinking water. Nevertheless, after pumping the groundwater into the water distribution system, water turned reddish to black in most of the sites due to the occurrence of iron and manganese in source groundwater (El Arabi 1999).

There is a great potential for the application of riverbank filtration (RBF) along the River Nile and main canals in Egypt. RBF is the naturally occurring inflow of surface water to the groundwater aquifers, via the bed and banks of the surface water body. During its passage, the water quality parameters improve due to microbial, physical and geochemical processes and by the mixing with ambient groundwater quality (Kuehn and Mueller 2000). Typical aquifers used for RBF in Europe consist of sand and gravel deposits that have hydraulic conductivity higher than 8.6 m/day (Grischek et al. 2002). The multi protective barrier concept, of bank filtration, including both natural and technical purification has proven to be a reliable method for drinking water production (Eckert et al. 2006). The natural attenuation of contaminants during bank filtration includes the elimination of suspended solids, particles, biodegradable compounds, bacteria, viruses, and parasites as well as the partial elimination of absorbable compounds (Hiscock and Grischek 2002). Bank filtrate is fairly biologically stable water with a lower disinfection dose thus long term application can decrease water treatment costs (Kuehn and Mueller 2000).

Hydrogeological conditions along the Nile River provide a promising application of RBF. Problems arise if thick clay layer exists at shallow depths that prevent seepage of surface water into groundwater and consume oxygen. Moreover, the redox conditions along the underground passage may lead to iron dissolution in addition to the ambient groundwater has iron concentrations >0.3 mg/L which is the case at many sites in Egypt. An Innovative approach is required to handle such constraints for simple RBF and removal of iron and manganese in the subsurface. A coupling of RBF and subsurface iron removal is used in this study. The objective of this work is to assess the efficiency of the proposed innovative design to enhance RBF and remove the iron and manganese within the aquifer as well as to evaluate the removal efficiency of bacteria and total organic carbon (TOC).

This study has been done in a new RBF site at Cairo along a main branch of River Nile called Al-Rayah Al-Naseri (Figure 14.1). Nevertheless, the hydrogeological conditions were not favorable for simple RBF design and the ambient groundwater quality has high iron (Fe) (0.97 mg/L) and manganese (Mn) (0.87 mg/L) contents. Moreover, surface water quality has high total coliform count (1,800 MPN/100 ml) and contains high amounts of total organic carbon (4.1 mg/L). The main aquifer that is of Pleistocene age and consists of sands and gravel has been targeted. The aquifer is overlaid by clay layer of average thickness (21 m) and upper thin aquifer that is directly connected with the surface water body.

The removal approach shown in this study differs from most other techniques used in Egypt that only investigated the removal using injection of enriched oxygenated water and recovery. It describes the results of the joint use of RBF and subsurface removal of iron and manganese in Nile Delta aquifer. The water quality of the RBF wells has been found suitable for drinking purposes except for Fe and Mn. It has high efficiency in removing coliform bacteria (100%) that is encouraging to apply RBF in similar countries that have high temperature or potential to have high temperature after climatic changes. Sand filters have facilitated the passage of oxygenated river water into the producing aquifer.

Figure 14.1. General location map for the study area (Birqash site along Al-Rayah Al-Naseri).

2. Iron and Manganese Distribution and Associated Problems in Egypt

Iron is found in natural fresh waters at levels ranging from 0.05 to 50 mg/L. No health-based guideline value is proposed from World Health Organization for iron (WHO 2006). The permissible limit according to Egyptian Drinking Water Standards (law no 458/2007) is 0.3 mg/L. At levels above 0.3 mg/L, iron stains laundry and plumbing fixtures. Anaerobic groundwater may contain ferrous iron at high concentrations without discoloration or turbidity in the water when directly pumped from a well. On exposure to the atmosphere, however, the ferrous iron oxidizes to ferric iron, giving an objectionable reddish-brown colour to the water due to the oxidation of the dissolved Fe(ll) to solid Fe-oxides. Iron also promotes the growth of iron bacteria, which derive their energy from the oxidation of

238 K.O. GHODEIF

ferrous iron to ferric iron and in the process deposit a slimy coating on the piping and distribution network.

Manganese is naturally occurring in many surface water and groundwater sources, particularly in anaerobic or low oxidation conditions. Occurrence levels in fresh water typically range from 0.001 to 0.2 mg/L, although levels as high as 10 mg/L in acidic groundwater have been reported (WHO 2006). The presence of manganese in drinking water, like that of iron, may lead to the accumulation of deposits in the distribution system (Kohl and Medlar 2007). Guideline value is 0.4 mg/L, concentrations at or below the health based guideline value may affect the appearance, taste or odor of the water, leading to consumer complaints. Upon aeration water turns gray-black due to the oxidation of the dissolved Mn(ll) to solid Mn-oxides. The permissible limit according to Egyptian Drinking Water Standards is 0.4 mg/L.

A baseline national water quality monitoring report (NWRC 2003) indicating a high level of iron and manganese in some observation wells in the Nile Delta and Western Desert aquifers. Iron concentrations exceeded the allowable standard for drinking water in about 51% of the monitored wells in the Nile valley aquifer. While in the Western Desert aquifer; concentrations of manganese exceeded drinking water limit in about 75% of the wells and iron exceeded drinking water standard in up to 100% of the wells (EEAA 2009). The high values of iron and manganese are related to the anaerobic conditions dominant in the aquifers. In a study around Cairo, the percentage of wells exceeding drinking water standard is 80% for manganese (Mn) and 20% for iron (Fe) where the concentrations of Mn were up to 1.5 mg/L and up to 1.0 mg/L for Fe (El Arabi 1999). In a study along the west side of the Nile near the intersection point between the Nile and its Delta, at Giza governorate, the range of Mn concentration is 0.058–2.78 mg/L and Fe concentration value lies within the range of 0.141–1.704 mg/L (Emara et al. 2007). The traditional technology for Fe and Mn removal includes an oxidation-precipitation of the soluble Fe(ll) and Mn(ll) to the insoluble Fe- and Mn-oxides followed by filtration. This can be achieved either above ground (conventional treatment) or in situ groundwater treatment, where Fe and Mn are precipitated within the aquifer.

3. Hydrogeological Conditions of the Nile Delta System

The main aquifer system in the Nile Delta consists of two water-bearing layers. The lower layer is formed of highly permeable alluvial sediments (sand and gravel) of Pleistocene age, and the upper one is formed of the Holocene clay-silt layer of relatively low and very low horizontal and vertical hydraulic conductivities, respectively. The Pliocene clay is the base of the main aquifer system (Idris and Nour 1990). The main aquifer in the Nile Delta is represented by Mit Ghamr Formation (El-Fayoumy 1968) that consists mainly from Pleistocene sand (Figure 14.2). The AQI average saturated thickness of the Nile delta aquifer ranges from about 200 m in the South near Qanatir el Qahiriya to about 800 m in the North, with a thickness reduction trend towards the delta fringes and southward to Cairo. The main aquifer is overlaid by a semi confining layer with a thickness ranges between 10 m in the South up to 30 m in the North (Dawoud et al. 2005). The groundwater aquifer underlying the Nile valley and Delta is recharged by seepage losses from the Nile, irrigation canals and deep percolation from irrigated lands (Figure 14.3) (RIGW 1989). Downward groundwater leakage occurs in the southern part of delta. The vertical water movement through the clay cap affects, to a great extent, the groundwater management and the drainage conditions. The existing rate of groundwater abstraction in the Nile valley and Delta regions is about 4.8 billion cubic meters (BCM)/year, which is still below the potential safe yield of the aquifer. Salinity of the Nile Delta aquifers is mainly below 1,500 mg/L. However, the average reported salinity of pumped groundwater lies in the range from 160 to 480 mg/L in the South Delta, 480 to 1,440 mg/L in the Middle Delta and >3,200 mg/L in the Northern Nile Delta (RIGW 1992).

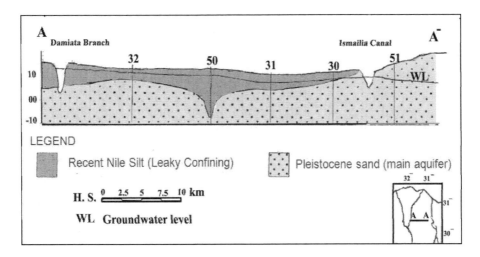

Figure 14.2. Hydrogeological cross-section in the eastern Nile Delta (modified after El-Fayoumy 1968).

4. Research Assumptions, Materials and Methods

Due to the occurrence of shallow clay layer under the surface water courses in the study area, an innovative approach is used. This has included inducing preferential flow through the clay layer and monitoring the quality changes in RBF wells. It is assumed that this innovative design will facilitate the passage of Nile water including oxygen into the producing aquifer and to improve the water quality and remove the Fe and Mn in subsurface. The approach has concentrated on inducing oxygen into the aquifer with bank filtrated water to prevent mobility of Fe and Mn. To test the efficiency of the proposed technique, a continuous water quality monitoring program, for a period of 12 months, was conducted. Water samples were collected from groundwater, surface water and from RBF wells for the determination of the physical, chemical and microbiological (fecal coliform) characteristics. Analyses for about 18 parameters include major ions, trace metals, nutrients, and other organic contaminants (TOC) were done in the laboratories of Geza Water Company and Suez Canal University that follow the American Public Health Association standard methods (APHA 2005). Hydrogeological data were collected for detail investigation. Hydrogeological environment is investigated using the conventional hydrogeological methods. Soil and aquifer materials were investigated and accordingly wells are designed.

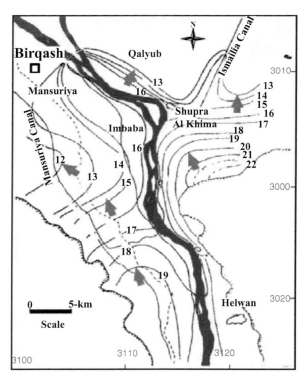

Figure 14.3. Groundwater level map and groundwater flow direction in the main aquifer in the south part of the Nile Delta (modified after RIGW 1989).

5. Design of the Site and Detailed Hydrogeology

The site has been selected in the old cultivated lands of the Nile Delta to be far away from residential areas where local sources of contamination predominate. The site locates along Al-Rayah Al-Naseri at AlSheikh Zayed City water intake that supplies drinking water to the rural residents of Birqash (Figure 14.4). The investigation of drilling cuttings deep to 98 m (Figure 14.5) has shown the presence of clay layer at shallow depth of average thickness 21 m. The upper layer is composed of fine sand and has thin thickness (average 7 m). It has shallow groundwater where water table exists at about 3 m below ground surface. This layer is in direct contact with surface water body. The main aquifer exists below the clay layer and is of Pleistocene age. It is composed mainly from sand and gravel and of average thickness 70 m and is underlain by dense Tertiary clay. Due to the specific constraints in the site that include the presence of thick clay layer at shallow depths and the presence of Fe and Mn in the ambient groundwater an innovative design has been proposed.

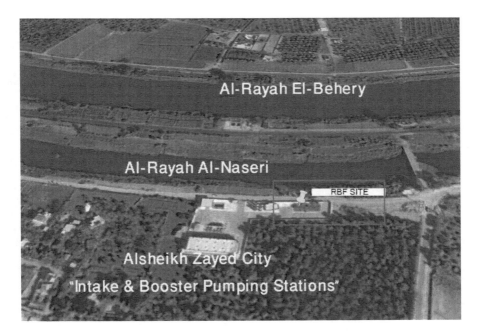

Figure 14.4. Detailed satellite image for the RBF site at Al-Rayah Al-Naseri.

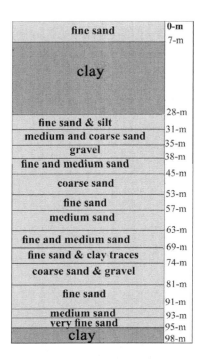

Figure 14.5. Geologic sections in the drilling site.

The site consists of two productive wells placed at 25 m from the surface water bank and two sand filters crossing the clay layer (Figure 14.6). The innovative RBF design has included inducing preferential flow through the clay layer. The two productive wells have been designed to withdraw groundwater from different depth. The well number one is abstracting groundwater from the upper part of the main aquifer (total depth 60 m). In this well, a preferential flow has been induced from the upper layer connected with surface water to the lower one through special design (Figure 14.7). The clay layer has penetrated through the outer annular space of the well and filled with graded sand. Screens were installed at depths range from 30 to 42 and 48 to 60 m below ground surface of total length 24 meters. The well number two is designed in traditional manner to withdraw groundwater from the bottom part of the main aquifer (total depth 90 m). Screens were installed at depths range from 54 to 60 and 66 to 90 m below ground surface of total length 30 m. Moreover, two sand filters crossing the clay layer has been constructed to facilitate the passage of river water and oxygen into the producing aquifer. The design of the sand filters took into consideration the natural geologic section (Figure 14.8). Filters were simply constructed directly on the river bank. Both of them have 0.56 m-inch diameter and 31 m length (crossing clay layer). The normal profile above the clay layer has been recovered and filling coarse sand and gravel instead of clay until reaching the main aquifer.

REMOVAL OF IRON AND MANGANESE WITHIN THE AQUIFER 243

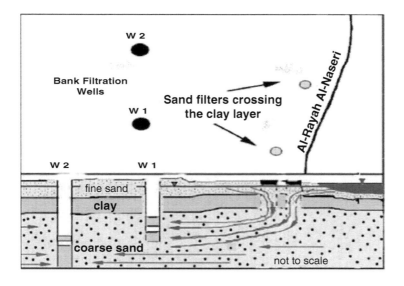

Figure 14.6. Site layout and design (site consists of two productive wells placed at 25 m from the surface water bank and two sand filters crossing the clay layer).

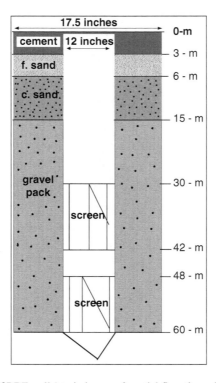

Figure 14.7. Design of RBF well 1 to induce preferential flow through the upper clay layer.

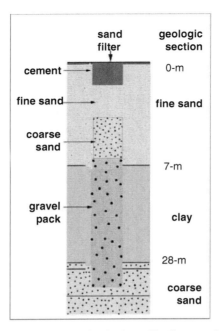

Figure 14.8. Design of sand filters crossing the clay layer (the diameter of the sand filter is 0.56 m).

6. Results and Discussion

6.1. Hydrogeology and RBF Wells Productivity

The main aquifer has high transmissivity where it is estimated about 10,500 m^2/d. It is calculated as the product of average aquifer thickness and hydraulic conductivity (60 m × 175 m/d). The company has targeted the main aquifer at the site to supply the surrounding communities with about 4,000 m^3/day from the two RBF wells. The productivity of the RBF wells (1 and 2) was high; both have average productivity of about 120 m^3/h. The theoretical volume of water that can move downward from the upper aquifer to the well screen through the innovative design that applied to RBF well (1) and the sand filters can be calculated by the Darcy's equation (Driscoll 1986):

$$Q = KIA \qquad (14.1)$$

where Q = vertical flow through the pack material, in m^3/day
K = hydraulic conductivity of filter pack, in m/day
I = Hydraulic gradient causing vertical flow in the filter pack
A = cross-sectional area of the filter pack, in m^2.

REMOVAL OF IRON AND MANGANESE WITHIN THE AQUIFER

The hydraulic gradient (I) is measured as the difference between the pumping level in the well and the static water level in the upper aquifer divided on the average distance through which the upper water must move, the distance from the midpoint of the upper aquifer to the top portion of the screen. In the case of the RBF well (1), the amount of water transmitted vertically is about 87 m^3/day. This amount is relatively small in comparison to the total amount of water pumped from the well. The volume of water that can be infiltrated through the two sand filters is about 520 m^3/day.

The proportion of bank filtrate in the RBF wells can be estimated using conservative tracer in the well and in the end-members (surface water and background groundwater). The chloride has been used because it is non-reactive and non-retarding and data about its distribution in the water environment is available. The differences in concentration of the two end-members should ideally be large and concentrations should be stable with time (Appelo and Postma 1993). The percentage of bank filtrate in the well is estimated according to the following equation:

$$X \ [\%]= [(C_w - C_{GW})/(C_{sw} - C_{GW})]*100 \qquad (14.2)$$

where C is the concentration of chloride in RBF well (C_W), groundwater (C_{GW}) and surface water (C_{WS}). The contribution of bank filtrate is estimated about 100% in the two wells. The area is surrounded by Nile water from all directions. These contributions may be from an old bank filtration. Isotope tracers are recommended to be used in this case.

6.2. Water Quality and Quality Changes

Results for specific quality indicators in surface water, RBF wells and background groundwater relative to Egyptian drinking water standards are shown in Table 14.1. The surface water source (El Rayeh Al Nasery) has fresh water quality where chloride content is ranging from 17 to 34 mg/L with median value 30 mg/L (Table 14.2). Sulfate content is ranging from 9.1 to 24.7 mg/L with median value 17.2 mg/L. The nitrogen species, ammonia, dominate in the source water while nitrate has completely disappeared. Ammonia content is ranging from 0.046 to 0.5 mg/L with median value 0.12 mg/L. The nitrite content is very low with median value 0.0035 mg/L. The average content of ammonia in Nile water at Giza was about 0.32 that comes mainly from domestic sewage (Ahmed et al. 1999). The surface water has relatively high content of TOC. The content of TOC is ranging from 3.9 to 4.9 mg/L with median value 3.9 mg/L while dissolved organic carbon (DOC) is ranging from 3.3 to 4.5 mg/L with median value 3.6 mg/L. The Surface water has relatively high potassium content of median value 5.7 mg/L. Turbidity is ranging from 3.14 to 5.1 NTU with median value 4.25 NTU. The total coliform bacteria were high where it ranges from 1,600 to 1,800 MPN/100 ml with median value 1,700 MPN/100 ml.

TABLE 14.1. Results of water analyses at the investigated site (March 2009) and average groundwater around the site.

	RBF Well 1 (upper part of the aquifer)	RBF Well 2 (bottom part of the aquifer)	El Rayeh Al Nasery	Average groundwater around the site	Egyptian standard (2007)
Odor	Nil	Nil	Nil	Nil	Nil
EC Micro (S/cm)	620	550	376	813	na
TDS (mg/L)	409.2	363	248.2	440	1000
Turbidity (NTU)	1.96	1.74	4.9	na	1.0
pH	7.5	7.5	8.3	7.5	6.5–8.5
NH_3 (mg/L)	0.06	Nil	0.14	Nil	0.5
NO_2 (mg/L)	0.004	Nil	Nil	Nil	0.2
NO_3 (mg/L)	Nil	Nil	Nil	Nil	45
Temperature (°C)	26.4	26	25.8	26	na
O_2 (mg/L)	na	na	10.2	na	na
TOC (mg/L)	1.8	2.0	4.1		na
Total alkalinity (CaCO3) (mg/L)	244	206	148	248	na
Total Hardness (CaCO3) (mg/L)	258	210	138	na	500
Ca^{++} (mg/L)	65.6	55.2	35.2	30	140
Mg^{++} (mg/L)	22.56	17.28	12	26	36
K^+ (mg/L)	na	na	5.7	na	na
Cl^- (mg/L)	24	26	28	70	250
SO_4^{-2} (mg/L)	24.2	32.8	24.69	50	250
Total Fe (mg/L)	0.43	0.5	0.1	1.7	0.3
Total Mn (mg/L)	0.51	0.56	Nil	0.91	0.4
Total coliform (MPN/100 ml)	Nil	Nil	1,600	Nil	<2 units/100 ml

na = not available

TABLE 14.2. Results of monitoring for specific indicators in the investigated site.

Parameter	Date	RBF Well 1	RBF Well 2	El Rayeh Al Nasery
Turbidity (NTU)	6/17/2008	5.09	12.2	5.1
	1/15/2009	1.58	1.7	3.14
	1/21/2009	1.82	1.6	3.6
	3/19/2009	1.96	1.74	4.9
	Median	1.89	1.72	4.25
NH_3 (mg/L)	6/17/2008	0.312	0.158	0.5
	1/15/2009	0.09	0.12	0.1
	1/21/2009	Nil	Nil	0.046
	3/19/2009	0.06	Nil	0.14
	Median	0.09	0.139	0.12
Chlorides Cl^- (mg/L)	6/17/2008	30	25	17
	1/15/2009	26	25	32
	1/21/2009	27	28	34
	3/19/2009	24	26	28
	Median	26.5	25.5	30
Total Iron (Fe) (mg/L)	6/17/2008	0.97	0.486	0.1
	1/15/2009	0.42	0.41	0.08
	1/21/2009	0.54	0.43	0.2
	3/19/2009	0.43	0.5	0.1
	Median	0.425	0.458	0.1
Total Manganese (Mn) (mg/L)	6/17/2008	0.824	0.713	Nil
	1/15/2009	0.64	0.63	0.21
	1/21/2009	0.46	0.41	Nil
	3/19/2009	0.51	0.56	Nil
	Median	0.575	0.595	0.21
Total coliform (MPN/100 ml)	6/17/2008	3	3	1800
	1/15/2009	6	9	1600
	1/21/2009	Nil	Nil	1800
	3/19/2009	Nil	Nil	1600
	Median	4.5	6	1700

The water quality of the RBF wells has been found suitable for drinking purposes except for Fe and Mn and turbidity. The coliform bacteria have disappeared completely from the RBF wells in the last two successive measurements (Table 14.2 and Figure 14.9). The permissible limit of total coliform bacteria according to Egyptian Drinking Water Standards is 2 cfu/100 ml and should not be detected in two successive measurements. It is clear that the applied RBF system has high efficiency in removing total coliform bacteria (100%) (Table 14.3). This result is encouraging to apply RBF in Egypt and similar countries that have relatively high temperature or potentially will have high temperature as expected within the climatic changes. The high removal efficiency of this technique for bacteria was mainly experienced in humid environment under average water temperature of about 10°C (KWB 2005) but the application under high temperature conditions is questionable. In this site, average water temperature is ranging from 15.6 in January to 29.6°C in August; such temperature variation may affect both the infiltration rate (Lin et al. 2003) and redox stage (Greskowiak 2006) and subsequent treatment processes (Hiscock and Grischek 2002). The higher temperature should induce a higher infiltration rate because of the higher hydraulic conductivity value (Eckert and Irmscher 2006). Moreover, under high temperature the redox stages changes and nitrate and manganese reducing conditions generally dominate below river bed (Greskowiak 2006). The passage of water underground can constitute an important treatment step to improve drinking water quality nevertheless it is not capable of removing all relevant contaminants nor is it applicable in all sites (Kuehn and Mueller 2000). The presence of clay layer is an important constraint to RBF.

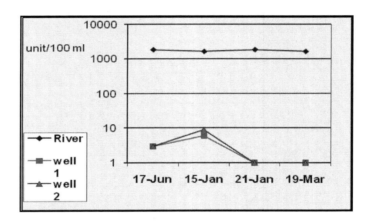

Figure 14.9. Distribution of coliform bacteria (unit/100 ml) in RBF wells and surface water.

TABLE 14.3. System removal efficiency (according to March 2009) for specific indicators.

Parameter	Surface water	Average of RBF wells	RE%
Fecal coliform (MPN/100 ml)	1800	0	100
Turbidity (NTU)	4.9	1.85	62.2
Ammonia NH_3 (mg/L)	0.14	0.03	78.57
Total organic carbon (TOC) (mg/L)	4.1	1.9	53.65

The distribution of ammonia in the RBF wells and surface water is shown in Table 14.2 and Figure 14.10. The removal efficiency of ammonia, turbidity and TOC were 79%, 62%, and 53%, successively (Table 14.3). The permissible limit for ammonia according to Egyptian Drinking Water Standards is 0.5 mg/L. The removal efficiency of ammonia depends on the healthy conditions of the river with respect to organic load and availability of oxygen in the river water. Removal of ammonia is important to keep oxygenated water for further treatment processes and removal of Fe and Mn from well water. RBF is efficient to protect from great variations of turbidity in surface water especially during flood seasons and engineering works in the Nile River. The high turbidity in the beginning is probably due to drilling residue and next almost constant. The TOC content in the surface water is about 4.1 mg/L; this amount is almost reduced by half in the RBF wells. This removal depends on the types (degradable/non-degradable) and prevailing oxic/anoxic/anaerobic conditions where rapid removal of TOC was observed, under oxic conditions whereas slower but continuing removal under both anoxic/anaerobic conditions (Schmidt et al. 2003; Grunheid et al. 2005). This result is of great concern in Egypt as chlorination is used in routine manner for disinfection. The decrease in TOC will reduce the formation of disinfection by-products such as trihalomethanes (THMs) species. The formation of THMs could be minimized by effective removal of organics from source water before chlorination (Reemtsma and Jekel 2006). The discharge of sewage effluent to the Nile River in Egypt should be stopped to recover the oxygen level in the Nile River water and minimize organic pollution.

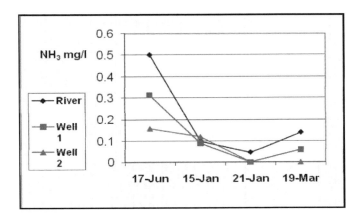

Figure 14.10. Distribution of Ammonia (NH$_3$ mg/L) in RBF wells and surface water.

The distribution of Fe and Mn in the RBF wells and surface water is shown in Table 14.2 and Figures 14.11 and 14.12. The mobility of Fe and Mn is connected with the redox conditions within the aquifer. After operating the wells under the effect of the new design, the removal efficiency of Fe and Mn were monitored continuously. Fe and Mn contents in the RBF wells range from 0.41 to 0.97 mg/L and 0.41 to 0.82 mg/L, respectively. Average concentrations of Fe and Mn are 0.46 and 0.53, respectively that are slightly above Egyptian drinking water standards. The permissible limit according to Egyptian Drinking Water Standards is 0.3 mg/L for Fe and 0.4 mg/L for Mn. WHO (2006) has no specific standards for iron content but guideline value for Mn is 0.4 mg/L. There is general trend for decrease in Fe content in the upper aquifer (well no. 1) but in the lower aquifer (well no. 2) the trend is not clear (Figure 14.11). The general changes in Mn content in both wells are decreasing (Figure 14.12). This can be explained by the effect of the infiltrated oxygenated water. The redox conditions below the innovative design in Egypt were largely dependent on temperature variations and oxygen content in the surface water. There is not significant difference between surface water temperature (25.8°C) and RBF wells temperature (the upper one has 26.4°C and the lower one has 26°C) in June 2008. The surface water oxygen (O_2) content is about 10.2 mg/L. Generally aerobic conditions prevail at low temperatures of the infiltrated water while anaerobic conditions dominate at high temperature. The Fe and Mn contents on average have been reduced with time. This can be explained by the migration of redox zones further in land. Taking into consideration the limited amounts of oxygenated surface water that could infiltrate through two sand filters of 22-inch diameter, the removal efficiency could be increased by installing two more sand filters. Most Fe removal systems operate on principal of oxidizing the Fe from ferrous, soluble form to ferric or insoluble form, followed by filtration. Coupling RBF and in situ removal of Fe and Mn differs from conventional treatment because removal of Fe and Mn takes place in the underground, in a wide precipitation zone far away from the extraction well. Detailed investigations of removal processes require long term monitoring of specific indicators and applying numerical flow and transport models.

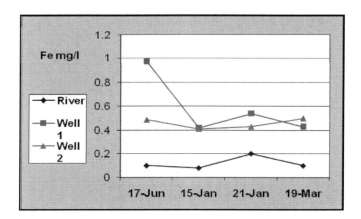

Figure 14.11. Distribution of iron (Fe mg/L) in RBF wells and surface water.

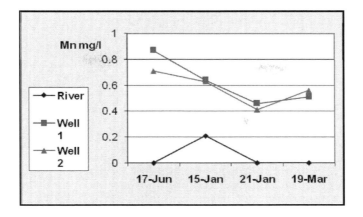

Figure 14.12. Distribution of manganese (Mn mg/L) in RBF wells and surface water.

7. Conclusion

The coupling of RBF technique and in situ removal of Fe and Mn is experimented in Egypt under unfavorable hydrogeological conditions and arid climatic conditions. An innovative RBF design has been installed to induce preferential flow through upper clay layer (21 m) using two sand filters. Sand filters have facilitated the passage of aerobic river water into the producing aquifer. Consequently, the Fe and Mn contents on average have been reduced within time. The mobility of iron and manganese is connected with the redox conditions within the aquifer. However, results suggest that the innovative idea of using sand filters through the clay layer worth further detail monitoring on the long run. The removal efficiency for coliform bacteria has reached 100% that is encouraging to apply RBF technique in similar arid climatic conditions. Total organic carbon has decreased by 53% which decrease the potential formation of chlorination by-products. The discharge of sewage effluent into the Nile River in Egypt should be stopped, otherwise oxygen decrease in the river water. High organic matter and low O_2 in the infiltrating river water would lead to anaerobic condition during RBF. Hence Fe and Mn would be mobilized also within the RBF-streamline, according to the German experience in 1960–1980 (Eckert and Irmscher 2006). For a sustainable use of RBF water protection is essential.

Acknowledgments The author would like to express deep appreciation for the valuable support of Giza Company for Drinking and Waste Water where the hydraulic installations and water analyses were done and also thanks extends to the nominated referees.

References

Abdel-Shafy HI, Aly RO (2002) Water issue in Egypt: Resources, pollution and protection endeavors. Cent Eur J Occup Environ Med 8:1–21

Ahmed S, Nader E, Mohamed E, Mahmoud H (1999) Analysis of Nile water pollution control strategies: A case study using the Decision Support System for water quality management. 2nd Inter-Regional Conference on Environment-Water, Cairo, Egypt

American Public Health Association (APHA) (2005) Standard Method for the Examination of Water & Wastewater. American Public Health Association, American Water Works Association, Water Pollution Control Facilities, Washington, DC

Appelo CAJ, Postma D (1993) Geochemistry, groundwater and pollution. Balkema, Amsterdam

Dawoud MA, Darwish MM, El-Kady MM (2005) GIS-based groundwater management model for Western Nile Delta. Water Resour Manag 19:585–604

Driscoll, FG (1986) Groundwater and Wells. 2nd. ed. Johnson Division, St. Paul, Minnesota

Eckert P, Irmscher R (2006) Over 130 years of experience with riverbank filtration in Düsseldorf, Germany. J Water Supply Res T 55:283–291

Eckert P, Rohns HP, Irmscher R (2006) Dynamic processes during bank filtration and their impact on raw water quality. Cited in UNESCO 2006: Recharge systems for protecting and enhancing groundwater resources, UNESCO, Paris. Proc. ISMAR5, Berlin, Germany: 17–22

Egyptian Environmental Affairs Agency (EEAA) (2009) Egypt State of Environment Report 2008 (Internal report Ministry of State for Environmental Affairs – Cairo – Egypt)

El Arabi N (1999) Problems of groundwater quality related to the urban environment in Greater Cairo. In: IAHS (ed) Impacts of urban growth on surface water and groundwater quality. Proceeding of IUGG 99 symposium HSS, Birmingham, IAHS Publ. No. 259

El-Fayoumy IF (1968) Geology of groundwater supplies in the region east of the Nile Delta and its extension in north Sinai. PhD Thesis, Fac. of Sci., Geol. Dept., Cairo Univ., Egypt

Emara MM, El Sabagh I, Kotb A et al. (2007) Evaluation of drinking groundwater for the rural areas adjacent to the nearby desert of Giza governorate of greater Cairo, Egypt. In: Linkov et al. (eds) Environmental Security in Harbors and Coastal Areas, 379–394. Springer

Greskowiak J (2006) Reactive transport processes in artificially recharged aquifers – Field and modelling studies. PhD dissertation, Humboldt-University – Berlin

Grischek T, Schoenheinz D, Worch E (2002) An overview of aquifer conditions and hydraulic controls. In: Dellion (ed) Management of aquifer recharge for sustainability. Swets & Zeitlinger B. V., Lisse, The Netherlands. Proc. ISAR-4. Adelaide, Australia: 485–488

Grunheid S, Amy G, Jekel M (2005) Removal of bulk dissolved organic carbon (DOC) and trace organic compounds by bank filtration and artificial recharge. Water Res 39:3219–3228

Hiscock KM, Grischek T (2002) Attenuation of groundwater pollution by bank filtration. J Hydrol 266:139–144

Idris H, Nour S (1990) Present groundwater status in Egypt and the environmental impacts. Environ Geol Water Sci 16(3):171–177, Springer-Verlag, New York

Kohl P and Medlar S (2007) Occurrence of manganese in drinking water and manganese control. American Water Works Association, Denver, USA, 184 pp

Kompetenz Wasser Berlin (KWB) (2005) Final report of NASRI project. Center of Competence for Water Berlin (KWB GmbH), Cicerostr.24, D-10709 Berlin, Germany

Kuehn W, Mueller U (2000) Riverbank filtration – An overview. J Am Water Works Assoc 12:60–69

Lin C, Greenwald D, Banin A (2003) Temperature dependence of infiltration rate during large-scale water recharge into soils. Soil Sci Soc Am J 67:487–493

Logsdon GS, Roger K, Solomon A, Shawn L (2002) Slow sand filtration for small water systems. J Environ Eng Sci 1:339–348.

National Water Research Center (NWRC) (2003) National Water Quality Monitoring Component 1000: National water quality and availability management (NAWQAM) project. National Water Research Center – Cairo – Egypt

Reemtsma T, Jekel M (2006) Organic pollutants in the water cycle. Wiley-VCH Verlag GmbH& Co. KgaA, Weinheim. 350 p

Research Institute for Groundwater (RIGW) (1989) Hydrogeological Map of Egypt. Research Institute for Groundwater, El Kanter El Khairia, Egypt

Research Institute for Groundwater (RIGW) (1992) Hydrogeological map for the Nile Delta area, Scale 1: 500000. Research Institute for Groundwater, El Kanter El Khairia, Egypt

Schmidt CK, Lange F, Brauch H, Kuhn W (2003) Experiences with riverbank filtration and infiltration in Germany. DVGW- Water Technology Center (TZW). Karlsruhe, Germany

World Health Organization (WHO) (2006) Guidelines for drinking-water quality: Incorporating first addendum. Vol. 1, Recommendations. – 3rd ed. WHO Press, World Health Organization, Geneva, Switzerland

Chapter 15 Riverbank Filtration as an Alternative Treatment Technology: AbuTieg Case Study, Egypt

Fathy A. Abdalla[1]* and Mohamed Shamrukh[2]

[1] Geology Dept., Faculty of Science, South Valley University, Qena, Egypt
[2] Dept. of Civil Engineering, Faculty of Engineering, Minia University, Minia, Egypt.
E-mail: mshamrukh@hotmail.com

Abstract For drinking water production, surface waters bodies are exposed to pollution as a result of discharging untreated liquid wastes into them from industrial, agricultural and domestic activities. Riverbank filtration (RBF) offers a cost-effective technique for producing drinking water by removing many suspended solids, pathogens, and micro-pollutants. The test site is the wellfield for water supply into AbuTieg residents (Assiut governorate, 370 km south to Cairo). It consists of seven municipal wells penetrating the Quaternary aquifer which is mainly fed by the bank filtrate. The results demonstrate the effectiveness of RBF for removing the pathogens and suspended solids. However, ammonium, dissolved iron and manganese are found in the bank filtrate at higher concentrations than those in the Nile water, but still within the allowable limits except for ammonia concentration in winter, the increased of ammonia concentration might be explained by less infiltration of oxygenated Nile water to the aquifer where the Nile received less precipitation in winter time (low-flow period) as well as the impact of sewerage system surrounding the wellfield. Results of the microbiological analysis showed that removal of total and fecal coliform (*E. coli*) bacteria (cfu/100 ml) is about 3.0 logs and total algal about 3.5 logs. Finally, all the examined abstracted water samples have concentrations below allowable limits for drinking water in Egypt.

Keywords: Riverbank filtration, Egypt, Nile Valley, groundwater, water supply, water quality, Assiut, AbuTieg

---

* Fathy A. Abdalla, Geology Dept., Faculty of Science, South Valley University, Qena, Egypt, e-mail: fathyhyd@yahoo.com

C. Ray and M. Shamrukh (eds.), *Riverbank Filtration for Water Security in Desert Countries*,
DOI 10.1007/978-94-007-0026-0_15, © Springer Science+Business Media B.V. 2011

1. Introduction

There are many water-related challenges facing Egypt. The first and most important challenge is related to water demand for public drinking water supply. Surface water bodies are exposed to pollution from many sources such as the discharging of untreated liquid wastes into them from industrial, agricultural and domestic activities. Compared with surface water, natural groundwater is well protected against most types of pollution, recognized as being free of pathogens and is of relatively good quality and constant/equilibrated temperature. The river loses water into the adjacent groundwater aquifer through the hydraulic interconnection, which is recognized as riverbank filtration (RBF) (Figure 15.1). Increased pumping action can also create a pressure head difference between the river and the adjacent aquifer and induces the river water to percolate through the riverbed and banks towards the pumping well. Therefore, wells extracts a mixture of groundwater originally present in the aquifer and infiltrated surface water from the river (Schoen 2006). The proportions of both kinds of water in the extracted one can vary depending on both extraction rate and river flow. The main natural purification processes result from mixing, biodegradation and sorption within two zones: active riverbed layer and along flowpath into abstraction wells. Intensive degradation and adsorption processes occur within the biologically active riverbed layer for short residence time. Along the flowpath into the abstraction wells, degradation rates and sorption capacities are lower and mixing processes greater. This is mainly determined to large extent by filtration and adsorption mechanisms and biological transformations during their underground flow path (Zullei-Seibert 1996, Sacher et al. 2001).

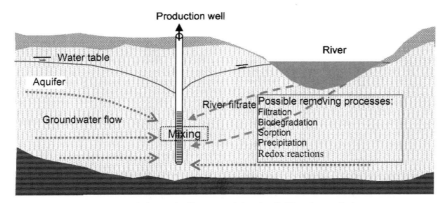

Figure 15.1. Schematic diagram of riverbank filtration technique.

This natural process holds promise as a relatively simple and low-cost way to remove particulates and microorganisms from surface water and make subsequent disinfection treatment easier. RBF has long been recognized in many countries worldwide (Ray et al. 2002). In Germany, it has been used for more than 130 years a long the Rhine and Elbe rivers (Sontheimer 1980, Doussan et al. 1997, Schmidt et al. 2003). In this method, production wells are placed in the flood plain/adjacent to the river bank, where they abstract river water that filtered through the alluvial sediments, which means clean bank filtrate is extracted (Figure 15.1) in the flood plain/adjacent to the river bank. Many factors controlling the travel time of river water along with pollutants should be taken into consideration while installing those wells, the most important are; the distance from the bank, well depth, discharge rate and the river conditions near the site. The effectiveness of RBF or the quality of bank-filtered water is also affected greatly by aquifer sediment properties, (e.g., grain size and distribution), the riverbed sediment, the infiltration velocity, and the residence time in the aquifer which varies greatly among sites (Literathy and Laszlo 1996). RBF has also long been recognized in many countries worldwide (Castany 1985, Doussan et al. 1997, Kim et al. 2003, Schmidt et al. 2003). In France, for instance, the proportion of bank filtrate amounts about 50% of the total drinking water production and in Germany about 16% of the drinking water (Schmidt et al. 2003). Further infiltration zones could be created by construction of artificial ditches and side channels. To improve the filtration effect of infiltration zones, a specific sand layer can be incorporated in percolation ditches, channels, and ponds. A further stage of development was finally the construction of recharge basins similar to those found in nearly all artificial groundwater recharge plants nowadays. In these recharge basins, raw water is passed through a filtering medium that consists of a layer of sand (Schmidt et al. 2003). Another adjective of the RBF is the ability of removing material of different character. The different removal mechanisms in the subsurface (e.g. sorption, biodegradation, filtration) combine to provide similar removal of the operationally defined hydrophilic and hydrophobic fractions of organic material upon ground passage. RBF may also serve as a significant barrier for the removal of microbial contaminants, including human pathogens (Weiss et al. 2003)

2. Statement of the Problem and Objectives

In Egypt, the River Nile is the main source of drinking water. As the result of varying human activities in and on the river body including tourist activities, industrial activities in addition to agricultural activities, the Nile receives wastewater discharge from 124 point sources from Aswan to Cairo (67 are agricultural drains, and the remainders are industrial drains). Therefore various pollutants have been detected in the Nile water such as trace metals and elevated COD (Abdo 2004a, b; Ibrahim and Tayel 2005). Algae, especially blue-green algae, are of interest to water treatment authorities because of their production of taste and odor compounds and natural toxins according to their exposure to some environmental conditions. Also, they interfere with certain water treatment processes for drinking water production (Shehata et al. 2008). Drinking water must meet specific criteria and standards

(WHO 1996, EHCW 2007) to ensure that water supplied to the public is safe and free-from pathogenic micro-organisms as well as hazardous compounds.

The traditional disinfectant methods (especially chlorine method) have been used widely to eliminate the risks of waterborne diseases such as typhoid fever, cholera and malaria. In chlorination method, if there is too much organic matter in water, it can react with the added chlorine to produce Trihalomethanes (THMs) as disinfectants by-products. The most commonly chlorine disinfection by-products species of THMs that may be found in drinking water are chloroform ($CHCl_2$) which considered as the major component (Geriesh et al. 2008, Rook 1974). Presence of THMs in drinking water is of carcinogenic effect and may damage the watering system by causing corrosion of distribution piping. Several studies (EPA 2003, IARC 1991) suggested that chlorination by-products may be linked to heart, lung, kidney, liver, reproductive problems including miscarriage rate, and central nervous system damage. For these reasons THM in public water supplies are limited by EPA (2003) to 0.08 mg/l.

According to the above mentioned problems, direct extraction of surface water for public-water supply should replace or supplement by artificial or natural subsoil passage of river water due to its efficiency in removing microorganisms, turbidity, organic matter, temperature deviations and other pollutants from the infiltrating surface water.

The main objective of this study is to evaluate the merits of RBF and its effectiveness for removing/controlling drinking water contaminants (e.g. microorganisms and turbidity). The results from this research work will improve our understanding of the effectiveness of riverbank filtration in contaminants removal and as pretreatment option that may help prevent outbreaks of waterborne disease and reduce compliance costs. The obtained research results provide a basis for decision making by the Egyptian holding company for drinking water and wastewater, Assiut branch. Furthermore, RBF has the ability to reduce the theoretical cancer risk due to reducing the formation of disinfection by-products formed upon chlorination of Nile water.

3. Examination Site

Riverbank filtration study site is located on the west bank of the River Nile at AbuTieg town, 20 km south of Assiut City. It contains seven vertical municipal wells penetrating the Quaternary aquifer. All wells are 160 m deep with 50 m screen length and installed at 20–80 m from the Nile bank (Figure 15.2). The quantity of water treated and supplied daily is about 10,000 m^3/day.

3.1. Nile Aquifer

From hydrogeological point of view, the study area constitutes a portion of the complex hydrogeological system dominating the Nile valley in Upper Egypt. It includes many aquifers, namely, Quaternary, Pliocene, Paleogene and Pre-Cenomanian aquifers. The Quaternary aquifer has the great importance in the Nile valley and Delta and it is our target in the present study. The following is a brief discussion about the Quaternary aquifer system in the study area.

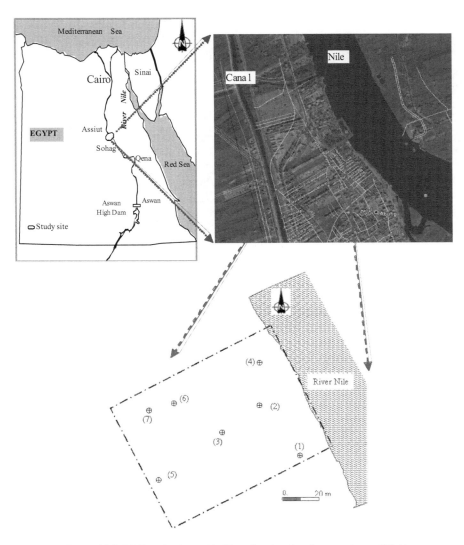

Figure 15.2. RBF study site at AbuTieg city showing the pumping wellfield.

3.2. Quaternary Aquifer

The Quaternary aquifer of the Nile Valley was studied by many authors such as: Attia (1985), Abd El-Moneim (1987), Abu El Ella (1989), Abd El-Bassier (1997), Kamel (2004), Abdalla et al. (2009). Based on these studies the groundwater system in the study area belongs to the regional Quaternary aquifer that extends along the Nile Valley. This aquifer can be categorized into two hydrogeological units with distinct hydraulic properties, Figure 15.3. These two units are sands and gravels at the base, and semi-permeable clay-silt layer at the top. The thickness of this aquifer as well as its width differs from one locality to another. At AbuTieg area, the thickness of the aquifer may reach around 200 m at the middle part of the flood plain and decreases gradually towards the edges of the plateau on both sides of the river (Figure 15.4). The horizontal and vertical permeability of the upper (Holocene) layer ranges from 0.4 to 1 m/day while the vertical hydraulic conductivity is low and increases with depth. The hydraulic conductivity of the Lower (Pleistocene) layer ranges from 60 to 100 m/day and transmissivity ranges from 2,000 to 6,000 m^2/day. The water in this aquifer is found under semi-confined conditions, and in other localities it is present under unconfined conditions where the Nile silt is absent. The main components of recharge of the aquifer in the study area are seepage from irrigation canals, subsurface drainage from the irrigated lands and upward leakage from the deep aquifers through fault planes. According to El Miligy (2004), potentiometric head of the Quaternary aquifer slopes gradually from south to north (+50 m at AbuTieg to +43 m at Dyirout). This means the general groundwater flows from south to north-east intersecting the River Nile. Lateral flow occurs from the River Nile to the aquifer is quite proved between AbuTieg and Assiut barrage (Mousa et al. 1994). Therefore, the study area could be considered as one of the most favorable area for RBF application.

El Miligy (2004) made a trial to estimate the direction and the amount of water flows from the River Nile to the aquifer using Darcy's equation:

$$Q = KbW \frac{dh}{dl} \tag{15.1}$$

where K is the hydraulic conductivity, b is the aquifer thickness at the midpoint between equipotential lines, W is the distance between two boundary flow lines, dh/dl is the hydraulic gradient. According to his study the area is directly recharged from the Nile water, therefore it is characterized by fresh water. The rate of natural flow (Q) of Nile water into the aquifer can be calculated based on the average transmissivity value T and applying Eq. (15.1).

4. Materials and Methods

To evaluate the efficiency of RBF site at AbuTieg, water samples from the two locations River Nile and the bank-filtered well waters were collected. Sampling was carried out at 3 months intervals (December 2008, February and May 2009). Three wells out of the total seven pumping wells (No. 4, No. 6, and No. 7), which are located at different distances from Nile, were sampled (Figure 15.2). Water samples analyzed to determine the physicochemical and bacteriological quality parameters. Turbidity, Alkalinity, Hardness (TH), Total Dissolved Solids (TDS), Temperature (T), pH, Electrical Conductivity (EC) of the water samples were measured immediately at the sampling sites. The water samples were analyzed for other major water quality parameters such as magnesium, calcium, chloride, nitrate, ammonia, iron and manganese, total algal, total coliform and *E. coli* bacteria according to standard methods of analysis (APHA 1998). The Atomic Adsorption Spectrophotometer (Model A Perken-Elmer 2380) was used for measuring the concentration of Mg Ca, Fe, and Mn. The spectrophotometer (SLLO Spectrophotometer) was used for measurement of NO_3 and NH_3 and Cl were determined by volumetric analysis. Bacteriological Measurements were conducted at the central laboratory of Ministry of Health in Cairo according to Egyptian standards and the procedures of the American Public Health Association (APHA 1998). The results were evaluated in accordance with the drinking water quality standards given by the World Health Organization (WHO 1996) and the Egyptian Higher Committee for Water (EHCW 2007).

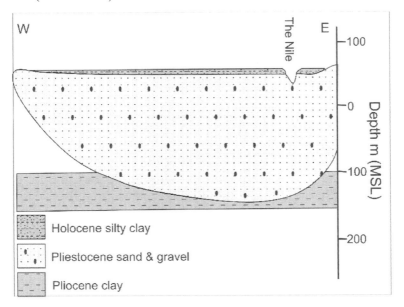

Figure 15.3. Hydrogeological cross-section at south of the study area (modified after Attia 1974, Mousa et al. 1994).

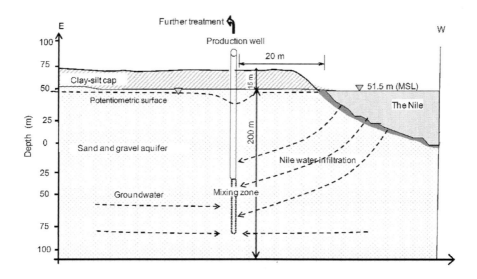

Figure 15.4. Schematic diagram of Nile and estimated aquifer flow lines at AbuTieg RBF site.

5. Results and Discussions

Measurements of water quality at the two locations, River Nile and the selected three pumping wells at three times from 2008 to 2009 are shown in Table 15.1. The comparison of water quality between Nile and the three wells are given in Figures 15.5 and 15.6.

TABLE 15.1. Water quality at the study site, AbuTieg, for Nile and three wells at 3 months.

Parameter	Nile	December 2008			February 2009			May 2009		
		W4	W6	W7	W4	W6	W7	W4	W6	W7
Temp	23–27	16.7	16.8	16.7	–	–	–	–	–	–
pH	7.6–7.7	7.7	7.6	7.7	7.6	7.7	7.7	7.7	7.7	7.8
Turbidity (NTU)	8–13	0.6	0.8	0.7	0.8	0.7	0.8	0.7	0.7	0.6
Conductivity (μS/cm^2)	220–278	370	418	610	–	–	–	–	–	–
TDS	190–210	264	298	335	250	313	342	218	302	311
Alkalinity	128–144	152	152	184	158	162	176	144	162	168
Total Hardness	118–135	145	130	211	136	152	200	136	148	194
Ca-Hardness	–	68	98	114	32	–	–	–	–	–
Mg-Hardness	–	77	32	97	16	–	–	–	–	–
Ca	23–30	–	–	–	32	44	48	34	42	51

Continued

AN ALTERNATIVE TREATMENT TECHNOLOGY

TABLE 15.1—*Continued.*

Parameter	Nile	December 2008 W4	W6	W7	February 2009 W4	W6	W7	May 2009 W4	W6	W7
Mg	13–19	–	–	–	16	21	21	16	20	22
Cl	18–20	16	20	29	21	22	26	21	21	24
SO$_4$	22–29	12	17	18	–	–	–	–	–	–
NO$_3$	1.7–2.6	1.0	0.9	1.1	1.0	1.2	1.2	1.0	1.1	1.3
NH$_3$	0.02–0.04	0.8	0.8	0.7	0.1	0.2	0.2	0.1	0.1	0.1
Fe	0.04–0.05	0.21	0.22	0.30	0.11	0.18	0.18	0.1	0.12	0.2
Mn	0.06–0.08	0.40	0.40	0.35	0.20	0.24	0.25	0.22	0.30	0.24
Total coliform (cfu/100 ml)	800–1,200	<1	<1	<1	<1	<1	<1	<1	<1	<1
E. coli (cfu/100 ml)	280–440	0	0	0	0	0	0	0	0	0
F. streptococci (cfu/100 ml)	–	<1	<1	<1	<1	<1	<1	<1	<1	<1
Total algal (unit/ml)	1,200–1,800	<1	<1	<1	–	–	–	–	–	–

Note: All parameters reported as mg/l or mentioned.

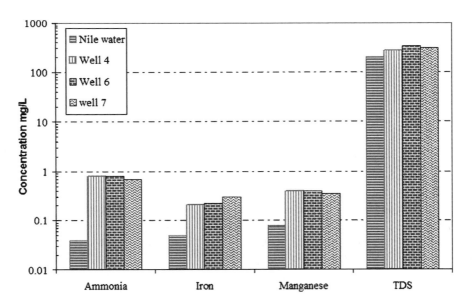

Figure 15.5. Comparison of water quality parameters in the Nile and bank filtered water, Dec. 2008.

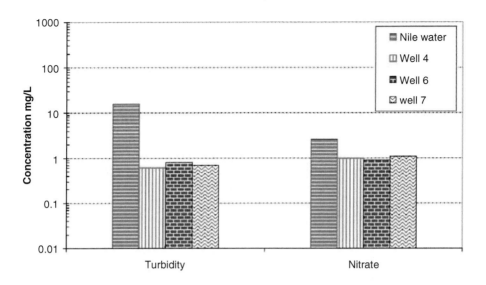

Figure 15.6. Comparison of turbidity and nitrate in the Nile and bank filtrate Dec. 2008.

5.1. Quality of Nile Water

Nile water qualities at the test site are monitored through three cycles sampling. The results showed the following: The Nile water temperature varies between 23°C and 27°C with an average of 25.2°C and pH values were slightly alkaline (7.6–7.7). Results have shown that the Nile water has moderate turbidity due to settling of suspended particulates upstream Aswan High Dam. Turbidity of Nile ranged between 8 and 13 NTU. However, Nile turbidity is more than the permissible limit of 1.0 NTU for drinking water (Table 15.1). Total Dissolved Solids (TDS) ranged between 190 and 210 mg/l while Total Hardness (TH) ranged between 118 and 135 mg/l whereas their permissible limits are 500 mg/l.

Iron and manganese concentrations in Nile water were much less than the permissible limits whereas their permissible limits are 0.3 and 0.4 mg/l, respectively. Results of Nile water have also shown that nutrients (nitrate and ammonia) concentrations were within permissible limits (45 and 0.5 mg/l, respectively) for all samples. Nitrate average concentrations ranged between 1.7 and 2.6 mg/l, where ammonia average concentrations ranged between 0.02 and 0.04 mg/l. The quality of Nile water in this study site is in good agreement with previous work carried out at Sidfa city, 30 km south to AbuTieg, by Shamrukh and Abdel-Wahab (2008).

5.2. Behavior of Biological Contaminants

For Nile water, results of the microbiological analysis showed that the total and fecal coliform (*E. coli*) bacteria (cfu/100 ml) as well as total algal counts at the three times have elevated values that are higher than normal occurring. This find may be indicates that Nile water is polluted by of sewage fecal contamination. This pollution might be due to agricultural drainage containing municipal wastewater discharges into the Nile. Figure 15.7 compares average total and fecal coliform (*E. coli*) bacteria (cfu/100 ml) as well as total algal count in the Nile and bank filtered water in the study area. Microbiological removal at this RBF site is more than 3.0 logs removal rate.

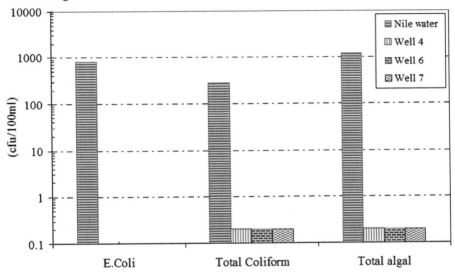

Figure 15.7. Comparison of total and fecal coliform (*E. coli*) bacteria as well as total algal in the Nile and bank filtrate Dec. 2008.

5.3. Quality of Bank Filtered Water

The results of the study revealed that the most physicochemical and bacteriological characteristics of infiltrated Nile water are different from those of the Nile. This is due to different removal mechanisms in the subsurface (e.g., sorption, biodegradation, filtration) as well as the expecting reducing conditions, particularly at winter seasons (probably due to the decrease of both discharges and level of Nile) and consequently less infiltration of oxygenated Nile water to the aquifer. The range of temperature of the bank filtrate is lower than in the Nile water with an average of 16.7°C. The obtained pH values did not show much difference from Nile water with mean values of 7.7 which reflects close association and relationship between them. The turbidity values of the infiltrated river water are less than the permissible limit in all samples (Table 15.1). This is likely because of the filtration effects of the underground passage.

Although iron and manganese concentrations are higher in infiltrated Nile water than those of the Nile, they are still under the permissible limits. This indicate the effectiveness of RBF technique in providing drinking water with Fe and Mn content less than the permissible limits. Rather, the significant positive correlation between iron and manganese during all seasons (Table 15.1) is shown to be a key indicator of the common source origin of the two elements during transportation and/or depositional reactions (Abdel-Satar and Elewa 2001).

Nitrate average concentrations ranged between 0.8 and 1.8 mg/l, where ammonia average concentrations ranged between 0.2 and 0.8 mg/l in winter samples, it has exceeded the permissible limits (0.5 mg/l). The decrease in nitrate concentration and the increase in ammonia concentration might be explained by the low-flow period where the flow rate and water level of Nile are low and in turn, decreases or less infiltration of oxygenated Nile water into the aquifer. However they were less than the permissible limit at the fall and spring sampling cycles where the concentration ranges between 0.1 and 0.2 mg/l. Furthermore, these elevated concentrations of ammonia might be due to impact of sewerage system surrounding the wells away from Nile bank with close distance less than 60 m to septic tanks of residents.

The present study showed a pronounced decrease in total and fecal coliform (*E. coli*) bacteria count (around three log removal rate) and total algal about three and half log removal rate as shown in Table 15.1 and Figure 15.7. This proven the effectiveness of RBF in microbiological treatment at AbuTieg site.

6. Conclusion

The application of riverbank filtration (RBF) as simple and natural treatment technique has proven very useful for removal of many contaminants and identifying opportunities for improving drinking water supplies management in AbuTieg area located in Nile valley. It can be concluded that removal of algae, bacteria and other suspended matter is very efficient through the studied RBF site. For iron and manganese which cause many problems in Nile groundwater, RBF supplied water with low concentrations than allowable limits. RBF technique might be used as stand alone or as pre-treatment in water supply system in Nile valley. Distance from Nile and depth of RBF abstraction wells are key aspects of RBF removal efficiency of Nile and aquifer contaminants. Moreover, protection of RBF wells from point-sources of contamination such as sewerage system is necessary. It is recommended to carry out further RBF investigation and measurements to give the final conclusion about its efficiency in Nile water treatment. On the other hand, it is recommended that Nile treatment plants integrate RBF to remove microbiological pollutants and to reduce chlorine and alum doses providing an aesthetically acceptable and biologically safe supply of water to the customers.

Acknowledgments   The authors would like to thank Assiut Company for Water and Wastewater for their support during the research work and to the anonymous reviewers for their valuable comments.

References

Abd El-Bassier M (1997) Hydrogeological and hydrochemical studies of the Quaternary aquifer in Qena Governorate. M.Sc. Thesis, Faculty of Science, Assiut University, Egypt

Abd El-Moneim AA (1987) Hydrogeology of the Nile basin in Sohage Province. MSc Thesis, Faculty of Science, Assiut University, Egypt

Abdalla FA, Ahmed A, Omer A (2009) Degradation of Groundwater Quality of Quaternary Aquifer at Qena, Egypt. J Environ Stud 1:18–30

Abdel-Satar AM, Elewa AA (2001) Water quality and environmental assessments of the River Nile at Rossetta Branch. The Second International Conference and Exhibition for Life and Environment, 3–5 April: 136–164

Abdo MH (2004a) Environmental studies on the River Nile at Damietta Branch region, Egypt. J Egypt Acad Soc Environ Dev 5(2):85–104

Abdo MH (2004b) Distribution of some chemical elements in the recent sediments of Damietta Branch, River Nile, Egypt. J Egypt Acad Soc Environ Dev 5(2):125–146

Abu El Ella EM (1989) Hydrogeochemistry of the Nile River in the area between Aswan and Assiut. PhD Thesis, Faculty of Science, Assiut University, Egypt

APHA AWWA, WEF (1998) Standard Methods for the Examination of Water and Wastewater 20th edn. American Public Health Association, American Water Work Association, Water Environment Federation, Washington, DC

Attia FA (1974) Parameter and characteristics of groundwater reservoir in Upper Egypt. MSc Thesis, Faculty of Engineering, Cairo University, Egypt

Attia FA (1985) Management of water systems in Upper Egypt. PhD Thesis, Faculty of Engineering, Cairo University, Egypt

Castany G (1985) Liaisons hydrauliques entre les aquifbres et les cours d'eau (in French). Stygologia. I:1–25

Doussan C, Poitevin G, Ledoux E, Detay M (1997) Riverbank filtration: Modeling of the changes in water chemistry with emphasis on nitrogen species. J Contam Hydrol 25:129–156

EHCW (2007) Egyptian standards for drinking and domestic uses. Egyptian Higher Committee for Water, Egyptian Governmental Press, Egypt

El Miligy E (2004) Groundwater resources evaluation of Assiut governorate. PhD Thesis, Faculty of Science, Assiut University, Egypt

EPA (2003) National interim primary drinking water regulations list of drinking water contaminants and their MCLs. U.S. Environmental Protection Agency, EPA Annual Report 816-F-03-016

Geriesh HM, Balke KD, El-Rayes AE (2008) Problems of drinking water treatment along Ismailia Canal Province, Egypt. J Zhejiang Univ Sci B 9(3):232–242

IARC (1991) Monographs on the evaluation of carcinogenic risks to humans: Chlorinated drinking water, Chlorination by-products; Some other halogenated compounds; Cobalt and cobalt compounds. International Agency for Research on Cancer, Lyon, France, 52:544

Ibrahim SA, Tayel SI (2005) Effect of heavy metals on gills of *Tilapia zillii* inhabiting the River Nile water (Damietta branch) and El-Rahawy drain. Egypt J Aquat Biol and Fish 9(2):111–128

Kamel R (2004) Geology of Luxor area and its relationship to groundwater uprising under the Pharaohs temples. MSc Thesis, Aswan Faculty of Science, South Valley University, Egypt

Kim S, Corapcioglu M, Kima D (2003) Effect of dissolved organic matter and bacteria on contaminant transport in riverbank filtration. J Contam Hydrol 66:1–23

Literathy P, Laszlo F (1996) Processes affecting the quality of bank-filtered water. In: Kivimaki AL, Suokko T, (eds) Proc. Int. Symposium on Artificial Recharge of Groundwater, NHP Report No. 38. (Nordic Hydrological Programme, Helsinki, Finland, 1996), pp. 53–64

Mousa SE, Attia FA, Abu ElFotouh AM (1994) Geological and hydrogeological study on the Quaternary aquifer in the Nile Valley between Assiut and Sohage Governorates, Egypt. Geol J Egypt 38:1–20

Ray C, Melin G, Linsky RB (2002) Riverbank Filtration: Improving Source Water Quality. Kluwer, The Netherlands

Rook J (1974) Formation of haloforms during chlorination of natural waters. J Water Treat Exam 23:234–-243

Sacher F, Brauch HJ, Kühn W (2001) Fate studies of organic micropollutants in riverbank filtration. Proc. Int. Riverbank Filtration Conference Rheinthemen, 4:139–148

Schmidt CK, Lange FT, Brauch HG, Kühn W (2003) Experiences with riverbank filtration and infiltration in Germany. DVGW-Water Technology Center, TZW, Germany

Schoen M (2006) Systematic comparison of riverbank filtration sites in Austria and India. MSc Thesis, University of Innsbruck, Austria

Shamrukh M, Abdel-Wahab A (2008) Riverbank filtration for sustainable water supply: Application to a large-scale facility on the Nile River. Clean Technol Environ Policy 10(4):351–358

Shehata AS, Ali HG, Wahba ZS (2008) Distribution pattern of Nile water algae with reference to its treatability in drinking water. J Appl Sci Res 4(6):722–730

Sontheimer H (1980) Experiences with riverbank filtration along the Rhine River. J Am Water Works Assoc 72:386–390

Weiss WJ, Bouwer EJ, Ball WP, O'Melia CR, LeChevallier MW, Arora A, Aboytes R, Speth TF (2003) Study of water quality improvements During riverbank filtration at three Midwestern united states drinking water utilities. Geophys Res Abstr 5, 04297

WHO (1996) Guidelines for drinking water quality, 2nd ed. World Health Organization, Geneva

Zullei-Seibert N (1996) Pesticides and artificial recharge of groundwater via slow sand filtration – elimination potential and limitations. International Symposium on Artificial Recharge of Groundwater, (Helsinki 1996) 247–253

Chapter 16 Quality of Riverbank Filtrated Water on the Base of Poznań City (Poland) Waterworks Experiences

Józef Górski*

Adam Mickiewicz University, Institute of Geology, 16 Maków Polnych Street, 61-606 Poznań, Poland.

Abstract   The article presents the results of 2-year investigation of riverbank filtrated water quality drawn from the wells situated at a three different distances to Warta River. The big influence on water quality abstracted from the drainage, horizontal well (with drains placed 5 m below the river bottom) and vertical wells situated 70–80 m from the river have been stated. This influence is visible in contamination of water with bacteria and planktonic organisms as well as nitrates and micropollutants. In the wells situated with far away distance from the river (250–1,100 m) and increasing of ground water portions in wells recharge, contaminations from the river water are almost not observed. However, the gradually increase of organic matter in the aquifer is visible. In the light of this investigation, the localization of wells at the distance of 150–250 m to the river is recommended.

Keywords: Riverbank filtration, water quality, river water quality changes in aquifers

1. Introduction

After World War II, many waterworks recharged by river bank filtration were built in Poland. These waterworks have different solutions concerning well locations in relation to a recharge river or rarely water reservoir. This situation exists on Mosina-Krajkowo Waterworks – the biggest source of water supply to Poznań City Community. There are many vertical wells situated at different distance to the river and one horizontal, drainage well with drains situated in the river bed. Water quality drawn from these wells is differentiated depending on well location in regards to the river and consequently the portions of ground water and river water in abstracted water. The article presents the results of long term observations of

---

* Józef Górski, Adam Mickiewicz University, Institute of Geology, 16 Maków Polnych Street, 61-606 Poznań, Poland. e-mail: gorski@amu.edu.pl

C. Ray and M. Shamrukh (eds.), *Riverbank Filtration for Water Security in Desert Countries*, DOI 10.1007/978-94-007-0026-0_16, © Springer Science+Business Media B.V. 2011

water quality drawn from these wells. Observations are mainly based on 2 year investigations of water quality from chosen wells situated at different distances to the river.

2. Characteristic of the Mosina-Krajkowo Waterworks

Mosina-Krajkowo waterworks, which produces about 130,000 m^3 per day, is situated in Warta River valley where the two main aquifers cross (Figure 16.1): buried valley and Warsaw-Berlin ice-marginal valley.

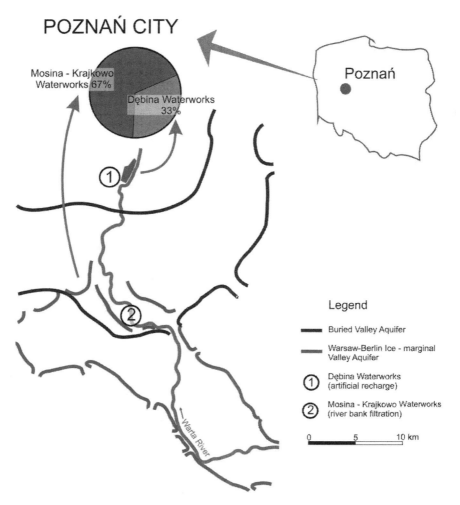

Figure 16.1. Localization of the Mosina-Krajkowo waterworks.

Bank filtrated water mixed with ground water is abstracted by two wells barrier (Figure 16.2). Barrier one is situated on a higher terrace at the distance of 480–1,100 m from the river, second barrier is located on the flood plain with the distance 70–80 m from the river. In addition to those two well barriers, there is one horizontally drainage well with eight drains, each 100 m long, placed in the river bed 5 m below bottom. This drainage well was built not as a typical Ranney well but by digging ditch in river bed and next installation of drains. Vertical wells tap the buried valley aquifer, which is connected with the upper aquifer (Figure 16.3). In some parts aquifers are divided by glacial tills. Depth of wells is mostly in the range of 30–40 m and hydraulic conductivity of the aquifer 24–96 m/day.

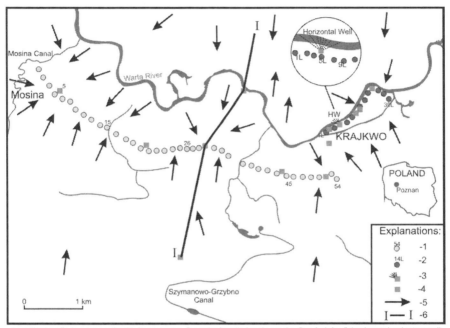

1-vertical wells barrier on the higher tarrace, 2-vertical wells barrie on the flood plain, 3-horizontal colector well 4-investigated well, 5-flow lines, 6-hydrogeological cross-section

Figure 16.2. Sketch of Mosina-Krajkowo wellfield.

3. Investigations

Recognition of water quality was done on the investigation sites included:
- Warta River
- Horizontal drainage well
- 4 VERTICAL wells from the barrier on the flood plain

- 4 vertical wells from the terrace barrier
- 1 separate well located at the distance of 250 m from the river (no. 78b/s)
- 1 observation well located at the distance of 30 m from the river (no. 86/b/2)

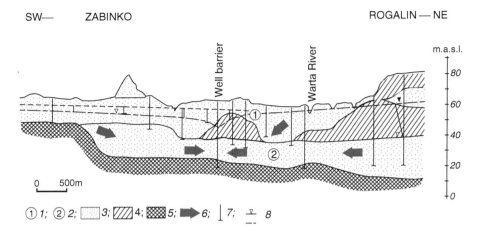

1-Warszawa-Berlin valley aquifer, 2-buried valley aquifer
3-sand and gravel, 4-glacial till, 5-clay, 6-flow lines,
7-screen intervals, 8-water level

Figure 16.3. Hydrogeological cross-section I-I (see Figure 16.2).

Detail characteristics of the investigation sites is presented in Table 16.1. The water quality analyses were done every 3 months within 2 years since October 1996 till June 1998. Investigations comprised:

- Physicochemical analyses in the range of temperature, smell, color, pH, NH_4, NO_2, NO_3, Fe, Mn, Cl, SO_4, hardness, COD_{Mn}, COD_{Cr}, alkalinity, O_2, BOD_5, TOC, PO_4, Na, K, Mg, Ca, N_{org}, H_2S, dry residue phenols, detergents Pb, Zn, Cd, Cr, Ni and for some analyses: dichloromethane, chloroform, 1,2-dichloroethane, carbon tetrachloride, trichloroethylene, 1,2-trichloroethane, tetrachloroetylene, hexachlorobenzene, lindan, heptachlor and its epoxide, DDT, metaxichlor, pentachlorophenol, PAH, aromatic and aliphatic hydrocarbons
- Microbiological analyses in the range of coliform bacteria and amount of bacteria culture at the temperature of 37°C and 20°C and
- Hydrobiological analyses encompassing determination of plankton

In addition, some detail chemical analyses for Warta River water, horizontal well and separate well on flood plain during the flood period are presented.

QUALITY OF RIVERBANK FILTRATED WATER 273

TABLE 16.1. Characteristic of the investigation sites.

Investigation sites	Morphology	Distance bank-well (m)	Depth of the well screen intervals (m)	Portion of river water in well recharge (%)	Residence time
Warta River	–	–	–	–	–
Horizontal well	Drains under river beds	0	5 m below to River bottom	~100	a few hours
Observation well no. 86 b2	By the river	30	6.0–10.0	90	a few weeks
Vertical wells barrier on the flood plain	Flood plain	70–80	15.8–35.8	75–85	1–3 months
Separate well no. 78 b/s	Higher terrace	250	18.0–28.0	60	6 months
Vertical well barrier on the higher terrace	Higher terrace	480–1100	23.0–47.1	50–60	5–15 years

4. Results of Investigations

Characteristic of the chosen water quality parameters in face of the analysed problem are presented in Table 16.2. The results show strict dependence between water quality in the Warta River, drainage well and observation well no. 86/b/2. Contamination level observed here in the range of nitrate compounds and detergents is similar to the one in waters of the Warta River. The symptom of the strict correlations with the river is also bacteriological contamination and plankton occurrence. The result of water infiltration through the sandy deposits of small thickness is a considerable reduction of bacteria and plankton. Lower reduction depends on TOC and oxidability.

TABLE 16.2. Chosen indicators of water quality (average and standard deviations).

A.

Investigation sites	TOC (mg/L)	COD_{Mn} (mg O_2/L)	COD_{Cr} (mg O_2/L)
Warta River	14.25(5.0)	10.4(1.0)	22.7(5.7)
Horizontal well	10.58(4.1)	6.2(1.1)	14.9(8.5)
Observation well no. 86 b2	13.03(6.1)	6.1(0.9)	23.4(8.4)
Vertical wells barrier on the flood plain	11.07(5.1)	5.5(1.0)	20.0(5.7)
Separate well no. 78 b/s	7.00(3.3)	4.7(0.3)	15.3(2.3)
Vertical well barrier on the higher terrace	8.75(4.6)	5.0(0.9)	14.6(5.3)

274 J. GÓRSKI

TABLE 16.2—*Continued.*

B.

Investigation sites	Fe^{2+} (mg/L)	Mn (mg/L)	NH_4 (mg/L)
Warta River	0.55(0.35)	0.14(0.09)	0.42(0.18)
Horizontal well	0.15(0,08)	0.18(0.11)	0.28(0.62)
Observation well no. 86 b2	0.12(0.06)	0.48(0.39)	0.28(0.40)
Vertical wells barrier on the flood plain	0.92(0.35)	0.52(0.10)	0.28(0.18)
Separate well no. 78 b/s	1.31(0.08)	0.61(0.05)	0.18(0.09)
Vertical well barrier on the higher terrace	4.00(0.91)	0.36(0.07)	0.45(0.14)

C.

Investigation sites	Detergents[c] (mg/L)		H_2S (mg/L)		Coliform bacteria*		Plankton**
	Max	%[a]	Max	Average	Max	%[a]	
Warta River	0.30	61	0	0	700,000	100	very numerous
Horizontal well	0.15	17	0.004	0.001	1,670	52	a few
Observation well no. 86 b2	0.20	33	0	0	398	75	14
Vertical wells barrier on the flood plain	0.20	50	0.008	0.001	5	40	a few to 64
Separate well no. 78 b/s	0	0	0.066	0.034	0	0	a few [b]
Vertical well barrier on the higher terrace	0.03	12	0.040	0.023	0	0	a few remains of organisms

[a] Percentage (%) of the analyses with detected contamination.

[b] Only after the flood.

[c] Anionics.

*Number of bacteria in 100 mL of water.

**Number of organisms in 1 mL of water.

Periodically in water from drainage well and observation well no. 86b/2 specific micropollutants were also detected (Table 16.3).

QUALITY OF RIVERBANK FILTRATED WATER

TABLE 16.3. Specific organic micropollutants investigation.

A. 1997–1998

Investigation sites	Dichloromethane (μg/L)	Chloroform (μg/L)
Warta River	n.d.	n.d.
Horizontal well	n.d.	n.d.
Observation well no. 86 b2	n.d.–9.71	n.d.–1.22
Vertical wells barrier on the flood plain	n.d.–11.67	n.d.–0.67
Separate well no. 78 b/s	n.d.	n.d.
Vertical well barrier on the higher terrace	n.d.	n.d.

B. 1992–1993

Investigation sites	Horizontal well*	Vertical well at the distance of 70 m to the Warta River*
Chlorobenzen (μg/L)	n.d.–20,0	n.d.–4.01
Heksachlorobenzen (μg/L)	n.d.–96	n.d.–0.52
Tetrachloroeten (μg/L)	n.d.–8.3	n.d.
Heptachlor (μg/L)	n.d.–0.03	n.d.–0.22
Gamma-HCH (μg/L)	n.d.	–

* Number of analyses = 6.
n.d. = not detected.

Significant influence on water quality from drainage well was observed during summer flood (Figure 16.4). The correlation with the Warta River quality is also visible in the case of the bank barrier wells located 70–80 m away from the river. The symptom of river influence is a periodical occurrence of the bacteriological contamination and plankton as well as specific micropollutants. This influence is also to be observed in case of nitrate, organic matter (TOC and COD - oxidability) and anionic detergent contents. In the range of the individual parameters, different reduction level in relation to river water was noticed. This indicates the intrinsic influence of the processes like ammonia sorption and heterotrophic denitrification. Waters of Warta River in the range of Mn, Fe and hardness are similar to groundwater due to minerals dissolution processes.

Water from the well no. 78 b/s, located 250 m from the river, differs distinctly from the wells discussed above. It shows no bacteriologic contamination, lower oxidability, lack of detergents and low concentration of nitrates. Some plankton organisms were found here, however, their occurrence can be explained by the flood influence as they were detected in the last two investigation series carried after the flood. This water presents the features typical for ground water and is therefore similar to water from the terrace wells. It contains higher concentrations of iron and hydrogen sulphide (H_2S) than water drawn from the bank wells. In relation to the terrace wells, a greater influence of the Warta River is singled out from the presence of nitrates.

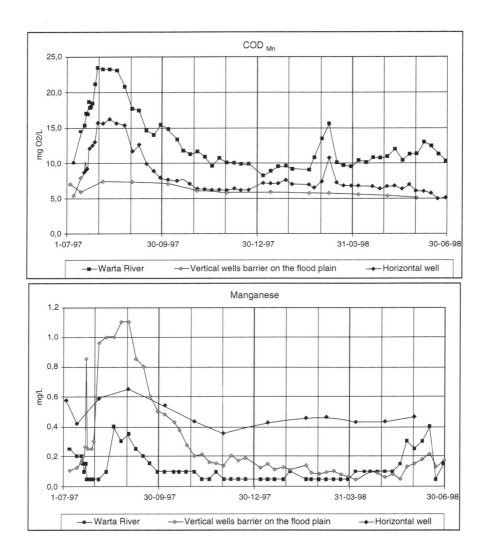

Figure 16.4. Contamination of Warta River and well water during flood.

Water exploited from the terrace wells has the composition typical for ground water. There is no distinct influence of the river within the contents of nitrates and micropollutants. Although some detergents were detected in water from well nos. 4G/3 and 50/3. Their presence is probably caused by local contamination. The influence of the Warta River on these waters is, however, also noticeable. It is reflected by the increased contents of organic matter and hydrogen sulphides. Besides, plankton remains were also discovered.

5. Discussion

The carried investigations have demonstrated that the changes of surface water quality during its passage into the drainage well and the well no. 86 b/2 are controlled mainly by physical processes, especially mechanical retention of suspension, partly of colloids as well as bacteria and plankton. Water quality depends strictly on surface water quality and is characterized by the increased concentrations of Mn, ammonia, color indicator and bacteria. Moreover, chemical micropollutants as well as plankton organisms typical for the river water occur in this pumped water. It becomes also enriched with Mn while filtrating through the layer of the alluvial sediments. The negative feature of water from the drainage well is temperature variability in the range from 0°C to 25°C. Water quality is subject to considerable fluctuations depending on water quality of the Warta River. The quality is particularly unfavorable in the period of melt water runoffs, during the periods when the river is frozen for longer time and during the floods. Water quality deterioration is also being observed after debris movements above well drains and concerns a specially bacteriology, color, and Mn (Górski and Przybyłek 1998). The only positive feature in comparison to other wells is the low content of Fe, which only incidentally exceeds drinking water quality standards.

Water quality of the bank barrier under the conditions of the contribution of ground water is also determined by the processes of mechanical suspension, colloids and plankton retention, as well as the influence of sorption, solution and red-ox processes which is visible. These processes contribute to the improvement of surface water quality not only in the range of turbidity, organic matter and bacteria but also in the range of ammonia and nitrates. Nevertheless, some of the wells are supplied with micropollutants, bacteria, and even plankton. Water temperature changes in the smaller range: 5–15°C and only during the summer flood in 1997 temperature increased up to 25°C. Water is getting enriched with Mn, Fe, Ca, and Mg in view of the solution and redox processes. Water quality is subject to the considerable fluctuation especially in the range of conservative pollutants during the flood periods. In comparison to the drainage well waters, the bank wells are less vulnerable to the influence of extreme hydrogeological climatic states (smaller influence of the flood and ice periods). However, after a long drought a significant deterioration of water quality was connected not only with a deterioration of Warta River quality but also with an increased development of hydrochemical changes related to sulphides and organic matter oxidation. These processes were mainly initiated by the growth of the depression cone on the right side of the Warta flood plain, in the zone of unfavorable geochemical environment enriched with sulphides and organic matter. The result of this situation was the increase of iron and sulphate contents. Beyond the drought period, no significant influence of hydrochemical changes was observed. This resulted from small depressions and a favorable geochemical environment on the left side of the flood plain in the wellfield region. A disadvantage of bank well barrier is a strong decrease of porosity of river bed sediments and a clogging of the hydrogeological environment particularly during the period of long-lasting low hydrological stages caused by drought and intensive water exploitation. Development of these processes during the drought period of 1989–1992 induced a growth of a depression cone on the right side of the Warta valley and clogging processes (Górski and Przybyłek 1991).

The influence of surface water on water quality of the terrace barrier is mainly noticeable within the conservative parameters, particularly chlorides. The amount of organic matter increases distinctly in the investigated waters and induces a decrease of redox potential as well as sulphides and hydrogen sulphide formation. Besides, the investigations revealed a penetration of plankton remains to the water. In case of other qualitative parameters, surface water is subject to various changes depending on the conditions of the hydrogeochemical environment and water quality of the wells is similar to the typical ground water. The observed water quality is thus a result of hydrogeochemical processes in the ground water environment. Oxidation processes in the aeration zone play the main role in the quality formation causing high contents of Fe, Mn, and sulphate. Denitrification process must be singled out as a process limiting a spread of nitrates occurring in significant concentration in surface water.

6. Conclusions

1. The investigations showed that in the conditions of significant surface water contamination, the bank wells close to a river (<50 m) and the collector well with horizontal drains located within bed sediments are inappropriate from the point of view of water quality. Wellfields of these types are also unfavorable because of the significant influence of the extreme hydrological stages (drought and flood) on water quality and the exploitation conditions (clogging).
2. Wells located at the distance of 70–80 m to the river showed the considerable influence of the river on the quality of the exploited water. Its symptoms are bacteria, plankton as well as micropollutants and nitrates. Water quality from the well no. 78 b/s located 250 m from the river is much better than water quality from the bank wells and is similar to the typical ground water.
3. The influence of the river in the range of nitrates and micropollutants is not observed in the case of the terrace wells located 480–1,100 m from the river. A composition of the exploited water is typical for ground water. However, soil and ground water medium are enriched with organic matter. The result of it is the increased oxidability and the occurrence of sulphides and plankton remains.
4. On the basis of the carried research it was found that the wells should be located 150–250 m from the river in order to obtain favorable quality of the exploited water. Such distance assures ground water residence time of at least 6 months.

QUALITY OF RIVERBANK FILTRATED WATER

Acknowledgments The author would like to thank Professor Jan Przybyłek from Adam Mickiewicz University in Poznań City for his cooperation in investigations, and the anonymous reviewers and editors for their help.

References

Górski J, Przybyłek J (1991) Clogging processes and their range in the Warta River beds along bank wellfield in Krajkowo. Proceedings of the Conference "Aspects of environmental protection." Akademy of Mining and Metallurgy, Kraków, Poland, pp. 217–224

Górski J, Przybyłek J (1998) The influence of climatic conditions on exploitation conditions and ground water quality. Proceedings of III International Conference "Municipal and rural water supply and water quality," Poznan, Poland, (Sozański M, editor), pp. 91–109

Chapter 17 Riverbank Filtration as an Alternative to Surface Water Abstraction for Safe Drinking Water Supply to the City of Khabarovsk, Russia

V.V. Kulakov[1], N.K. Fisher[1]*, L.M. Kondratjeva[1], and T. Grischek[2]

[1] Institute of Water and Ecological Problems, Far Eastern Branch of the Russian Academy of Sciences, 680000 Khabarovsk, Russia. E-mail: vvkulakov@mail.kht.ru, kondratjeva@ivep.as.khb.ru

[2] University of Applied Sciences Dresden, 01069 Dresden, Germany. E-mail: grischek@htw-dresden.de

Abstract This paper shows the pollution of the Amur River, which is the water supply source for the city of Khabarovsk. The Amur River is polluted by aromatic and polyaromatic hydrocarbons, phthalates, chlororganic compounds, metals, etc. Major pollution of the Amur River comes from the Sungari River, which is located in the Chinese territory. An alternative source of water for Khabarovsk is groundwater. It is 10 km west of Khabarovsk on the opposite side of the Amur River. The capacity of the Tungus site is estimated to 500,000 m^3/day. The distance between the abstraction wells and the river is about 1.8 km. Calculations show that the first portion of bank filtrate will approach the nearest well after about 2 years.

Keywords: Riverbank filtration, Khabarovsk, Amur River, Sungari River, water pollution

1. Introduction

The city of Khabarovsk is located in the Far East of Russia on the right bank of the Amur River (Figure 17.1). Khabarovsk has about 600,000 inhabitants. At present the drinking water supply is based on surface water abstraction from the Amur River.

The first water supply system in the Far East of Russia was built in Khabarovsk. In 1907 the first waterworks with a capacity of 1,200 "buckets" per day (~15,000 L) was constructed. This waterworks was renovated many times and is today no longer in operation (Chaikovski and Soshnikov 2003). A new waterworks was

---

* N.K. Fisher, Institute of Water and Ecological Problems, Far Eastern Branch of the Russian Academy of Sciences, Khabarovsk, Russia, e-mail: fisher@ivep.as.khb.ru

C. Ray and M. Shamrukh (eds.), *Riverbank Filtration for Water Security in Desert Countries*, DOI 10.1007/978-94-007-0026-0_17, © Springer Science+Business Media B.V. 2011

built, today having a capacity to treat 374.000 m^3/day surface water. The main intake of surface water from the Amur River is located more than 50 m from the right bank of the Amur River where the city is located. The average water consumption in Khabarovsk is 220 L/capita/day.

Figure 17.1. Amur River map (arrow indicates flow direction).

The first stage of the main treatment facilities began operation in 1974. Two treatment steps are used for river water treatment: sedimentation and filtration. From the water intake the water comes to four vertical mixing vessels and further to flocculation chambers. Then, suspended solids are allowed to settle in nine horizontal sedimentation basins for about 3 h. Afterwards, the water flows through ten rapid sand filters. After filtration the water is disinfected using chlorine and stored in three storage tanks having a total volume of 17,000 m^3 (Chaikovski and Soshnikov 2003).

Chlorine is added at two different times: before mixing and after filtration to destroy persistent substances and to disinfect the treated water, respectively. To destabilize suspended and colloidal substances a coagulant (aluminum sulfate or aluminum chloride) is added before mixing. To enhance the formation of flocs during coagulation and sedimentation, and to improve sorption during filtration, Praestol is added as a flocculent (Chaikovski and Soshnikov 2003).

2. Hydrology of the Amur River

The Amur River is among the ten largest world rivers and ranks third in the Russian Federation in terms of length and fourth in terms of drainage area. The Amur River basin area is 1,856,000 km^2, the total length is 4,444 km, and the length from the confluence of the rivers Shilka and Argun is 2,824 km. The basin lies in the territories of three countries – Russia, China, and Mongolia (54, 44, and 2% of the basin area, respectively). The Amur River is the largest transboundary river in Russia. There are about 5 million inhabitants in the Russian part of the basin and about 100 million inhabitants in the Chinese part.

The average discharge of the Amur River to the sea is 10,900 m^3/s and at Khabarovsk 8,300 m^3/s (Chaikovski and Soshnikov 2003). Precipitation falling on Russian territory accounts for 72% of the annual discharge. The maximum discharge registered at Khabarovsk was 40,000 m^3/s in September 1897 and the minimum discharge was 153 m^3/s in March 1922. Throughout the year the discharge is highly variable. In the cold period from November to March the discharge is low. Snowmelt in the spring causes rising water levels. The lowest water level in non-freezing periods is usually recorded at the beginning of summer. Maximum levels and discharge are observed in summer and autumn from July to October because of frequent rain storms brought by typhoons and cyclones from East-Asian seas (Mahinov et al. 2006). The Amur River receives water from surface water runoff after precipitation (75–80%), snow melt (15–20%), and groundwater (5–10%) (Voronov 2004). The average amount of precipitation at Khabarovsk is 672 mm/year. During the summer, precipitation account for up to 90% of annual total (Chaikovski and Soshnikov 2003).

The river flow velocity in winter is 30–50 km/day, and in summer 70–80 km/day (Mahinov et al. 2006). The river width changes considerably along its course. At Khabarovsk, the width of the river is about 2 km, compared to 8 km near the mouth. Depths of the Amur also vary from shallow to 10–15 m. The average river depth at Khabarovsk is about 10 m.

The Amur River basin spans approximately 1,000 km from north to south and 2,000 km from east to west. The Amur River flows mainly in a latitudinal direction and crosses regions with different environmental and climatic characteristics. In the headwaters, the climate is predominantly continental with strong temperature fluctuations during winter and summer in addition to early frosts. The average annual air temperature is above zero only in the middle part of the Amur River. The average annual temperature at Khabarovsk is 1.4°C, the average temperature in winter is −23°C, and in summer 25.7°C. The period with sub-zero temperatures lasts about 162 days on average. The Amur River at Khabarovsk usually freezes at the beginning of November, and thawing begins at the end of April. The duration of a freeze-up is 178 days on average, and the mean thickness of ice is 118 cm. In winter, the monthly average temperature of water is 0°C and in July it is 22.5°C. In winter, the depth of frost penetration into the ground at Khabarovsk is 2.69 m (Chaikovski and Soshnikov2003).

Bad conditions for water supply and treatment are brought on by floods, which frequently last a long time (up to 70 days). During floods the water quality of the Amur River water decreases considerably. As a result, the quantity of reagents needed in the water treatment process increases substantially. Floods usually occur in July and August.

The river water is characterized by low total dissolved solids (TDS) (40 mg/L), dark color due to high concentrations of humic acids (3.9–4.5 mg C/L or 42–47% C_{total}) and a high concentration of Fe (0.38–1.12 mg/L at Khabarovsk) (Levshina 2007, Nagao et al. 2006, Shesterkin et al. 2007). The content of dissolved organic compounds, measured by a chemical oxygen demand, at Khabarovsk in summer 2006 was up to 38 mg O_2/L (Levshina 2007).

The Amur River water has a high content of various pollutants. Major pollution of the Amur River comes from the Sungari River (Songhua in Chinese) which flows in on the right bank of the Amur River and whose water tends to stay near that side. The Sungari River contributes approximately 25% of the total flow of the Lower Amur, but the pollution load during the last few years represented more than 80% of the total pollution (Novorotsky 2008).

3. Pollution of the Sungari River

Along the banks of the Sungari River, there are more than 100 factories whose activities involve petrochemicals, pulp and paper, machinery, and intensive agriculture. The Chinese banks of the Amur River are also highly developed.

Insufficiently developed infrastructure of wastewater treatment facilities, increasing application of fertilizers, and increasing soil erosion due to excessive logging of trees all strongly contribute to pollution in the Sungari River (Bettinger et al. 2007, Xiang et al. 2007).

From 1960 to 1990 the main pollutants in the Sungari River were methylated mercury and phenolic compounds (Xiang et al. 2007). From 1958 to 1982 the Sungari was seriously contaminated by mercury from the acetic acid plant of the Jilin Chemical Company. The plant discharged 113.2 tons of total Hg and 5.4 tons of methylated Hg to the Sungari River, which constituted 69.8% and 99.3% of total anthropogenic Hg and methylated Hg input to this river, respectively. In addition, some small plants distributed inside the drainage area of the Sungari River also used Hg and discharged it into the river (Li et al. 2009). After 24 years as the largest source of mercury, the acetic acid plant of Jilin Chemical Company completely terminated discharge of Hg to the Sungari River in 1982. Total Hg and methylated Hg concentrations in river water and total Hg concentrations in sediments decreased significantly. In the sediments at effluent discharge sites mercury concentrations decreased from 16.8 mg/kg in 1974 to 0.09 mg/kg in 2005 (Lin et al. 2007).

The Sungari River is heavily polluted by nitrogen-containing substances, especially during flood periods (Liu and Yu 1999, Yu et al. 2003). The concentration of NH_4^+ is up to 3 mg/L, NO_3^- up to 5.35 mg/L, and NO_2^- up to 0.46 mg/L (Shesterkin et al. 2007). It is supposed that the main source of nitrogen is soil erosion and surface runoff from farmland enriched by fertilizers and pesticides (Xing et al. 2005).

Water and riverbed sediments are also polluted by aromatic and polycyclic aromatic hydrocarbons (PAHs). Concentrations of PAHs and polychlorinated biphenyls in riverbed sediments are up to 15 µg/g and 36.8 ng/g, respectively (Guo et al. 2007, Zhang and Wang 2006). The level of PAHs in the sediments of the Sungari River is higher in comparison with values reported from other rivers and marine systems around the world. The most contaminated sediment samples were found in the upstream area of the Sungari River located near the Jilin petrochemical industrial company in both flood and icebound seasons. Sediments are dominated by 4–6 ring PAHs due to their higher persistence (Guo et al. 2007). The concentration of polychlorinated biphenyls in water is up to 0.088 ng/L (Zhang and Wang 2006).

Sungari riverbed sediments are also polluted by heavy metals such as Co, Cu, Cr, Ni, Pb and Zn with concentrations up to 14.7, 79, 75, 29, 124, and 403 mg/kg, respectively (Lin et al. 2006). Maximum concentrations of heavy metals were observed near the industrial centre of the city of Jilin (Guo et al. 2006, Lin et al. 2008).

Research in 2002–2003 showed that organic extracts of water samples have genotoxic activity and a risk of carcinogenic potential to human health. At present the mutagenicity of water samples is elevated compared to the results in 1994–1995 (Liu et al. 2009). Due to intensive pollution of the Sungari River by mercury in the last century, a high content of mercury was determined in the hair of residents (0.16–199 mg/kg) (Zhang and Wang 2006).

Thus, the Sungari River contributes various pollutants to the Amur River (Kondratjeva 2005, Shesterkin 2008). The pollution affects not only the water quality at the intake of the city of Khabarovsk but also the entire Amur Estuary further downstream.

4. Pollution in the Amur River

In Khabarovsk, there was no equipment for measurements of a wide spectrum of organic trace substances in the Amur River. The major part of the analysis consisted of qualitative analyses and determination of sum or group parameters such as biological oxygen demand (BOD) and chemical oxygen demand (COD).

During 2001–2002, an investigation of PAHs in Amur River water near the water intake of Khabarovsk showed that the concentration of total PAHs was 9.8 ng/L in winter. Principal types of PAHs were anthracene (1.8 ng/l) and pyrene

(3.3 ng/L). The concentration of benzo(a)pyrene was 0.5 ng/L (Kondratjeva et al. 2007). In summer, the total PAH concentration in the Amur River water near Khabarovsk was up to 10 times higher than in winter.

In different seasons, phthalates (C7–C9; C15–C20) were detected repeatedly at Khabarovsk, of which dibutylphthalate was most prevalent. Among volatile substances highly toxic 2-chloromethyl-N-(2-ethyl-6-methylphenyl)-acetamide, 1-tert-butyl-2-methyl-1,3-propandiol-ether of isobutyric acid, tert-butyl-ether of palmitic acid, 2,6-di-tert-butyl-4-methylphenol and 3,5-di-tert-butyl-4-hydroxy-benzaldehyde were also identified in river water (Kondratjeva 2008, Rapoport and Kondratjeva 2008).

Water analysis for metals in Amur River water showed higher concentrations at the right river bank (the Sungari River's influence zone) than at the left bank (Kondratjeva et al. 2006, Mahinov et al. 2006). For example, concentrations of Cd, Pb and Co at the left bank were 0.02, 0.88, and 0.27 µg/L, respectively, and at the right bank concentration were 0.71, 2.73, and 0.57 µg/L, respectively (Table 17.1). Hg was detected only near the right bank with a concentration of 0.17 µg/L (Mahinov et al. 2006).

TABLE 17.1. Maximum concentrations of heavy metals (µg/L) in Amur River water at Khabarovsk in September 2006 (Mahinov et al. 2006).

Cr	Cu	Zn	Cd	Hg	Pb	Mn	Fe	Co	Ni	Sn	Sb
2.37	72.7	85.6	0.71	0.17	2.73	75.8	1773	0.57	3.57	1.23	0.72

Besides water a health risk to the population of Khabarovsk are the fish caught in the Amur River. In 2002, pesticides DDT and hexachlorocyclohexane, trimethylamine, heavy metals, aldehydes and butyric acid ethers were identified in fish (Kondratjeva and Rapoport 2008).

5. Accident in the Sungari River Basin in 2005

The transboundary pollution of the Amur River due to an accident in the Jilin province (Figure 17.1) in November 2005 aggravated the poor environmental situation in the Lower Amur (Kondratjeva et al. 2009). At the Jilin Chemical Plant explosions occurred in the nitration unit for aniline production. A T-102 tower jammed up but was not handled properly (People 2005). The result of the explosion was the input into the river of about 100 tons of nitrobenzene with admixtures of benzene, aniline, xylene, and toluene.

The water quality monitoring in the Amur River was started on November 24, 2005 and finished on January 13, 2006. In total, more than 3,500 water samples were taken and up to 38 parameters measured. Water sampling was conducted at

14 cross-sections at the right and left banks and in the middle of the river, and included sampling of surface water directly under the ice cover and near to the riverbed. Sampling was carried out up to 8 times per day. When the nitrobenzene pollution passed Khabarovsk, the water in the river and in the water intakes was analyzed in more than 170 samples per day, and after passing through treatment facilities the water was analyzed every 2 h. Experts from all organizations involved worked round the clock (Andrienko et al. 2006). The on-site analysis of nitrobenzene in water was accomplished by two laboratories. One of them was an existing laboratory in Khabarovsk, and a second, additional mobile laboratory was established by the Ministry for Emergency Measures on the Amur River bank.

Four days after the accident, the nitrobenzene concentration in the Sungari River was 0.804 mg/L (Figure 17.2) (Simonov and Dahmer 2008). Almost a month after the accident, the pollution reached the Amur River. The maximum concentration of nitrobenzene in the Amur River at the right Chinese bank was 0.209 mg/L on December 18 (Berdnikov et al. 2006). Upstream of Khabarovsk, the zone of nitrobenzene pollution gradually dispersed across the width of the river. On December 22 the pollution reached Khabarovsk and passed during a 7-day period (Figure 17.3) with a maximum concentration of 0.05 mg/L. The nitrobenzene concentration in the Amur River water near the water intake of Khabarovsk was <0.003 mg/L. The latest analyzed water samples were taken on January 13, 2006, in the lower reaches of the Amur River near the village Bogorodskoe. Nitrobenzene was detected in only one sample, where its concentration was 0.001 mg/L; nitrobenzene concentration in other samples was below the detection limit (<0.001 mg/L).

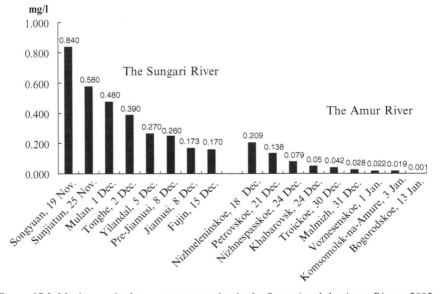

Figure 17.2. Maximum nitrobenzene concentration in the Sungari and the Amur Rivers, 2005–2006 (Berdnikov et al. 2006, Simonov and Dahmer 2008).

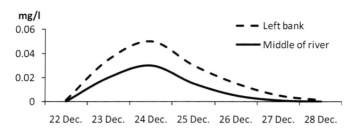

Figure 17.3. Nitrobenzene concentration in the Amur River at Khabarovsk, 2005 (Berdnikov et al. 2006).

Calculations showed that about 11 tons of nitrobenzene passed at Khabarovsk during 7 days, based on the discharge of 1,847 m^3/s of the Amur River in winter at Khabarovsk (Shesterkin 2008) and observed nitrobenzene concentrations during the 7 days (Berdnikov et al. 2006).

In order to dilute the pollutant concentration in the Sungari River, water discharge from the Fengman hydroelectric power station into the river was temporarily increased (UNEP 2005).

When the polluted waters of the Sungari River came into the Amur River, many chlororganic compounds were found in the water (chloroform, tetrachloromethane, chlorophenols, chlorobenzenes) (Andrienko et al. 2006, Kondratjeva and Fisher 2009). Aside from nitrobenzene and other chlororganic compounds, benzene derivatives, isobutylphthalate, dibutylphthalate, biphenyls, pyridine derivatives, benzopyrene, benzofluoranthene, atrazine, and cyclohexane derivatives were identified in polluted Amur River water.

The accident posed a real threat to the water supply of Khabarovsk. To prevent intake of the polluted river water at Khabarovsk, a dam was built in the river channel where the water intake is located. Additional drinking water treatment with powdered activated carbon provided the population with potable water according to the requirements and as a result kept the people healthy. To further protect the people's health, the Khabarovsk government ordered a ban on fishing in the Amur River and in the Amur estuary and in summer prohibited swimming in the river.

After the accident, China donated equipment and materials—including six pieces of monitoring equipment, 150 tons of activated carbon and six air compressors for hydrochemical laboratories—to assist Russia in responding to the potential damage and risks (UNEP 2005).

6. Amur and Sungari Rivers Pollution After the Accident

During nitrobenzene pollution the surface water of the Sungari River was sometimes frozen, and some of the nitrobenzene was captured in the ice. Preliminary tests by the Chinese Academy of Environmental Sciences indicated that the concentration of nitrobenzene in tested ice sample was one quarter that in river water (UNEP 2005). Investigations by Russian scientists showed that different pollutants were captured in ice. The most polluted ice of the Sungari River was found near the city of Jiamusi close to the right bank (Table 17.2). This ice contained many clay particles and had a strong smell. Concentration profiles of ice of the Sungari River indicated shock loads of pollutants during the winter. Research results showed that icebound pollutants during an ice drift entered the Sungari and the Amur River water and had a negative influence on self-purification of the Amur River (Fisher 2007).

TABLE. 17.2. Icebound pollutants in the Sungari River ice near the city of Jiamusi close to the right bank (Kondratjeva 2008).

Substances	c ($\mu g/L$)
Benzene and alkylbenzenes	0.1–0.3
m, p-cresoles	16
2-brom-4-tertbutylphenol	0.8
Tertbutylphenol	0.6
Benzoic acid	0.5
Benzoacetic acid	7.5
Naphthalene	3.0
Phenanthrene	0.131
Benz(a)pyrene	0.008
11 PAHs	0.348

Three months after the accident it was found that water and bottom sediments in the Sungari and the Amur Rivers were still contaminated by different pollutants. High concentrations of PAHs such as pyrene (31.4 $\mu g/kg$), perylene (26.6 $\mu g/kg$), benz(a)pyrene (24.4 $\mu g/kg$), chrysene (13.6 $\mu g/kg$), and naphthalene (8.6 $\mu g/kg$) were revealed in riverbed sediments of the Sungari River (Kondratjeva and Stukova 2008). In Amur riverbed sediments collected in the zone of Sungari impact, PAHs concentrations were even higher. For example, perylene concentration was 128 $\mu g/kg$ and benz(b+j)fluoranthene concentration was 112 $\mu g/kg$. Furthermore, high concentrations of chloroform, carbontetrachloride, heavy metals, benzene, ethylbenzene, toluene, xylene, and phthalates were found in riverbed sediments of the both rivers. Nitrobenzene was not found in sediments (Berdnikov et al. 2006). In Amur River water, the concentration of total PAHs was as high as 0.036 $\mu g/L$ (Kondratjeva and Stukova 2008).

In Sungari River water and transboundary water of the Amur River, concentrations of Fe, Zn, Cu, Mn, Pb and Ni exceeded permissible levels. High concentrations of dichlorophenol (30 times higher than the permissible level), trichlorophenol (up to 30 times the permissible level), bromophenol (permissible level is not established), and chlororganic pesticides (five times the permissible level) were also found (Andrienko et al. 2006).

In spring 2006, benzene (up to 3.13 µg/L) and xylene (up to 22 µg/L) were measured in the Amur River water (Levshina 2007).

Table 17.3 shows concentrations of selected ions in the Sungari and Amur River water in spring 2006.

TABLE 17.3. Selected ion concentrations (mg/L) in Sungari and Amur river waters in spring 2006 (Shesterkin et al. 2007).

	Ca^{2+}	Cl^-	SO_4^{2-}	NH_4^+	NO_3^-	NO_2^-	HPO_4^{2-}
Sungari River	23.8	13.7	34	3.0	5.4	0.46	0.28
Amur River	23.8	12.8	32	3.1	3.5	0.20	0.14

In July 2006, research show that pollutants other than benzene derivatives accumulated in riverbed sediments, fish and mollusks in the Amur River (Table 17.4). Phthalates, naphthalene, highly toxic anisole and benzothiazole, acetochlorine and triazine pesticides as well as new-generation fluorine and phosphor containing pesticides were identified for the first time (Kondratjeva and Fisher 2009). Pollutant concentrations frequently exceeded the permissible levels.

TABLE 17.4. Toxic substances in the Amur River after accident (Kondratjeva and Fisher 2009).

	Freeze-up period (2005–2006)
Water	Nitrobenzene, benzene, toluene, xylene, ethylbenzene, chloroform, dichloromethane, dichlorobenzene, PAH, heavy metals
Bed sediments	Chloroform, tetrachloromethane, chlorobenzenes, PAH, heavy metals
Fish	Nitrobenzene, benzene, toluene, xylene, ethylbenzene, heavy metals, chlorine containing pesticides
	May–June 2006
Water	Phthalates, dibutylphthalates, chlorophenols, pesticides, acetochlorine, atrazine, PAH, heavy metals
Bed sediments	Diethylphthalate, dioctylphthalate, methylphenol, benzo(b)fluoranthene, perylene, benzene, xylene, toluene, chlorophenols, methylbenzene
Fish	Benzothiazole, anisole, naphthalene, phthalates, xylene, toluene, ethylbenzene, heavy metals, phosphororganic pesticides

According to news sources in China, accidents often occur that lead to water pollution. For example, from November 13, 2005 till September 2006 in China more than 130 water pollution accidents were reported across the country, averaging one every two or three days (Xinhua 2006a). During this period, there were three major accidents: the release of toxic wastewater into the Beijiang River (a branch of the Zhu River) in December 2005; the release of cadmium-containing wastewater into the Xiangjiang River (a branch of Yangtze River); and a spill of diesel oil into the Huang River in January 2006 (WB 2007).

In the Sungari River basin, a few more accidents happened after the accident in Jilin. On April 6, 2006, the Xiguang Chemical Plant near Harbin suffered an accident at the site just 4 km away from the Sungari River. Tanks containing raw materials for producing diluting agents exploded and four tons of dimethylbenzene and cinnamene were destroyed. Environmental safety officials said that pollutants were not discharged into the Sungari River (Xinhua 2006b).

On August 20, 2006, the Changbaishan Jingxi Chemical Company dumped about 10 m^3 of petrol-blending chemical xylidine into the Mangniu River, tributary of the Sungari River, resulting in a 5 km reddish-brown slick (ENC 2006). To prevent an inflow of pollutants into the Sungari River a pollution interception dam and two dams with activated carbon were built to adsorb the pollutants. Through filtration, the stretch of polluted water had bypassed the three dams and flowed into the Sungari River (CGOWP 2006).

On 5 June 2008, during cutting of a gas bags which arrived for recycling in the city of Cicikar near Harbin, a leak of phosgene gas occurred causing three deaths and poisoning eight people. According to the news, there was no pollution of the Nen River, a tributary of the Sungari River.

Four months after the initial accident in Jilin, the Chinese Government developed a long-term water pollution control plan for the Sungari River basin beyond 2010, creating a provision for policy inputs to the 11th Five-Year Plan and drafting the Sungari River Basin Pollution Control Master Plan for reducing the discharge of key pollutants by 10% within a 5-year time frame. Despite quality assurance of water and the construction of wastewater treatment facilities in China, the quality of the Amur River water remains unsatisfactory for drinking water purposes. The transboundary pollution of the Amur River by different organic compounds represents a real danger for population health and makes the water treatment process considerably more difficult and expensive.

7. An Alternative Means of Drinking Water Production

An alternative source of water for Khabarovsk is groundwater. The question of changing the drinking water supply of Khabarovsk from surface water to groundwater abstraction was already considered in the beginning of the 1960s. During nitrobenzene pollution in 2005, the possibility of building groundwater intakes for water supply was raised again, and the Tungus site was explored for water supply potential. It is located near Khabarovsk between the Amur River and the Tunguska River, 10 km west of Khabarovsk on the opposite side of the Amur River.

Full-scale hydrogeochemical and microbiological investigations into groundwater abstraction at the Tungus site, drilling of test wells and the creation of special experimental installations were carried out from 1993 to 1998 (Kulakov and Kondratjeva 2008) and continued from 2007 to 2009 (Kondratjeva et al. 2008). The intent was to determine the possibility of using its groundwater resources for water supply to Khabarovsk and to carry out in-situ removal of iron and manganese in the aquifer. The capacity of the Tungus site is estimated to 500,000 m^3/day. At the initial stage of site development it will be 100,000 m^3/day. Groundwater reserves are recharged on the order of 65% by surface water from the Amur River. The distance between the abstraction wells and the river is about 1.8 km. Calculations show that the first portion of bank filtrate will approach the nearest well after about 2 years (Figure 17.4).

Worldwide studies and long-time investigations have demonstrated the efficiency of riverbank filtration (RBF) for the removal of many organic compounds. RBF provides significant removal of pollutants, natural organic matter, heavy metals, microorganisms, and turbidity (Hiscock and Grischek 2002). Physical, chemical and biological processes are involved in natural water purification in the aquifer.

Sustainable removal of organic compounds during RBF depends on filtration, adsorption and biodegradation. The effectiveness of RBF for water quality improvement depends on a number of variables such as the characteristics and composition of the alluvial aquifer materials, river water quality, groundwater dilution, filtration velocity and the distance of the well(s) from the river, temperature of river water, pumping rate, and sediment characteristics at the river–aquifer interface (Hiscock and Grischek 2002, Ray et al. 2002).

Data in Table 17.5 demonstrate the removal efficiency of RBF for pollutants relevant to the Amur River even at 13–100 m from a river. Thus, at the Tungus site, after a flow path length of more than 1,800 m in the aquifer, the bank filtrate should by no means contain these pollutants. During RBF surface water qualities will be fully transformed by filtration through the thick sandy-argillaceous mass providing a high sorption capacity. The composition of bank filtrate from the Amur River will be similar to that of groundwater formed under natural conditions. Due to the long travel time of the bank filtrate, 50% of all wells will not have received any bank filtrate by the end of their standard operation lifetime of 25 years (Figure 17.4). They will only abstract natural groundwater.

RIVERBANK FILTRATION AS AN ALTERNATIVE TO SURFACE WATER 293

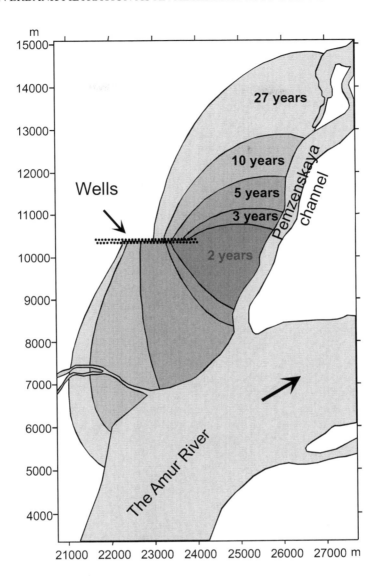

Figure 17.4. Calculated travel time of bank filtrate (Kulakov and Shtengelov 2009).

TABLE 17.5. Removal efficiency of riverbank filtration for pollutants relevant for the Amur River.

Location of site	Distance well-river (m)	Flow time (days)	Pollutant	Removal rate (%)	Reference
USA, Nebraska, City of Lincoln, Platte River	25	6	Triazines	up to 84	a
			Acetamides	up to 76	
Germany, Heilbronn, Neckar River	100	58	Toluene	80	b
			o-Xylene	78	
			p-Xylene	67	
			Benzene	66	
			Ethylbenzene	79	
			Naphthalene	89	
The Netherlands, Rhine River	45	not available	Aniline	81	c
			Nitrobenzene	81	
Switzerland, Glatt River	13	18–54	Nonylphenol	Up to 100	d

Note: a = Verstraeten et al. 2002, b = Jüttner 1999, c = Noordsij et al. 1985, d = Ahel et al. 1996.

Figure 17.5 shows simulated piezometric contour lines in the middle part of the aquifer, flow path lines, and wellhead protection zones (WPZ). WPZs can be established around the water intake or based on protective properties of the aeration zone and low-permeability horizons that may be located within rocks (water permeable and waterproof) on the intake facility site.

The size of WPZ I is estimated based on engineering requirements. The radius of the WPZ I belt around the abstraction well is calculated to be within the range 30–50 m. Belt sizes are determined based on requirements to exclude microbial (belt II) and chemical (belt III) pollution of water during the estimated time period. In Russia, the time period for belt II depends on bacteria lifetime in the aquifer (400 days), and for belt III it is equal to the facility production term (usually over 25 years or 10,000 days).

The main problems with using groundwater for drinking water supply are high concentrations of iron (12–28 mg/L), manganese (0.38–1.86 mg/L), and silica (11.2–19.3 mg/L). The Russian drinking water guideline sets upper limits of 0.3 mg/L for Fe, 0.1 mg/L for Mn, and 10 mg/L for Si. The groundwater at the Tungus site contains about 200 mg/L TDS and 1.5 mg/L dissolved organic compounds. Abnormal factors are a low pH (5.8–6.1) and a high concentration of dissolved CO_2 (220–250 mg/L).

To build a groundwater abstraction scheme for Khabarovsk, the German technology SUBTERRA was chosen. Pilot wells went into production in November 2007. In January 2008, microbiological and chemical analyses for in-situ iron and manganese removal were started. Companies from Germany were involved in the design and construction of specific wells used both for abstraction and infiltration.

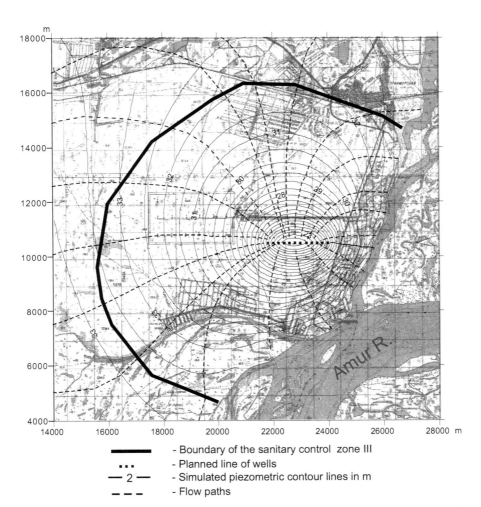

Figure 17.5. Simulated piezometric contour lines and boundary of wellhead protection zone III around planned wells (Kulakov and Shtengelov 2009).

8. Conclusion

Riverbank filtration is a promising alternative to direct surface water abstraction from the polluted Amur River for water supply in Khabarovsk. One of the main benefits in using riverbank filtration is protection against shock loads. Input concentrations of different compounds in river water are equilibrated along passage through the aquifer. Riverbank filtration contributes to a higher drinking water quality and provides safe drinking water during periods of accidental spills.

Because of low concentrations of dissolved organic compounds, formation of disinfection byproducts is expected to be low.

In comparison with treatment of surface water, the natural treatment achieved through RBF obviates the construction of additional facilities and decreases the use of reagents. Furthermore, using subsurface iron and manganese removal techniques, environmental pollution by sludge and backwash water can be avoided. In general, using RBF results in cost savings in water treatment.

References

Ahel M, Schaffner C, Giger W (1996) Behavior of alkylphenol polyethoxylate surfactants in the aquatic environment – III. Occurrence and elimination of their persistent metabolites during infiltration of river water to groundwater. Water Res 30:37–46

Andrienko SN, Bardyuk VV, Veselovskaya OV (2006) On the implementation of integrated measures for ensuring environmental safety of the population in the Russian Amur Region. Proc. Int. Scientific Conf. in the Field of Ecology and Life Safety. Komsomol'sk-na-Amure: GOUVPO "KnAGTU". 10–15 (in Russian)

Berdnikov NV, Rapoport VL, Rybas OV et al. (2006) Monitoring of pollution of the Amur ecosystem resulting from an accident at Jilin petrochemical plant. Tikhookean Geol 25:94–103 (in Russian)

Bettinger R L, Barton L, Richerson PJ et al. (2007) The transition to agriculture in northwestern China. Dev Quat Sci 9:83–101

CGOWP (2006) Jilin: River chemical pollution under control. Chinese Government's Official Web Portal, http://www.gov.cn/english/2006-08/23/content_368626.htm

Chaikovski GP, Soshnikov EV (2003) The Amur River is the source of water supply. DVGUPS. Khabarovsk. 83 p. (in Russian)

ENC (2006) China ups tempo on pollution prosecutions. Environmental news center, http://www.edie.net/news/news_story.asp?id=11983&channel=0

Fisher NK (2007) Microbiological research of the components of the Amur River ecosystem after accident in the Peoples Republic of China (Jilin). In: Science - to Khabarovsk Territory: Proc. IX Regional Competition of Young Scientists. TOGU. Khabarovsk. 55–70 (in Russian)

Guo S, Wang X, Li Y et al. (2006) Investigation on Fe, Mn, Zn, Cu, Pb and Cd fractions in the natural surface coating samples and surficial sediments in the Songhua River, China. J Environ Sci 18:1193–1198

Guo W, He M, Yang Z et al. (2007) Comparison of polycyclic aromatic hydrocarbons in sediments from the Songhuajiang River (China) during different sampling seasons. J Environ Sci Heal A 42:119–127

Hiscock KM, Grischek T (2002) Attenuation of groundwater pollution by bank filtration. J Hydrol 266:139–144

Jüttner F (1999) Efficacy of bank filtration for the removal of fragrance compounds and aromatic hydrocarbons. Water Sci Technol 40:123–128

Kondratjeva LM (2005) Environmental risk of aquatic ecosystem pollution. Dalnauka. Vladivostok, 299 p (in Russian)

Kondratjeva LM (2008) Transboundary pollution of the Amur River with stable toxic organic substances. In: Riabinin NA (ed) Pure Amur – A Long Life: Proc. Int. Conf. IWEP FEB RAS. Khabarovsk, 11–17

Kondratjeva L, Fisher N (2009) Estimation of ecological risk of transboundary pollution of the Amur River. In: Jones JA, Vardanian T, Hakopian C (eds) Threats to Global Water Security. Springer Science-Business Media B V., 385–388

Kondratjeva LM, Rapoport VL (2008) The analysis of the content of toxic organic substances in fish of the Amur River. In: Fresh-Water Ecosystems of the Amur River Basin. Dalnauka. Vladivostok, 249–256 (in Russian)

Kondratjeva LM, Stukova OY (2008) Biogenic studies of polycyclic carbons discharge from the Amur River into the Far Eastern Seas. Report on Amur-Okhotsk Project 5., Japan

Kondratjeva LM, Fisher NK, Berdnikov NV (2009) Microbiological estimate of water quality in the Amur and Sungari Rivers after a technogenic accident in china in 2005. Water Resour 36:548–559

Kondratjeva LM., Fisher NK, Stukova OY et al. (2007) Amur River pollution by polyaromatic hydrocarbons. Vestnik FEB RAS 4:17–26 (in Russian)

Kondratjeva LM, Kanciber VS, Zazulina VE (2006) Influence of main inflows on the heavy metals content in water and bed sediment of the Amur River. Tihookean Geol 25:103–114 (in Russian)

Kondratjeva LM, Kulakov VV, Fisher NK (2008) Biogeochemical research of ferriferous groundwater at the Tungus Site. In: Proc. Int. Scientific Conf. in the Field of Ecology and Life Safety. GOUVPO "KnAGTU". Komsomol'sk-na-Amure, 338-341 (in Russian)

Kulakov VV, Kondratjeva LM (2008) Biogeochemical aspects of groundwater treatment in Priamurje. Tihookean Geol 27:109–118 (in Russian)

Kulakov VV, Shtengelov PS (2009) Estimation of supply of fresh groundwater in river Priamur'e Valleys. Proc. XIX Meeting on Groundwater of Siberia and Far East, 254–257 (in Russian)

Levshina SI (2007) Content and dynamics of organic substances in water of the Amur and Sungari Rivers. Geogr and Nat Resour 2:44–51 (in Russian)

Li P, Feng XB, Qiu GL et al. (2009) Mercury pollution in Asia: A review of the contaminated sites. J Hazard Mater 168:591–601

Lin C, He M, Yan B et al. (2006) Heavy metal contamination in the sediment of the Second Songhua River. Chinese J Geochem 25:118–119

Lin C, He M, Zhou Y (2007) Mercury contamination and dynamics in the sediment of the Second Songhua River, China. Soil Sediment Contam 16:397–411

Lin C, He M, Zhou Y et al. (2008) Distribution and contamination assessment of heavy metals in sediment of the Second Songhua River, China. Environ Monit Assess 137:329–342

Liu J, Yu J (1999) Dynamic variation of nitrogen content in the Second Songhua River. Chinese Geogr Sci 9:368–372

Liu J-R, Dong H-W, Tang X-L et al. (2009) Genotoxicity of water from the Songhua River, China, in 1994–1995 and 2002–2003: Potential risks for human health. Environ Pollut 157:357–364

Mahinov AN, Kim VI, Kuznetsov AM et al. (2006) Assessment of the discharge of some chemical substances from the Amur into the Seas of Japan and Okhotsk. Report on Amur-Okhotsk Project 5:49–56

Nagao S, Terashima M, Kodama H et al. (2006) Migration behavior of Fe in the Amur River basin. Report on Amur-Okhotsk Project 5:37–48

Noordsij A, Puyker LM, van der Gaag MA (1985) The quality of drinking water prepared from bank-filtered river water in the Netherlands. Sci Total Environ 47:273–292

Novorotsky PV (2008) Fluctuations of the Sungari discharge. In: Environmental problems of the large rivers basins – 4. Theses of reports of the international conference. IEWB RAN. Toliyatti, 120 (in Russian)

People (2005) Five dead, one missing, nearly 70 injured after chemical plant blasts. People's Daily Online, http://english.people.com.cn/200511/15/eng20051115_221428.html. Accessed 15 November 2005

Rapoport VL, Kondratjeva LM (2008) Pollution of the Amur River by anthropogenic and natural organic substances. Contemp Probl Ecol 1:377–386

Ray C, Melin G, Linsky RB (2002) Riverbank filtration: Improving source-water quality. Kluwer Academic Publishers, Dordrecht, 366 p

Shesterkin VP (2008) The basic stages of formation of a chemical compound of the Low Amur River water. In: Freshwater ecosystem of the Amur River basin. Dalnauka. Vladivostok, 11–17 (in Russian)

Shesterkin VP, Shesterkina NM, Forina Yu A et al. (2007) Transboundary pollution of the Amur River in winter during low water 2005-2006 (in Russian). Geogr Nat Resour 2:40–44

Simonov EA, Dahmer TD (2008) Amur-Heilong River basin. Ecosistems Ltd. Hong Kong

UNEP (2005) The Songhua River spill, China. Field mission report. United Nations, http://www.unep.org/PDF/China_Songhua_River_Spill_draft_7_301205.pdf

Verstraeten IM, Thurman EM, Lindsey ME et al. (2002) Changes in concentrations of triazine and acetamide herbicides by bank filtration, ozonation, and chlorination in a public water supply. J Hydrol 266:190–208

Voronov BA (2004) Complex surveys of the Amur River basin. Report of Amur-Okhotsk Project 2:35–39

WB (2007) Water pollution emergencies in China. Prevention and response. Sustainable development department. East Asia and Pacific Region. The World Bank (2007) Washington, D.C. 30 p

Xiang P, Zhou Y, Huang H et al. (2007) Discussion on the green tax stimulation measure of nitrogen fertilizer non-point source pollution control – taking the Dongting Lake area in China as a case. Agr Sci China 6:732–741

Xing Y, Lu Y, Dawson RW, Shi Y et al. (2005) A spatial temporal assessment of pollution from PCBs in China. Chemosphere 60:731–739

Xinhua (2006a) China sees one water pollution accident in 2–3 days. Xinhua News Agency, September 11, 2006, http://en.invest.china.cn/english/environment/180736.htm

Xinhua (2006b) Environment gets all-clear after chemical plant blast. Xinhua News Agency, April 17, 2006, http://china.org.cn/english/environment/165819.htm

Yu S, Shang J, Zhao J et al. (2003) Factor analysis and dynamics of water quality of the Songhua River, Northeast China. Water Air Soil Poll 144:159–169

Zhang L, Wang Q (2006) Preliminary study on health risk from mercury exposure to residents of Wujiazhan town on the Di'er Songhua River, Northeast China. Environ Geochem Health 28:67–71

INDEX

A

Abiotic process, 133
AbuTieg, 23, 255–267
Accident, 25, 73–76, 78, 282, 286–291, 295–296
Activity coefficient, 116
Agricultural chemicals, 73
Agrochemicals, 11, 16
AHD. *See* Aswan High Dam
Ahmedabad, 3
Albany region, 29, 31
Aldehydes, 286
Algae, 17, 64, 65, 98, 257, 266
Alluvial aquifer, 1, 84, 129, 169, 172–173, 175, 178, 182, 186, 292
Al-Rayah Al-Naseri, 236, 237, 241, 243
Ambient groundwater, 12, 16, 17, 23, 24, 33, 37, 53, 236, 241
Amur River, 281–295
Anoxic environment, 4, 112, 117
Anthropogenic, 55, 56, 61, 152, 153, 163–165, 220, 284
Aquifer storage recovery (ASR), 3, 29–47, 98, 99, 101, 109–112, 120, 122
Aquifer Storage Transfer Recovery (ASTR), 98, 99, 101, 109–111, 113, 120, 122
Arsenic (As), 12, 32, 33, 41, 45–47, 217
Artificial recharge (AR), 97–125, 138
ASR. *See* Aquifer storage recovery
Assiut, 7, 8, 16, 18, 25, 258–260, 267
Assiut Water Company, 25, 258, 267
ASTR. *See* Aquifer Storage Transfer Recovery
Aswan High Dam (AHD), 8, 11, 18, 19, 259, 264
Attenuation, 1–2, 17, 31, 45, 72, 75, 84, 85, 88, 89, 103, 138, 195, 236

B

Bacteria, 7, 16, 31, 32, 61, 71, 82, 84–86, 92, 93, 130, 153, 157, 174, 197, 220, 236–238, 245–248, 251, 261, 265, 266, 272–274, 277, 278, 294
Bacteriophage, 86, 92
Bank filtration (BF), 17, 22, 47, 51–65, 69, 82, 130, 138, 139, 141, 143, 152, 154, 162, 166, 203–221, 236, 243, 245, 269
Belgrade, 2
Benzene, 122, 158, 272, 288–291, 294
BF. *See* Bank filtration

BHC, 10, 16
Biodegradable organic carbon, 8, 24, 69, 71
Biodegradation, 24, 31, 69, 138, 143, 145–146, 148, 161, 256, 257, 265, 292
Biodegradation zone, 24, 256
Biogeochemical, 4, 17, 32
Biogeochemical equilibrium, 63
Biological clogging, 32
Biological oxygen demand (BOD), 7, 10, 19, 23, 131, 272, 285
Biotic process, 133, 145
BOD. *See* Biological oxygen demand
Bonn, 2
Bratislava, 2
Breakthrough, 32, 63, 64, 74, 76–78, 87, 89, 92, 131, 133
    curve, 75, 84, 86, 88, 134
Budapest, 2

C

Caisson, 175, 182–190, 193, 196, 200, 229–231
Calibration, 32, 54, 74, 114, 120, 165, 166, 178, 179, 218, 219
California, 30, 32, 169, 172, 173, 176–177, 183, 184, 197
Cape Cod, 84–86, 90, 91
Cedar Rapids, 2
Cedar River, 2
Central Ground Water Board (CGWB), 206–209, 212
Central siphon point, 225
Changing boundary conditions, 51–52, 56–57, 60, 64
Chattahoochee River, 33
CHC. *See* Chlorinated hydrocarbons
Chem3D, 141, 142
Chemical oxygen demand (COD), 10, 23, 108, 131, 257, 272, 273, 275, 276, 284, 285
Chemical spills, 69–78
Chloracetophenon case, 76–77
Chlorinated hydrocarbons (CHC), 107, 125, 158
Chlorine, 13, 15, 25, 82–83, 196, 258, 266–267, 282, 290
Chronic toxicity, 15
Circulating flow regime, 57–59, 62, 63, 155–157, 159
Clean Ottawa sand, 131, 132

C. Ray and M. Shamrukh (eds.), *Riverbank Filtration for Water Security in Desert Countries*,
DOI 10.1007/978-94-007-0026-0, © Springer Science+Business Media B.V. 2011

300 INDEX

Climate change, 4, 51, 52, 64, 98, 152
COD. *See* Chemical oxygen demand
Coliform removal, 220, 237, 247
Colloids, 1–2, 17, 82, 86, 88–90, 277, 282
Column experiment, 57, 58, 60, 62, 132, 145, 154–156, 158, 159, 164
Conceptual model, 33, 153, 161–166
Conservative solute, 83–84, 90
Conservative tracer, 84, 89, 90, 245
Construction access shaft, 226, 232, 233
Contact time, 58, 60, 62–64, 156, 158–159, 162, 164
Cryptosporidium, 82–83, 86–91, 224
Cryptosporidium parvum, 81–93
Cyanobacterial, 22

D

Damietta Branch, 10
DBPs. *See* Disinfection by-products
DDT, 16, 272, 286
Degradation concept, 162
Degradation kinetics, 40, 58, 62, 63, 159–161, 165
Delta of Nile, 7–11, 13, 235, 237–241, 259
Denitrification, 32, 40, 42–45, 275, 278
Desert environment, 1–4
Des Moines, 2
Diara formations, 206
Disinfection, 13, 25, 64, 83, 130, 173, 175, 196, 220, 229, 235, 236, 249, 256, 258, 282
Disinfection by-products (DBPs), 15, 17, 25, 32, 47, 130, 153, 224, 249, 258, 296
Dispersion coefficient, 74
Dissociation constant, 116
Dissolved organic carbon (DOC), 32, 33, 39–47, 51–65, 98, 104, 108, 117, 119–122, 125, 151–166, 212, 217, 245
   removal efficiency, 59, 63, 154
Dissolved oxygen, 4, 8, 17, 19, 37, 39–43, 138, 217
DOC. *See* Dissolved organic carbon
DRAGON, 141, 142
Drawdown, 1, 76, 187, 193, 199, 214, 225, 226
Dresden, 2, 3, 52, 56, 64–65
Drilling, 178, 182, 186, 188–193, 201, 207, 208, 229, 230, 241, 242, 249, 292
Drinking water, 7–16, 29–47, 203–221, 281–296
Drought period, 277, 278
Düsseldorf, 2, 52, 69, 70, 72, 75, 76, 212

E

Easy-Leacher (EL), 122–124
EDCs. *See* Endocrine disrupting compounds
EDTA, 55
Egypt, 5–25, 235–238, 245–251, 255–267
EL. *See* Easy-Leacher
Elbe River, 2, 17, 52, 53, 55, 56, 58, 65, 140, 166, 213, 256–257
Electron trapping capacity (ETC), 24
Endocrine disrupting compounds (EDCs), 139, 143, 198
Endosulfan, 73
EPI suite, 141, 142
Escherichia coli, 84, 85, 261, 263, 265, 266
ETC. *See* Electron trapping capacity
EXCEL spreadsheet, 113, 115, 121, 122, 125
Explosive chemicals (TNT, DNT, RDX, HMX), 129–135

F

Fatwa formations, 206, 207
FCM. *See* Fluorescent carboxylated microspheres
Fecal coliform, 10, 23, 240, 248, 265, 266
Fecal streptococci, 262–263
Fence line, 223–234
Fe-oxides, 237
Fertilizers, 8, 9, 11, 12, 16, 25, 56, 82, 284, 285
Flash flood, 4, 8, 25
Flehe waterworks, 71
Flint River, 31, 33–35, 37, 39, 43
Floridan aquifer, 30, 33–35, 37, 39, 43
Flow through regime, 57–59, 61, 159
Fluorescent carboxylated microspheres (FCM), 82–84, 86–90, 92, 93
Fluorescent microspheres, 81–93

G

Ganga River, 53, 203–221
Geochemical reactions, 32, 34–35, 117–118, 121
Geza Water Company, 240
Glasgow Waterworks, 82
Groundwater, 7, 8, 10–13, 15–17, 23, 24, 30, 33, 35, 37, 53, 64, 65, 76, 82–84, 86, 87, 92, 98, 99, 104–106, 108–111, 114, 116–117, 119–121, 130, 151–166, 173–175, 178, 179, 192, 198, 201, 203–207, 209–220, 223–234, 236–242, 245, 246, 256, 257, 260, 266, 275, 283, 291–292, 294
Groundwater protist, 86
Groundwater recharge, 51–52, 54, 152, 165
Gujrat, 3

INDEX

H
Hawaii, 131
Hazardous, 25, 63, 74, 257–258
HCWW. *See* Holding Company for Water and Wastewater
Heavy metals, 8, 10, 15, 17, 31, 34, 104, 105, 112, 125, 217, 285, 286, 289, 290, 292
Heidelberg, 2
Hexachlorocyclohexane, 286
Holding Company for Water and Wastewater (HCWW), 13, 258
Hollandsch Diep site, 122
Honolulu, 131
Horizontal collector, 2, 3, 71, 132, 224–226, 278
Hydraulic conductivity, 21, 39, 54, 56, 60, 90, 173, 177, 188, 207, 209, 213, 236, 238, 244, 248, 260, 271

I
Immobilization, 84–86, 88, 92
India, 3, 17, 24, 30, 53, 64–65, 203–221
Inorganic compounds, 17, 64, 107
Inverse modeling, 113
Ion exchange, 32
Ionic strength, 87, 88, 107, 116
Iowa, 2
Iron (Fe), 3, 7, 11, 12, 16, 17, 19, 25, 32, 90, 105, 106, 112, 117, 122, 217, 235–251, 261, 264, 266, 275, 277, 292, 294, 296
oxyhydroxides, 24
reduction, 4, 33, 40, 42, 43, 45
Isotope analyses, 212, 219–220

J
Jacob's Straight Line Method, 207, 209
Jilin province, 286

K
Khabarovsk city, 281–296
Köln, 2

L
Lag coefficient, 74
Laterals, 34, 90, 132, 175, 182–201, 208, 229, 231, 260
Leverkusen accident, 75
Lincoln, 2
Linear isotherms, 40
Louisville Water Company (LWC), 223–224

M
Managed aquifer recharge (MAR), 98, 99, 105, 107, 117, 122, 137–148
Managed aquifer recharge and recovery, 137–148
Manganese (Mn), 3, 4, 7, 11, 12, 16, 17, 19, 24, 25, 32, 72, 73, 106, 217, 235–251, 261, 264, 266, 276, 292, 294, 296
Manganese oxyhydroxides, 24
MAR. *See* Managed aquifer recharge
Mass balance, 113–122, 125
Maximum contamination level (MCL), 2, 15
Mercury (Hg), 284, 285
Methanogenesis, 122
Microbacterium strain, 84
Microbial degradation, 17, 56
Microcystins, 17, 22
Microorganism, 8, 47, 69, 84, 85, 92, 93, 104, 125, 158–159, 165, 235, 256–258, 292
Micropollutants, 15, 17, 25, 112, 274–278
Migration velocity, 57, 59–62
Milwaukee, 83
Mineralization, 40, 104, 164, 215
Mirror Lake, 84
Mit Ghamr Formation, 238
MLR. *See* Multiple linear regression
Mn-oxides, 238
Modeling, 31, 33, 40, 41, 45–47, 152, 153, 159–165, 178, 183, 193, 195, 198, 201, 205, 210, 213, 218–219
MODFLOW model, 34–35, 40, 165, 179, 183, 193, 196, 198, 212
Molecular descriptors, 141–145, 148
Monsoon, 30, 205, 212–215, 217–220
Morphology, 10, 84, 92, 103, 214–215, 272–273
Mosina-Krajkowo waterworks, 269–271
MT3DMS, 34–35, 40, 122
Multiple linear regression (MLR), 142, 143

N
Naga Hamadi, 7, 18–25
Natural treatment systems, 25, 138, 267
Nebraska, 2, 294
Netherlands, 17, 32, 73, 98–100, 108–110, 113, 118, 294
Nile aquifer, 10–12, 19–22, 24, 25, 237–239, 259, 260, 265
Nile flow rate, 7–9, 18, 260, 262
Nile valley, 6–8, 10, 11, 13, 18–25, 235, 238, 239, 259, 260, 266
Nitrobenzene, 286–291, 294
North Rhine-Westphalia, 72

O

Ohio, 89
Ohio River, 53, 88, 223–224
Oocysts, 81–93
Organic micropollutants (OMPs), 8, 15, 104–105, 107, 113, 125, 137–148, 275
Organic compounds, 4, 8, 17, 24, 55, 56, 60, 63, 64, 107, 152–154, 156, 158–162, 165, 284, 291, 292, 294, 296
Organic matter, 10, 15, 17, 23, 40, 104, 130, 141, 198, 251, 258, 275–278, 292
Organochlorine pesticides, 10
Oxic environment, 117
Oxidation capacity, 72
Oxidation reduction potential (ORP), 105
Oxygen demand, 72

P

PAH. See Polycyclic aromatic hydrocarbons
Particulates, 17, 84, 256, 264
Pathogens, 1–3, 8, 12, 14–17, 22–23, 32, 47, 63, 64, 71, 81–93, 104–105, 112, 125, 130, 198, 256–258
Patna, 203–221
Perennial river, 4, 205
Persistent compounds, 63, 64, 147, 152
Pesticides, 2, 8, 9, 11, 12, 15, 16, 31, 56, 72, 73, 125, 143, 197, 224, 285, 286, 290
Pharmaceuticals, 4, 32, 47, 55, 72, 107, 125, 130, 138, 139, 141, 144, 198
PHREEQC-2, 34–35, 37, 40, 122
PHT3D, 34–35, 40, 122
Physicochemical condition, 7, 82, 88, 261, 265
Plankton, 272–278
Platte River, 2, 294
Pleistocene aquifer, 108
PMWIN, 213, 214
Poland, 269–279
Pollutant plume, 76
Polycyclic aromatic hydrocarbons (PAH), 15, 104, 112, 158, 272, 285–286, 289, 290
Potable water, 4, 6, 17, 22, 32, 45, 130, 154, 288
Poznań City, 269–279
Praestol, 282
Protozoa, 82–84, 86, 89, 90, 92
Pumice stone, 57, 58, 60
Pumping efficiency, 193, 195
Pumping rate, 2, 17, 59, 138, 187, 195, 199, 292
Pump station, 169, 224–226, 228, 232–234
Pyrite-bearing aquifer, 32
Pyrite oxidation, 32, 40, 41, 45–46, 104, 105, 109, 119, 120, 122

Q

Qena, 7, 16, 18, 259
Quantitative structure activity relationship (QSAR), 138–139, 141–148
Quaternary, 10, 20, 71, 206–207, 210, 258–260

R

Raccoon River, 2
Radionuclide, 104–105, 125
Radio-opaque substances, 55
Ranney well, 271
RDX. See Research Department Explosive
REACTIONS+6, 113–121
Reactive transport modeling, 40, 98, 122–125
Recovery, 3, 25, 30–33, 39, 47, 82–86, 88–93, 101, 104, 105, 109–112, 114, 120
Redox, 4, 32, 41, 99, 104–113, 120, 125, 157, 158, 236, 248–250, 277
Redox potential, 33, 105–106, 278
Redox-sensitive chemicals, 3, 32, 40, 105
Reducing conditions, 4, 24, 33, 45, 248, 265
Research Department Explosive (RDX), 130–135
Retardation, 84, 86, 92, 112, 139
Retention time, 57, 59–60, 62–64, 71, 82, 155–159, 162, 164–166
Rhine River, 2, 17, 52, 53, 56, 69–78, 108, 121, 140, 157, 162, 256–257, 294
Risk assessment, 69–78
Riverbank filtration (RBF), 1–25, 29–47, 69–70, 81–93, 97–125, 129–135, 141, 152, 175, 204, 220–221, 223–251, 255–267, 281–296
River Nile, 4–25, 236, 249, 251, 257–258, 260–262
River water quality changes in aquifers, 31–32
Rosetta Branch, 9, 10
Russia, 166, 281–296
Russian River, 30, 88, 90, 91, 170–175, 180, 198, 283

S

Sabarmati River, 3
Sand filters, 87, 104, 235, 237, 242–245, 250, 251, 282
SANDOZ, 73
SAT. See Soil Aquifer Treatment
Scatter plots, 143–145
SCWA. See Sonoma County Water Agency
Security risks, 223–234
Seine River, 17

INDEX

Seismic refraction, 177–178, 180, 186, 205–206
Sewage room, 12, 16
Sidfa, 18, 23, 264
Soil Aquifer Treatment (SAT), 99, 138, 152
Soil Organic Material (SOM), 105, 109–111, 117, 121, 122
SOM. *See* Soil Organic Material
Sonoma County Water Agency (SCWA), 30, 169–201
Sorption equilibrium, 133
Specific yield, 177, 188
Spumella guttula, 85, 86
Stadtwerke Duesseldorf, 212
Steady-state calibration, 218
Storativity, 207
Sulfate reduction, 4, 40, 42, 43, 45
Sungari River, 282, 284–291
Surface complexation reactions, 32
Surrogates, 81–93, 152
Sustainable operations, 4, 198
Switzerland, 73, 84–86, 294

T

Temperature, 4, 7, 17, 19, 32, 39, 40, 43, 52–62, 64, 107, 109, 114, 116, 130, 131, 158, 166, 178, 179, 197, 198, 237, 246–248, 250, 256, 258, 261, 265, 272, 277, 283
Testfilter, 154, 155
Theis Curve Matching Method, 207
THM. *See* Trihalomethanes
TNT. *See* Trinitrotoluene
Toluene, 286, 289, 290, 294
Tracer test, 74, 86, 92
Transmissivity, 11, 39, 173, 177, 188, 207, 209, 244, 260
Travel distance, 63, 89, 138, 139, 143, 144, 148
Travel time, 2–4, 17, 74, 109, 110, 114, 116–117, 138, 139, 143–145, 148, 163–165, 204, 257, 292, 293
Trihalomethanes (THM), 15, 32–33, 130, 249, 258
Trinitrotoluene (TNT), 130–134
Tunnel, 182, 183, 223–234
Turbidity, 19, 23, 31, 69, 130, 187, 193, 197–198, 220, 237, 245–249, 258, 261, 262, 264, 266, 277, 292

U

Ultraviolet absorption (UVA) coefficient, 58
Uniformity coefficient, 60
United States Environmental Protection Agency (USEPA), 132
United States Geological Survey (USGS), 35, 37, 91, 171
UVA coefficient. *See* Ultraviolet absorption coefficient

V

Vacuum-siphon extraction systems, 225
Vertical wells, 2, 18, 20, 25, 37, 39, 69, 71, 132, 173, 175, 210, 224–226, 245, 258, 269, 271–276
Viruses, 31, 71, 82, 84–86, 93, 236

W

Wadi, 3, 4
Warning and Alarm Plan (WAP), 74
Warta River, 270–278
Water pollution, 5–25, 69, 71–78, 152, 235, 249, 256, 265, 284–291, 294, 296
Water quality, 3, 4, 6–10, 12–19, 22–25, 30–32, 35, 37, 47, 52, 53, 64, 65, 69, 71–74, 82, 98, 101, 102, 106, 110, 117, 120–122, 125, 138, 152, 153, 173–177, 183, 184, 187, 196–198, 200, 201, 204, 211, 212, 215–217, 220, 223–224, 235–240, 245–251, 256, 257, 261–266, 269–279, 284–287, 291, 292, 295
Water scarcity, 30, 31, 46
Water supply, 1, 2, 5–25, 30, 31, 33, 34, 43, 51, 52, 64, 65, 69, 70, 73–76, 78, 82, 83, 90, 93, 98, 99, 130, 169–171, 173–178, 197, 198, 203–221, 223–224, 236, 241, 256–258, 266–267, 269, 277, 281–296
Water temperature, 4, 19, 37, 39, 40, 43, 52–58, 61, 130, 177–179, 187, 196, 198, 217, 237, 246, 248, 250, 256, 258, 261, 264, 265, 277, 283, 292
Well management, 75–78
Well screen, 105, 110, 113, 134, 186, 187, 209, 229, 242–245, 258, 273
World Health Organization (WHO), 15, 16, 237, 238, 249–250, 257–258, 261

Z

Zonation pattern, 108–112